MW00669214

MODELING:
Gateway to the Unknown

A Work by Rom Harré

Rom Harré

STUDIES IN MULTIDISCIPLINARITY VOLUME 1

MODELING: Gateway to the Unknown

A Work by Rom Harré

EDITED BY

Daniel Rothbart

George Mason University
Fairfax, Virginia, USA

2004

ELSEVIER

Amsterdam – Boston – Heidelberg – London – New York – Oxford
Paris – San Diego – San Francisco – Singapore – Sydney – Tokyo

ELSEVIER B.V.	ELSEVIER Inc.	ELSEVIER Ltd	ELSEVIER Ltd
Sara Burgerhartstraat 25	525 B Street, Suite 1900	The Boulevard, Langford Lane	84 Theobalds Road
P.O. Box 211, 1000 AE	San Diego, CA 92101-4495	Kidlington, Oxford OX5 1GB	London WC1X 8RR
Amsterdam, The Netherlands	USA	UK	UK

First edition 2004

Library of Congress Cataloging in Publication Data
A catalog record is available from the Library of Congress.

British Library Cataloguing in Publication Data
A catalogue record is available from the British Library.

ISBN: 0-444-51464-3

♾ The paper used in this publication meets the requirements of ANSI/NISO Z39.48-1992 (Permanence of Paper).
Printed in The Netherlands.

CONTENTS

SECTION FOUR
A METAPHYSICS FOR PHYSICS

SECTION FIVE
A METAPHYSICS FOR EXPERIMENTS

SECTION SIX
THE ETHOGENIC POINT OF VIEW IN SOCIAL SCIENCE

GENERAL INTRODUCTION

GENERAL INTRODUCTION

1. WHY MODELING?

The unknown has always been alluring. Since the time of the ancients, scientists have refined techniques, apparatus, and methodologies for disclosing things and events that lie beyond the senses. Some of the greatest discoveries of science occur in areas that transcend the here and now, exposing a world that is bizarre in relation to everyday material bodies. Such discoveries reveal alien beings that challenge our capacities of imagination. How can these discoveries be explained? For many twentieth century philosophers, principles of logic provide pathways to truth about the known and unknown. If a scientific account violated certain patterns of reasoning, it is dismissed as "unscientific." But the problems and paradoxes of rationalist programs are legendary. Logical positivists were plagued by serious challenges to their inductivist program, and Popperian proposals for a falsificationist methodology were also challenged for their hidden inductivist underpinnings.

The storm of criticism against rationalist philosophies of science brought with it serious challenges to the mission of scientists and the privileged status they enjoy in society. For example, relativists committed to the Strong Programme in the Sociology of Scientific Knowledge understood research activities through social and political influences, showing in particular the vulnerabilities of scientists to institutional pressures. The honorific terms "truth," "knowledge," and "evidence" were said to have nothing to do with the way the world is, but were deployed as propaganda devices in campaigns for glory or power. Scientists are presumably committing a kind of intellectual fraud when they declare triumphantly to have discovered real-world properties, independent of human interests.

How can scientific discoveries be explained without denigrating the mission of scientists, and without succumbing to the mistakes of rationalist philosophy? In a corpus of writings that spans four decades and continues to grow, Rom Harré has produced a philosophy of science that is comprehensive in scope and revolutionary in comparison to earlier programs. For Harré, the central mistake of positivists is the unceasing reliance on discursive language as the unit of knowledge, and the privileged status given to logic as a basis for scientific thought. In his Copernican Revolution in philosophy of science, Harré identifies modeling as an essential aspect of inquiry. As the frontiers of science are continually pushed back, and the distance between experimenter and the world widens, the intelligibility of the world demands the construction and manipulation of models. Scientific discourse is often used to convey the information from well-grounded models. Scientific thinking is inescapably modeling and intimately involved with inquiry.

A model can be as simple as a line drawn in the sand to indicate landmarks or as complex as cosmological models of a multidimensional universe. Through modeling, scientists manipulate symbols with meanings to represent an environment with structure. Such manipulations take place to fulfill a human need, solve a problem, or create a product. When constructing a model, one works in the cognitive space of ideas. Models are used to encapsulate, highlight, replicate or represent patterns of events and the structures of things. Of course, no model provides an exact duplication of the subject matter being modeled. Details are hidden, features are skewed, and certain properties are emphasized. Models are abstract and idealized. As an abstraction, a model omits some features of the subject matter, while retaining only significant properties. As an idealization, a model depicts a subject's properties in a more perfect form. The model citizen of a a nation has certain character traits that are more perfect than those of any actual citizen. Some of these traits are idealized, never materializing exactly the way in which the models operate.

Harré's writings had the effect of destabilizing many central doctrines of logical empiricism. Consider for example the claim that sensory experience is, in principle, separable from cognition. Appealing to important finding of experimental psychology, he demonstrates in elegant detail how sensory experience requires modeling. In many settings, a viewer faces a difficult perceptual problem--how to distinguish between the "real" movement of material bodies in an environment and the "imaginary" aspects caused by a viewer's own movements. The solution demands of viewers complex skills of gathering information, apprehending patterns of spatial relations among material bodies, comparing present experiences with the past, and anticipating experiences to come. A viewer is constantly moving back and forth from the actual perceptions to ideas that emerge in continuous testing, comparing, and revising of models.

He shows how modeling does not require that one leave the sensible realm. In particular models used to explain events are constructed around icons of experience. An iconic model offers a conceptual picture, typically regarding concrete images, of certain patterns of events and structures of things. Ideas are marked by images, and conveyed by pictorial symbols. Sensory images are used to represent abstract properties. An icon is an idea, usually pictorial, that places constraints upon, and opportunities for, model-construction. The mind forms visual diagrams in the imagination, producing schematic maps of information. Such imaginative capacities are necessary to replicate processes that extend beyond possible experiences. For example, the 1953 Watson-Crick theoretical model of DNA replicates the double helical structure, not as a summary of isolated "facts" but as an abstract simulation of biochemical structures. Through iconic modeling, biochemists can visualize DNA replication, geophysicists can replicate tectonic plates, and astrophysicists can conceptualize processes of distant galaxies.

2. REALISM RESCUED

For Harré, a commitment to real-world activities and entities is required for experimental research. A researcher's skillful action mediates between experimental phenomena and causal capacities of nature. Researchers are expected to know how certain properties of materials can be exploited in the construction of instruments, how human abilities for manipulation and control are enhanced through the power of instruments, and how capacities of substances under investigation are revealed during an experiment.

Harré endorses the view that science discovers because it produces artifacts. In contemporary research, the old distinction between pure science, as the discovery of a given-reality, and technology, as the deliberate production of artifacts, collapses. The properties of experimental phenomena cannot be cleansed of the pragmatics associated with instrumental detection. The character of laboratory skills cannot be reduced, explained, or understood through inductive argument patterns. An experimental researcher is a skilled agent who exploits both natural and artificial resources for retrieving information. The validity of data is inseparable from the skillful use of the apparatus needed to obtain such data. Like the ancient artisan, a researcher starts a process, removes obstacles, and exploits concealed capacities of materials. These capacities hidden from immediate detection from the naked senses, but revealed indirectly through the methods of detection.

Scientific research rests on commitment to three kinds of models, corresponding to three realms of existence. Realm 1 type models represent perceivable entities, properties and processes, revealing a world of common experience. Realm 2 type models replicate perceivable entities that can be experienced only with technological assistance, when obstacles are overcome and underlying properties are exploited. Realm 3 models resemble unobservable entities, properties, and processes, all of which are inaccessible to possible experience even with technological assistance. These three realms of models are needed to explain the instrumental techniques. Accounts laboratory methods demand an appeal to causal powers. Procedural knowledge about how research should take place rests on declarative knowledge about the causal mechanisms that are responsible for experimental phenomena. During research, experimenters provoke nature to exercise its powers, released when apparatus is properly used. An apparatus is a modified version of processes found in the wild, enabling experimenters to domesticate nature. For example, an Atwood's machine is a domesticated version of a cliff. Both the machine and the cliff are governed by principles of gravity, among others.

But these three realms are not immune to change. The designation of an entity as Realm 2 or Realm 3 depends on the detection methods at a particular time. For example, atoms were unobservable before the modern instruments of detection, prior to the so-called instrumentation revolution of the twentieth century. Atoms became detectable (and in a loose sense observable) only after 1981 with advances in electron microscopy. The observable/unobservable distinction requires a moving boundary, but a boundary nonetheless.

For Harré, real world properties of a causal mechanism are capacities of certain entities to exhibit detectable effects under certain conditions.[1] A cannon-ball has a tendency to fall toward the earth, just as an electron has a tendency to accelerate toward a positively charged plate. Earthquakes are explained by the causal actions of tectonic plates, extending principles of mechanics to geology by analogy. Such capacities are revealed empirically under the appropriate releasing conditions, and the suppression of obstructing influences. Of course, real world properties do not lose their capacities when conditions are not met. He writes: "A *particular Being* has a *Tendency* which if *Released*, in a certain type of situation, is manifested in some observable *Action* but when *Blocked* has no observable effect." (Harré's emphasis.) (1986, p. 284).

Explanatory models are replicas of real-world processes, highlighting structures that are analogous to those they represent. To explain the occurrence of observed events, scientists resort to an analytical analogue of a productive mechanism. An analogue of the observed process can be imagined to be caused by an unknown productive process. The conceptual content for this analogue can be found in an appropriate source model. Harré introduces the notion of a paramorph to identify the kind of analogical model that is essential to scientific explanations. A paramorphic model is one in which the source of the model is distinct from the subject matter. Of course, in the case of the double helical model, the source is found in concepts of information processes and the subject of biomolecular replication. Similar examples abound throughout the

[1] Nancy Cartwright argues that the physical world is a world of nomological machines (1999, p. 52). Her work draws heavily on Harré's earlier writings on the commitment to the existence of causal mechanisms. Although not immediately observable, such a mechanism is understood through various analogical associations to known systems. The ontological underpinnings of causal events with effects are the capacities of particular entities that emanate from a generative mechanism (Harré and Madden 1977, p. 11).

history of science.[2] The concepts of light waves, electrical flow, flow of time, and information retrieval during protein synthesis were born from paramorphs.

A paramorph-shift may precipitate a crisis in literal vocabulary. A scientist's commitment to a new paramorph may strain the capacity of the conventional vocabulary to express innovative ideas. What was impossible, inconceivable, and incoherent based on literal vocabulary becomes possible, conceivable, and coherent through metaphoric redescription. Combinations of terms that were incoherent, in relation to the conventional rules of meaning, become meaningful. Metaphoric description arises from a momentary suspension of the rules for literal vocabulary. The semantics of a metaphor convey an alternative realm of conceptual possibilities, through a new set of possible attributes. Of course, not all scientific language is metaphoric. But when unexpected empirical findings raise serious doubts about a familiar scientific theory, a satisfactory resolution occur through the use of metaphoric vocabulary. In some cases metaphor offers the only avenue for expressing promising but unexplored modes of thought for a particular area of inquiry (Rothbart 1997, Chapter 2).

3. MODELING MINDS

Harré's major contributions to the field of social psychology center on the complexity of action and identity. Action involves intentionality, in contrast to behavior, which does not. Winks are actions and blinks are behaviors, to borrow Clifford Geertz's neat example. The difference between a wink and a blink rests on ability of social agents to construct models. Through modeling, agents represent actions of close relatives, remote acquaintances, and distant strangers. An agent can review someone's action, anticipate the effects on a social setting, and plan a possible response.

Harré's arguments draw on Erving Goffman's dramaturgical perspective of self as actor. In Goffman's work, the self/actor relies on a script to act on a stage in front of an audience and critics. Before the performance, each actor imagines performing the roles, directing a scene, and possibly reviewing a play. An actor's conception of self demands skills of playwright, director, and critic in a monodrama of performance. Management of the self/actor requires a private conversation, as if the separate skill sets are engaged in an intimate discussion about what should be done. But the self is not reducible to mere introspection. Each social encounter provides an expression of plays within plays. A monodrama of one actor can be included within the monodrama of another actor, leading to an ascending order of metamonodrama (the play that surrounds the play).

In his social psychology, Harré understands a social agent through intentions to influence other agents and critical reflection on what took place. An agent's intention functions as a template or a preformed plan constructed prior to action. In constructing a template, an agent is guided by local conventions for certain roles that are appropriate for particular settings. But he warns against attributing any causal power to such templates. Just as a grammatical rule guide

[2]Analogical modeling contributed to important developments in Newtonian mechanics (J. North 1980), Maxwell's electromagnetism (R. Kargon 1969; M. Hesse 1974), and subatomic physics (Miller 1986). Episodes of analogical modeling in biology are well known (G. Canguilhem 1963; T. Hall 1968). In some episodes of scientific development, analogical models are used to generate entirely new disciplines. The nineteenth century unification of biology and chemistry was advanced by the exploration of analogical associations (J. Brooke 1980, 1987).

linguistic behavior without being causally responsible for action, a preformed plan provides instructions for action. A preformed plan constitutes a *formal* cause, that is, a cognitive template for human agency.

In his later writings, Harré abandons the conception of intention as a template for action in favor of a narratological perspective. A narrative is not a template superimposed on action, and not a preformed script written before the performance. Narratives are private conversations that bring past performances to bear on present and future encounters. This conversation produces a story line in which the salient features of a particular episode are named and the context for the episode is framed. To construct a narrative, one constructs an idealized scene, pictured in the mind's eye. Narratives provide guidance for one's own performance and standards for adequate responses to conversations that unfold during a social encounter. A sense of self emerges from reflection on, and management of, one's own narratives, that is a nested hierarchy of patterns of private conversations. A social self is revealed through one's social and moral position in relation to others.

A narrative is similar to a model in three ways. First, narratives, like models, are conceptual constructions under the control of a story teller. Second, a narrative replicates some aspects of past experiences, recalling events that are at least temporally remote, and in most cases far away. Here's the present teller, close to the reader or listener, and there at a distance is the tale. Third, a narrative has a projective dimension. Reflection on past activity leads to planning and projection of future activity, so that the story teller anticipates encounters yet to occur. The projective aspect of narratives, and models, is essential for revealing unobserved, but observable, events.

An actor imagines how the course of events could change, and how the expected state of affairs could be modified through manipulation. Past patterns of interaction are deployed as instruments for imagining new possibilities. Adjusting to the exigencies of emerging situations, an actor shifts attention away from the present and toward future possibilities. The self-as-actor constructs images of where he or she thinks they are going and how to get there, resulting in a life trajectory of events that gives shape and direction to performances. Of course, the patterns of interaction can change as one acquires more experiences. Through reflection on present encounters, an agent can reconstruct narratives of the past, adding and readjusting repertoires of life experiences. In responding to the challenges of social encounters, actors can distance themselves from the scripts and traditions that constrain their performance.

As people tell stories about their social interaction, their relative positions emerge. In each life story, a person is "located" within a social encounter. The self is an abstraction of the speaker's various positions in a social encounter. For Harré, each story not only reveals a speaker's location in past encounters, but also builds a sense of personal identity as stories are told. Personal identity can be maintained through the use of pronouns. The use of "I" fixes a speaker's utterance to a spatio-temporal location from one story, and establishes a continuity across many stories.

In sum, an enlightened understanding of both physical and social phenomena is possible through modeling, as various stages of inquiry. Just as real world models are inseparable from experiences of the empirical world, a narrative is thoroughly implicated in a social encounter. Actors resort to narratives as they respond to the movements of others and project possibilities for future encounters. An actor becomes a virtual witness to an idealized scene, drawing upon past encounters to construct a picture of future, hypothetical events. As social encounters take place, the actor is transported from the specific circumstances of the local environment to a realm of possibilities. Similarly, the properties of physical events demands attention to the entire empirical system in which the event occurs. In laboratory research using complex apparatus, the fleeting experimental events are made intelligible within an entire instrumental system. Just as narratives are used for control and management of action, iconic models of the instrumental system are needed to render experimental phenomena intelligible. The world is explained through iconic models. Returning to the former example, an actor uses iconic modeling when he or she envisions episodes that are not actually present, blending the immediacy of the moment with an awareness of unseen possibilities.

A philosophy of great scope and depth unfolds in these chapters. Many of his insights from his early years are now standard fare in current literatures. His contributions to social psychology, ethnography, linguistics, and cognitive psychology transcend traditional disciplinary boundaries, and reinforce the power of scientific modeling as a technique for representing the known and as a gateway to the unknown.

This work offers for the first time under a single volume Harré's comprehensive and celebrated ideas on modeling as essential to inquiry. The essays listed below originally appeared in other places.

1970. "Models in Theories", *The Principles of Scientific Thinking*, Chicago, Chicago University Press, Chapter 2, pp. 33-62.
1976. "Images of the world and societal icons", in K. D. Knorr, H. Strasser and H. G. Zilian, eds., *Determinants and Controls of Scientific Development*, Dordrecht: Reidel, pp. 257-283.
1976: "The constructive role of models", in R. Collins, ed., *The Use of Models in the Social Sciences,* London, Tavistock, pp. 25-43.
1979. "The Uses of Models II: Social Action as Drama and as Work," *Social Being*, Oxford, Blackwell, pp. 146-147 and 161-185.
1981. "Creativity in Science", in D. Dutton and M. Krausz, eds., *The Concept of Creativity in Science and Art,* The Hague, Martinus Nijhoff, pp. 19-46.
1982. "Metaphor in Science", with J. Martin Soskice, in D. Miall, ed., *Metaphor: Problems and Perspective,* Brighton, Harvester, pp. 89-105.
1986, "Theory-Families and the Concept of Plausibility", *Varieties of Realism*, Oxford, Blackwell, Chapter 11, pp. 201-221.
1986. "Deeper into Real 3: Quantum Field Theory", *Varieties of Realism*, Oxford, Blackwell, Chapter 14, pp. 261-280.
1986. "Explanation", *The Philosophies of Science,* Oxford, Oxford University Press, 2nd edition Chapter 6, pp. 168-183.
1990. "Exploring the Human Umwelt", in R. Bhaskar, ed., *Harré and his Critics,* Oxford, Blackwell, pp. 297-319.
1990. "Life as conversation: updating the role-rule model and the act-action distinction", in K. Grawe, et al., eds., Uber die richtige Art, Psychologie zu bettreiben. Gottingen: Hografe
1990. "Tracks and Affordances; the Sources of a Physical Ontology", *International Studies in the Philosophy of Science, 4,* 149-157.
1990. "Vygotsky and artificial intelligence", *Midwestern Studies in Philosophy,* vol. XV, pp. 389-399.
1992. "Rethinking the laboratory experiment",with F. Moghaddam, *American Behavioral Scientist,* vol. 36, pp. 22-38.

1994. "Some Proposals for The Formal Analysis of the Use of Models in Science" with J. Aronson and E. Way, *Realism Rescued,* London, Duckworth, Chapter Four, pp. 72-87.
1995. "Realism and an ontology of powerful particulars",.*International Studies in the Philosophy of Science*, vol. 9, pp. 285-300.
1996. "From observability to manipulability: the inductive arguments for realism", *Synthese*, vol. 108, pp. 137-155.
1997. "The ontological duality of space-time variables", *International Studies in Philosophy of Science*, vol. 11, pp. 83-96.
1998. "Rhetorical Uses of Science", *Greenspeak: a Study of Environmental Discourse,* Los Angeles and London, Sage, Chapter 3, pp. 51-68.
1998. "Recovering the experiment", *Philosophy,* vol. *73,* pp. 353-377.
2003. "The Materiality of Instruments in a Metaphysics for Experiments" in H. Radder, ed., *The Philosophy of Scientific Experimentation*, Pittsburgh: University of Pittsburgh Press, pp. 19-38.

DANIEL ROTHBART

SECTION ONE

THE STRUCTURE AND FUNCTION OF MODELS

SECTION ONE

THE STRUCTURE AND FUNCTION OF MODELS

INTRODUCTION

During the two hundred years of the transformation, we call the renaissance the focus of philosophical reflection began to change from its once exclusive emphasis on the interpretation and justification of religion. The scientific method of enquiry into the material and social worlds began to crystallize out as a distinctive method for acquiring reliable knowledge. What gave it this apparently inexhaustible power?

Since then philosophers of science have followed two rather different paths guided by different conceptions of the task of philosophy. For some, philosophers reveal the most general abstract forms of knowledge making. Philosophy of science is, in the end, a province in the domain of logic. It is easy to cite a good many 'logicists'. In the nineteenth century, John Stuart Mill looked for a formal account of the genesis of scientific knowledge, parallel to the Aristotelian account of the powers of deduction. The twentieth century, for a time was dominated by figures such as Bertrand Russell, Rudolph Camap, Karl Popper and many more. For another dynasty of philosophers it seemed entirely obvious that the rationality of science was scarcely captured at all by the abstraction of logical forms. Indeed it has proved only to easy to declare the sciences irrational from that standpoint. In the nineteenth century William Whewell, and in the twentieth N. R. Campbell and many who followed his lead in the second half of the century the principles of scientific thinking were grounded in another kind of cognition, concrete representations of reality rather than abstract forms of the discourses of scientists, or what were taken to be their discourses.

The logicists tended to be positivists, relying only on the verisimilitude of observations. The others tended to be realists, convinced that carefully crafted representations of the world beyond the bounds of sense made legitimate contributions to knowledge.

There were several terminologies in use to describe the methodology of the Whewellian school. The word that has survived throughout, enlarging its meaning and in some circles losing meaning altogether, is the word 'model'. Thinking with models is likewise called 'modeling'. We make and use 'models', concrete representations of the central material entities, structures and processes of a domain into which a scientist might be enquiring. The focus shifts from discourse, talking and writing about nature, the most general theory of which is logic, to modeling, reproducing and representing nature materially, the most general theory of which is analogy. Analogy, we will come to see, rests on something yet more fundamental, the sorting of things into representatives of types and kinds.

Once one takes modeling seriously as the core of scientific method, the issue of rational constraints on model making loomed large. Thinking of models as entities of the same type as that which they represent is one way of bringing this issue into focus. Such a constraint requires another limitation. The source of a model must also be of the same generic type as both the model and its subject, that which it represents. In the most interesting cases model making is used to represent beings, processes and structures with which scientists are not yet acquainted. This prompts the general question: Is the common type under which source, model and subject all fall, represent a plausible ontology for the unknown beings for which the model is a stand in? The same question can be posed in terms of natural kinds. Are the natural kinds presupposed by the use of a model to represent that which has yet to be observed plausible for the domain in question?

For those cases in which the subject of a model is such that with technical advances it could be observed, for instance Pasteur's 'microorganism' model of the vectors of disease, natural kinds are conserved. The ontology of the observable is roughly comprised of space-time locatable substances and their properties and changes therein. However, there are many situations in physics and in psychology that involve building models of realms of beings which could never be observed. The most important of these are powers and potentials. A certain location in space and time is ascribed a gravitational potential. What would it be like to observe the potential itself? The effect of that being on a test body is readily observed, but not the causal power itself.

The full story of model making in the sciences is intimately linked with the fundamental ontological questions concerning the investigation of an ontology of dispositions and causal powers, that is deeper than that of substances and their attributes and relations, the ontology of the readily observable beings of common experience. The first stages of physics and chemistry extended this ontology into the knowable but unknown realms of nature. The later stages of physics and chemistry require the acknowledgment of a more fundamental ontology to characterize the way of being of the realms of reality that lie outside the bounds of sense.

To sum up: The conceptual foundations of physics are made up of a variety of dynamicist concepts. Mechanistic concepts are used only for modeling of entities nearest to the perceptible world of the unassisted senses. Studies of modeling in the human sciences show that there too dynamicist models are required to represent the most fundamental beings, though these beings are among the most easily perceptible of our world. In sociology, psychology, history. ethology and so on, higher organisms are treated as the sources of activity in the relevant domains.

A thorough study of modeling as the core of scientific method cannot be divorced from reflections on the ontology of the sciences. Such reflections lead us quickly to the insight that dynamicist concepts must dominate mechanistic concepts in scientific thinking in general.

CHAPTER ONE

ICONIC MODELS

1. INTRODUCTION

I begin my positive account of the nature of scientific thinking by setting out a new view of theories in which they are to be seen as essentially concerned with the mechanisms of nature, and only derivatively with the patterns of phenomena. I shall be considering only one component of those intellectual constructions which I have used the metaphor of 'statement-picture complex' to describe, and which are actual theories *in vivo.* Scientists, in much of their theoretical activity are trying to form a picture of the mechanisms of nature which are responsible for the phenomena we observe. The chief means by which this is done is by the making or imagining of models. Since enduring structures are at least as important a feature of nature as the flux of events, there is always the chance that some models can be supposed to be hypothetical mechanisms, and that these hypothetical mechanisms are identical with real natural structures.

Theories are seen as solutions to a peculiar style of problem: namely, 'Why is it that the patterns of phenomena are the way they are?'. A theory answers this question by supplying an account of the constitution and behaviour of those things whose interactions with each other are responsible for the manifested patterns of behaviour. This might be a microexplanation like the explanations of chemical reactions in terms of the behaviour of chemical atoms. It might be a macroexplanation like the explanation of the tides in terms of the behaviour and powers of the heavenly bodies. To achieve this a theory must very often fill in gaps in our knowledge of the structures and constitutions of things. This it does, in ways to be detailed, by conceiving of a model for the presently unknown mechanism of nature. Such a model is, in the first instance, no more than a putative analogue for the real mechanism. The model is itself modelled on things and materials and processes which we do understand. In a creative piece of theory construction, the relation between the model of the unknown mechanism and what it is modelled on is also a relation of analogy. Thus, at the heart of a theory are various modelling relations which are types of analogy.

The rational construction of models will be shown to proceed under the canons of a theory of models, to which the old logic of analogy is a crude approximation. In developing this theory we shall be looking at the way scientists train themselves in the skilful judgement of likeness and unlikeness, plausibility, and the like. But we can hold no hope of successful formalization of the principles of this art. At this stage we describe them.

A theory cannot, therefore, be a single deductive structure. It consists, as we shall see, of at least three sets of sentences, the successful deductive organization of any of which is quite fortuitous. There will be one for the description of the phenomena for which the theory is devised. There will be another for the description of the central model, and one or more describing the material upon which the central model is based. These sets of sentences are tied together by various relations of analogy, that is by further sets of sentences whose extent cannot be discovered a *priori.*

Finally, it will be shown that this is the structure which generates the most crucial of all the kinds of scientific hypotheses, namely existential ones. This happens by the step of coming to treat the central model as a hypothetical mechanism. All sorts of consequences follow from this view of theories; consequences which tie together the scientific enterprise in a rational structure which no simpler theory of theories, however elegant, can approximate.

I hope it will become that I do not in the least intend anything specifically mechanical by the word 'mechanisms'. Clockwork is a mechanism, Faraday's strained space is a mechanism, electron quantum jumps is a mechanism, and so on. Some mechanisms are mechanical, others are not. I choose the word 'mechanism' for this use largely because it is the word most usually used for this purpose. We talk of 'the mechanism of a chemical reaction', 'the mechanism of bodily temperature control', 'the mechanism of star creation', and so on.

I turn now to a detailed exposition of the nature, use and structure of models. . . . My use of the word 'model' will not be precisely that of everyday, though not unconnected with it. I use it as a technical term, and the precise technical usage I intend can be grasped, only after my exposition of the way I intend to use the term is complete. In the next section I give a general definition of model, but it is rather a sketch of certain broad conditions under which I intend to use it than a precise definition. Just to show that I do know what I am about, I introduce at this point a sketch of the ordinary usages of the term, not all of which will be reproduced in the technical usage.

1.1. Participle and Preposition

(a) The Ghanaian Assembly is *modelled on* the British Parliament.' Here the British Parliament is the source of certain features of the Ghanaian Assembly, but it would be incorrect to say that that assembly is a model of the British Parliament, or that it is a model Parliament.

(b) 'The gold lamé cocktail dress was *modelled by* Samantha.' This is a very derivative usage, harking back to Samantha being a model, and that too is derivative, harking back to Samantha being, in some sense, ideal.

(c) 'The head was *modelled in* clay.' Here, by transferred epithet, the process of making a model is called 'modelling', and what it is modelled in, asks of what substance has it been made.

(d) 'She made a living *modelling for* the art class.' She might be an ideal shape for Venus, or for a grandmother, but she might just be something to draw. In this sense, though not in that of *b,* it is necessary for her to be an ideal for her to be a model.

1.2. The Verb 'to model'

Two common usages that I ignore are 'to wear in an exemplary fashion' and 'to make something with the fingers out of a plastic material'.

1.3. Models as Types

Then there is the use of the word 'model' to designate a type. A new version of a car is a 'new model'. Sometimes this slides over, in careless talk, into reference to the token as a token representative of the type, in such expressions as 'I bought this year's model' where I mean to indicate that I bought one of this year's type. . . .

2. MODELS AND SYMBOLS

I now turn to a more exact account of how I intend the word 'model' to be used. First the difference between model and symbol; I shall use 'model' so as to include pictures. This distinction has to be stated with some care, since both models and symbols depend upon a convention or set of conventions by which they become vehicles for thought about a certain subject matter. I am going to distinguish projective conventions for models from arbitrary conventions for symbols. It might be noticed in passing that Wittgenstein in the *Tractatus* (1922) tried to assimilate arbitrary conventions to projective conventions. A model can become a symbol when its source of projection is lost or forgotten as, for instance, the bull's head became *alpha,* but a symbol cannot become a model. This last clause is slightly stipulative, since some people still talk of equations as models of motions and processes. At that rate every vehicle for thought would become a model, and a valuable and interesting distinction would be lost. If anyone insists that that is just how he uses the word 'model', or that that is how the word model is used in his culture-circle, I will give him the word 'model' gladly, and mark my distinction by means of another word, let us say 'modella'. It's well to remember the old saying, if our eyes were made of green glass then *nothing* would be green.

What is the difference between a projective and an arbitrary convention? Projective conventions involve the use of the source of the model or picture in the act of construction of the model or picture. The reason why *A* is a model of, or for, *B* in one kind of case, is that *A* was constructed from *B,* so that *A* and *B* have certain likenesses, and it is these that make *A* a model of *B*. But because *A* and *B* must be unlike in certain ways (or else *A* is just a replica of *B),* and just how unlike and in what respects is at the will, and under the guide of the purpose of the model-builder, I call the projection of *B* into *A,* a convention. Arbitrary conventions have, *inter alia,* the following *differentium:* whatever may have been the source of the symbol β, no likeness or unlikeness it may bear to α its subject matter counts as a reason why it is a symbol for, or of, α. Any sign would do as well, provided a symbol convention had been agreed for it. Hieroglyphs provide an interesting case, since the marks seem to function both as models and as symbols. I understand from Egyptologists that this appearance is misleading, and that they make a clear distinction themselves between the comic strip era, and the syllabary. . . .

A cardinal distinction, of the first importance in understanding the way models work, is that between the source of a model and the subject of a model. The *subject* is, of course, whatever it is that the model represents: that it is a model of, or model for. There are several different relations between models and their subjects. The *source* is whatever it is the model is based upon. There are again several possible relations between a model and its source. Broadly speaking models belong in two great categories, depending on whether the subject of the model is also the source of the model, or whether the subject and the source differ. For example a toy car is a model for which source and subject are identical, the toy being modelled on a car, and being a model of the very same car. However, even with toy cars, the matter is not quite so simple as it seems at first glance. Is the toy car modelled on a particular instance of the 1963 Mini-Cooper, say the one which won the Monte Carlo Rally, or is it modelled on the 1963 Mini-Cooper type? Both cases occur. Models of the particular 1963 Mini-Cooper which won the 1963 Monte Carlo Rally are sold, and models of the 1963 Mini-Cooper are also sold. The only difference that I can see is that while all models of the particular 1963 Mini-Cooper must be, in all respects in which they *are* models of *the* car, the same, the models of the 1963 Mini-Cooper can be different, in the same kind of respects as those in which 1963 Mini-Coopers differ, namely in colour, radiator grille, presence or absence of spotlights, and so on. If someone produced a model of the 1963 Buick Riviera, and declared that he wished to treat it as a model of a 1963 Mini-Cooper, we would, quite properly, have to tell him that he *could not do this,* despite the common possession of four wheels, two doors and so on. Again this shows how models differ from words. Only a certain latitude is permissible in projective conventions, and is determined by the need to eliminate certain possibilities of being misled, in particular cases. Of course, if it *is* a model, as contrasted with a replica, then there will always be *some* way of being misled, just because a model is not exactly like the subject in all respects.

With Lord Kelvin's model of gyroscopes and elastic thread for a particle of the luminiferous ether, . . . source and subject are different. The subject is the unknown mechanism of the transmission of light (supposing light is transmitted!), the source is the known mechanical behaviour of gyroscopes and of elastic threads. Of course, a model's subject and source may differ even when the subject is some known mechanism, since models may be made, or imagined, for such conveniences as ready intelligibility, or any one of a number of reasons which would suggest that the mechanism of the thing or process, etc. modelled, should be modeled by some other kind of mechanism having suitable characteristics.

Part of the great importance of models in science derives from their role as the progenitors of hypothetical mechanisms. It often happens that the antecedents of an effect are well known, but the casual mechanism by which the antecedents bring about the effect is not. Consider, for instance, the long history of the use of catalysts, and of antibiotics, before the way they worked was found out. Then, in the imagination, we make a model for the unknown mechanism. Whatever *is* in the black box, one might say, could be like this. This process differs fundamentally from making a model of a subject which is already well known. I shall sharpen up the terminology a bit by using '*A* is a model for *B*' where *B* is quite unknown, and '*A* is a model of *B*', where *B* is known. Generally speaking, making models for unknown mechanisms is the creative process in science, by which potential advances are initiated, while the making of models of known things and processes has, generally speaking a more heuristic value. . . .

An object, real or imagined, is not a model in itself. But it functions as a model when it is viewed as being in certain relationships to other things. So the classification of models is ultimately a classification of the ways things and processes can function as models. Fashion models are not a special category of humans, but fashion modelling is something that some people can do.

2.1. Homoemorphs

I shall begin a detailed analysis with those models for which source and subject are the same. . . . Depending on just how models are related to their source-subject, there are *idealizations, abstractions*, and *typifications*. Suppose that the task is to model an object which has distinguishable properties $p_1 \ldots p_n$, relevant to the enquiry or task.

2.1.1. Idealizations

Then a model of it may be made in one of two ways: an idealization is created either by building a model, or selecting something to serve as a model, having properties $p_1 \ldots p_n$, where each property of the source-subject is matched by a property of the model, but the model properties are distinguished by being, according to some scale of values, more perfect than the source-subject's properties. A fashion model, for instance, has all the characteristics of an ordinary man or woman, but has them in more perfect form, according to some currently accepted scale of values. A glance at an old fashion magazine shows immediately that what shape distinguishes an ideal man or woman at one time may not meet the standards of another age. A model pupil presents some difficulty of classification, since, while he or she does all the things a pupil should do better than his peers, and does not do some of the things that they do, when functioning as a model, the pupil may be, in my sense, an abstraction as well. Another point of importance emerges here too, and will be discussed later. It is not so much any particular woman or any particular pupil that serves as the source-subject of the model, though the model must be particular. Rather it is some average or type which is the source subject. An ideal pupil is an idealization of all pupils, and a fashion model is similarly related logically to all humans. I shall return to this point.

2.1.2. Abstractions

The other way of making . . . a model is by abstraction. If the source-subject has properties $p_1 . . . p_n$ and here again the source-subject may be a particular, representing the type of all members of some class of things, or it may be a kind of average. An abstraction has properties $p_j . . . p_k$, $l < j < k < n$, that is fewer than its source-subject. Which properties are chosen depends largely on the purposes for which the model is created. This is because the properties which are not modelled are those which are irrelevant, and relevance and irrelevance are relative to purposes. . . . Scale models differ from their subject, which is also their source, only in dimensions. They are created by applying to a suitable characterization of their source-subject some such instruction as 'Reduce (Increase) the main linear dimensions by 1/64.' A solid model of an airliner used for wind-tunnel testing some important features of the source-subject are omitted, like the seats in the passenger cabin, and the pressurization system. Similarly, fibreglass models of viruses, millionfold enlargements do not replicate every structure of the source-subject. The rules of transformation from source-subject to model and back again may not be uniform for all features of the subject. In model aeroplane making, the control surfaces must be scaled down by a lesser factor than the major linear dimensions, if the stability and maneuverability of the model is to match that of the plane. The exact scale factors for any features other than the major linear dimensions may have to be discovered by experiment, if the model is going to be able not only to represent the physical structure, of the plane but also to behave like it. This is generally true of *pilot plants.* To model both behaviour and structure, some compromise on the details of structure is required.

Typical abstractions include those structures in coloured wires which model the blood vascular systems of animals, and which represent some relations of connectivity and relative length and size among the blood vessels, but leave out the valves, the relations of elasticity, the capillaries and of course the blood! Since the model itself is a wire and plastic structure it has many features which are irrelevant to its functioning as a model, such as the chemical composition of the wires and plastic coating. . . .

Another abstraction . . . is the gas molecule of the kinetic theory of gases. Molecules have the mass and velocity of material objects, but in the earliest form of the theory lacked the volume and imperfect elasticity of perceived things. In later theories molecules still lack some of the properties of perceived things: they have no warmth, for instance, since this characteristic is defined only for aggregates of molecules.

So far no distinction has been made between real models and conceptual models, between everything from *objects trouvés* to things built as models in workshops, and imaginative constructions about which and with which one only thinks. . . . Theory can fruitfully be looked upon as the imaginative construction of models, according to well-chosen principles, and that, in many ways, the theory of 'Ideas', in Whewell's sense, is more helpful in the theory of theories and scientific method generally, than the logic of statements. . . .

2.1.3. Typifications

As a third sort of model for which source and subject are identical, typification is a model that represents a class, typically generated by some mathematical equation. . . . The typification has just as many properties, in the sense of determinables, as a typical member of the class it represents, but the determinate properties of a typification are arrived at by an averaging or similar mathematical operation on the set of determinates under each determinable, possessed by all the members of the class to be represented. . . . So, the typical family which consists of 2.63 . . . children, has characteristics which no actual human family could have. . . .

[Abstractions] can also represent classes. An abstraction does so because what may be omitted from the set of properties of the individual, from which it is modelled may be just the individual differences. If the purpose of the maker of the . . . model is to represent the class from which his course-subject is drawn, then he may simply omit those properties which differ from individual to individual, either by dropping out a whole determinable which fluctuates too widely, or by arbitrarily choosing a mean determinate, as [above]. [In starting with the source-subject identity], the model-maker has passed to a new situation in which his source remains an individual, but his subject has now become a class, of which that individual is a member. Since abstraction always involves some loss of properties from source to model, there will always be the possibility, with an abstraction, that the subject may be switched form the individual modelled to the class of which it is a member.

2.2. Paramorphs

The second main category of models [I will call] 'paramorphs. A paramorph models a subject, but its form and mode of operation are drawn from a source different from its subject. . . . The relation of a paramorphic model to its subject may be of three different kinds. To differentiate them, it is necessary to notice that paramorphs are usually constructed with the ultimate aim of modelling processes, and this is often achieved by actually modelling equipment, or naturally occurring things, in which the processes one wishes ultimately to model are occurring. The paramorph is needed, because, to operate with or study the real process raises difficulties of various sorts. This may be because the process in which the interest lies occurs in nature by some path which is currently unknown, as for instance when genetic factors were introduced as a model for the unknown process of the inheritance of Mendelian characters. Sometimes it is because the equipment which one wishes to use is awkward to work with, so one builds a model [for example] an electrical network . . . to study a hydraulic system. Sometimes it is because the mechanism of the process being modelled is complex, as in heat transfer across phase boundaries. Sometimes a model is resorted to because it is not at all clear where to look for the mechanisms of the process, for instance in the theory of human learning. Sometimes it is too dangerous to work with the real thing. Usually some event or state initiates some process which has some other event or state as outcome. The dimensions of variety of paramorphs arise because there are a number of different relations which may exist between the model and its subject. (1) The initial and final states of model and subject may be exactly alike, and the processes by which they were reached differ, as for instance in computer simulation of arithmetical calculation by a human being. (2) The initial and final states may only be similar, and that similarity may indeed extend only to the similarity between tables of numbers, as for instance in electrical simulation of hydraulic networks where what is similar is the structure of tables of numerical results derived from measurements of operations with each network. Let us call these models partial paramorphic analogues. The processes occurring in the model and subject are different as in case (1).

This sort of paramorphic analogue is exemplified by Bohr's atom. The process by which it stores energy, and later releases it, is modelled on both electro-magnetic and mechanical processes, that is, our conception of it is built up by drawing upon the concepts of electromagnetism and mechanics. But since it contains a mechanical impossibility: the 'motion' of electrons from orbit to orbit with loss or gain of energy, in no time at all, the process is, at best, an analogue of whatever process does occur in radiating atoms. However, input and output for Bohr atoms are identical with input and output for real atoms, namely electromagnetic radiation. The classical studies on the tube-boiler are a case of a kind of model, which I shall call a 'complete paramorphic analogue', where *both* model-process and model input and output are analogues of subject-process and subject input and output. The model is built by using sticks of sugar to represent the hot tubes, and the process of mass-transfer to circulating water, models the process of heat-transfer in the boiler. Finally, by a suitable transformation, the profiles of the sugar sticks, after running the model, can be transformed into the heat profiles of the tubes.

Though the processes in a paramorphic model and its subject, i.e. their manner of working, would usually be different, in those cases where the initial and outcome states are analogues but not identical, there could be the case of what I shall call a paramorphic *homologue.* The model and subject would have an identical manner of working but only analogous inputs and outputs. The idea of using living nerve cells for building computers would be a case in point, since the input and output would be only analogous to giving information and receiving reports, but the process by which the cellular computer and the brain found the answer would be the same, whatever that is.

The relation of paramorph to source is similarly threefold. I distinguish singly connected, multiply connected and semi-connected paramorphs. A paramorph is constructed or imagined as operating according to certain principles, commonly drawn from known sciences and technology, though sometimes, in works of the imagination, fudged up for the occasion, and vitally important this last is for science as we shall see. A paramorph, like the corpuscular theory of gases is singly connected, because the principles of only one science, mechanics, supply the definitions of the entities and the laws of the processes which constitute the model. Bohr's atom is multiply connected, since to construct it in the imagination, one must draw on the sciences of mechanics and electromagnetism, and the principles of these sciences are not explicable, one in terms of the other. Freud's 'psychic energy' mind model is semiconnected, because.. . . to construct it, one [not only draws] on some principles of energetics from physics, [but also from] processes occurring according to principles unknown to energetics, or any other Science. This does not, of course, count against the Freudian theory, *a priori,* because sometimes semi-connected paramorphs are just what give us a new scientific development, by suggesting the idea of a new kind of entity, or process.

A difficulty here concerns what is to count as <u>one</u> source. Only if that is clear can singly and multiply connected paramorphs be distinguished. I mentioned a weak criterion, immediately above, that the principles of the one science have not, as yet, been explained in terms of the other. There is a stronger criterion with which I propose to settle any question as to whether the entities and processes dealt with by two sciences count as two sources for a paramorph. It is simply to ask whether two individuals, one defined in terms of the one science, and the other in terms of the other, can be allowed to occupy the same place at the same time, and the one not be a part of the other. This would make electromagnetism and mechanics different sciences as sources for paramorphs, physiology and psychology different sciences, but chemistry and quantum mechanics branches of the same science.

3. THE PROBLEM OF REAL MODELS

Consider again the case of a [model] created by abstraction, and the example of a wire structure as a model of the vascular system. There is a case for saying that the wire structure itself is not the model but represents the model: that, as I put it in introducing this problem above, the model is the meaning of the wire structure. If the model is an abstraction from the subject-source then, in the interesting case, whole determinables are not represented in the model. For instance, the wire model does not represent any of the tensile qualities of the vascular system, but it has tensile qualities of its own. We are to ignore the tensile qualities of the wire model, because the projective conventions which together constitute the modelling transform do not give them a role, in the modelling. One may well now ask, 'What is it that lacks tensile qualities?' It certainly is not the wire structure. The spatial organization of the wire structure models the spatial arrangements of the vascular system, and that is all. So the model lacks colour, temperature, volume, and so on: all qualities both of the wire structure and the vascular system. This tempts us to ask, 'Is the wire structure really the model after all?' Perhaps it only represents the model, which is something abstract and mysterious. This would be a mistake. There is no other object involved but the wire structure, but it has to be understood in a certain way. It must be read, like a sentence, so that it is treated as having a certain spatial configuration, and nothing else, for the purpose in hand. But what makes things, unlike words, models, is a projective convention, that is, what characteristics are read off the object as the model serve their function through physical similarity to the characteristics of the object modelled. What makes a thing a word is an arbitrary convention.

4. THEORIES AND MODELS

The Copernican Revolution in the philosophy of science consists in the following ideas: theory construction is primarily model building, in particular imagining paramorphs. Imagined paramorphs involve imaginary processes among real or imagined entities. The crucial point for the understanding of science is that in either case they may invite existential questions, since unlike homoeomorphs, they introduce additional entities other than the given, provided it seems plausible to treat these as a casual mechanism. It is through imagined paramorphs and their connection with their sources, multiple, single, semi or fragmentary, that theoretical terms gain part, and a vital part, of their meaning. It is by being associated with a paramorphic model, in ways that I will go into later, that many laws of nature get their additional strength of connection among the predicates that they associate, that distinguishes them from accidental generalizations. A scientific explanation of a process or pattern among phenomena is provided by a theory constructed in this way.

Models, analogies and metaphors are closely related, though not identical tools for rational thought. At this point I want to draw a distinction, in passing, between models merely as the source of picturesque terminology, and models as the source of genuine science-extending existential hypotheses. A model for something, be it thing or process, can be described in the language of simile as a thing or process analogous to that of which it is a model. Thus electricity can be described as like a flow of something, indicating that electrical circuits have some likeness to hydraulic networks, that is, are analogous in certain respects to them. The fluid model provides an existential hypothesis, since it made sense, once, to ask, 'Is there an electric fluid or fluids?' In a case like this, we are not redescribing electrical phenomena in the language of hydraulics, we are offering the fluid as a causal mechanism to explain the electrical phenomena. This should be compared with an example like economic cycles, with its picturesque terminology of inflation and deflation, depression and boom. Somewhere in the background of this system of metaphors lies a model of the economy, and of trade, in which transactions are treated as the parts of some substance which can expand and contract, a model which no longer has an explanatory function. The model offers us nothing by way of explanation, and no existential hypotheses, but it does provide, in the system of metaphors, a picturesque terminology. Many metaphors are indeed just this, the terminological debris of a dead model. There has been a great deal of muddle about these distinctions, and some writers, particularly Poincaré, have not succeeded in keeping the cases distinct. At this point I am concerned solely with the use of models to provide hypothetical mechanisms, if they are paramorphs; or readily understandable analogues, if they are homoeomorphs. I want to turn now to a more detailed examination of the way paramorphs prompt existential questions.

Even for a singly connected paramorph the problem is not so simple. For the sake of exposition I shall take a straightforward example, the mechanical corpuscle as it figures, say, in nineteenth century models for chemistry, physics and other studies. Mechanics is the science of material things, in certain kinds of mutual interactions. Mechanics does give an account of the relations between velocities, masses, and so on of interacting bodies, but does not give an account of the interactions of their colours or temperatures. The colour and the temperature, for instance, simply do not figure among the mechanical parameters. So if a model is devised in accordance with the science of mechanics, the properties with which its entities will be endowed will be the mechanical parameters, and, in the first instance, and for a paramorph singly connected with mechanics, nothing else. But the ordinary things in the world of whose existence we can be sure have rather more properties than the mechanical parameters, though they do have those, too. The metaphysical question which has to be dealt with right at this point in the use of a single science for model-source, is this: what properties constitute the minimal set for us to be able to say that what we have is a thing existing in the world, *and forming one of the parts of which an ordinary thing is the whole, or a whole of which ordinary things are the parts?* Do the parts of a coloured thing have to be coloured? Does the whole, of which the parts bear structural relations to each other, which exhibit the relations of the law of entropy, itself exhibit that law? One answer to this question was provided in the seventeenth century with the doctrine of primary qualities, that the essential properties of things were their bulk, figure, texture (arrangement of parts) and motion, and that the properties of the whole were simply the summation of the properties of the parts. All other characteristics which things seemed to have were nothing but different ways in which the essential properties affected us. Other solutions have been given. But what we need to notice now is that without some metaphysical doctrine, even if only tacit, we cannot use the simplest kind of paramorph for its proper scientific purposes.

For multiply connected paramorphs, the problems raised by existential hypotheses are even more profound. If we suppose our imaginary model is perhaps a real mechanism, process, or thing, we have to ask whether nature, as we know it, can admit of such a thing. Sometimes this can be settled, in the negative only, *a priori.* Treated as a work of the imagination a centaur is a multiply connected model devised by drawing on both horses and men, but not drawing upon the characteristics of either completely. But a moment's thought on the parts of each that are conjoined, will show that a centaur is an anatomical impossibility. Where we do not have this easy way out we depend on what I shall call 'plausibility control'. This can be exercised through such questions as: 'Would existential affirmation of the entities and processes of the multiply connected paramorph introduce something into the world which, while not impossible, is in sharp conflict with the ideas currently held as to what there is?' The unpopular minority were right who said, of the Bohr-Rutherford atom, that it was a jerry-built contraption, on the basis of its incidentally requiring a massy particle, as the electron was then understood, to change position without traversing the intervening space or perhaps worse, to traverse it in no time at all.

This can be looked at in another way. For singly connected paramorphs there are only two cases. Either there is a full transference of properties from source to model, and this leads to philosophically uninteresting questions like 'Is there a planet between Mercury and the Sun whose influence explains the anomalous behaviour of Mercury?' Or there may be a partial transference. Here the existential questions raise more interesting issues, because the transference may omit only so many determinables without an entity losing substantiality. At some point we say that no existential question can be sensibly raised in the context, as for instance one might be tempted to say about point-atoms. In either case, we tacitly accept and operate with the standard received conception of an object, and in the latter, as I pointed out above, with the distinction between primary and secondary qualities.

Multiply connected paramorphs can lead to changes in these conceptions. The Bohr atom at least raised the question of the absoluteness of Newtonian criteria of existence. Perhaps entities of a new kind should be admitted to exist alongside those presently conceived possible, entities which did not obey the received principle, hitherto categorical, that one thing cannot be in two places at the same time. Part of the origin of the difficulties over electrons can be traced to the fact that they were multiply connected paramorphs, with, as it emerged, only *partial* transference of what seemed essential properties, from each source, the mechanical and electromagnetic respectively. The hypothesis of their existence would involve existential novelty. It would not be merely the addition of further items to a kind of entity whose existence was accepted on all hands only differing from standard examples of that kind in non-essential ways, typically only differing in size. Even though the acceptance of a multiply connected paramorph as a candidate for existence does raise problems they are nevertheless sometimes admitted. Electromagnetic radiation, viruses, quasars, electrons, the benzene ring, infant sexuality each had its conceptual origin as a multiply connected paramorph. Who would be so bold as to deny that there are such things?

Part of the difficulty of doing philosophy of science in terms of received logic, that is by concentrating on a particular aspect of the language of science, is that the principle of non-contradiction, or some similar principle or principles, serves as our only plausibility control, when we construct hypothetical mechanisms, that is in the language *genre,* define theoretical terms. Nearly all the difficulties of classical philosophy of science can be put down to this, and its associated bit of nonsense, that the sense of theoretical terms is a product of the sense of the observation terms to which they are, à *la* Carnap, reducible, plus their logical role as calculating devices. In short, plausibility control is exhausted, in this way of thinking, by the kind of observations already made, and the principle of non-contradiction. But in the new way of thinking, we can ask, 'How plausible is the model as a hypothetical mechanism?' even before the question of the reality of that hypothetical mechanism is raised. The plausibility question cannot be answered by any routine enquiry. Plausibility for a model is determined partly by the slowly changing general assumptions of the scientific community as to what the world is really like, partly by the way the model fits in to the particular circumstances for which it was created. . . .

4.1. Model Making Illustrated

N. Sutherland (1963) found that an octopus could distinguish between rectangles presented to it horizontally and rectangles presented obliquely. A triangle with apex up was treated by the octopus just like a diamond with apex up, while both were distinguished from squares and rectangles presented with two sides vertical and two horizontal. . . . The octopus cannot distinguish one oblique rectangle lying along a left to right diagonal from one lying along the opposite diagonal. . . .

The next step is to construct (in the imagination first) an [explanatory model] in terms of electrical circuitry, as a model of whatever neural mechanism might exist in the octopus, capable of performing the projections envisaged in the protomorph. The first of several such paramorphs, which the investigators invented, involved a square matrix of photosensitive cells, with a master circuit which collected and compared the total number of cells firing by rows, and by columns. In this example we have a two-step development of the paramorph through the prior invention of a geometrization. Once the paramorph is constructed, if it has existential plausibility, it is open to the investigators to suppose that it is the hypothetical mechanism, and then look within the octopus for something like this mechanism. In this particular case, the investigators did not find a mechanism like that supposed in the paramorph. They found something quite different, which banished both the paramorph and the protomorph from the explanation of the phenomena. The mechanism hypothesized simply did not exist.

4.2. Model and Subject

I have discussed, so far, how models are related to their sources, that is how models are created. But the function of a model is to form the basis of a theory, and a theory is invented to explain some phenomena. It has a subject. I turn now to the relation between model and subject, in greater detail. An iconic model stands in for the real mechanism of nature, of which we happen to be ignorant. That is its function and that is its importance. A model, considered from the point of view of its subject phenomena, beginning merely as an analogue of the real mechanism which is supposed unknown, can become, by a switch in the attitude of its users to its status, none other than a hypothetical mechanism, and, as such, models bear hypothetically to their subject matter, just exactly the relations that real mechanisms really bear to the phenomena for which they are responsible. These can be of two kinds.

1. Some state of the mechanism is causally responsible for some state of things which can be observed. For instance a particular set up of springs, gears, and so on inside a clock is responsible for the way the hands are oriented on the face, their position from time to time and their relative motions. Or a particular arrangement of molecules in the genetic material is responsible for a particular shape or colour or chemical composition of a structure of an adult organism. Or a particular energy transaction in an atom is responsible for the colour of the light emitted. Or a complex of repressed experience is responsible for a certain pattern of hysterical behaviour. Model states are linked to phenomena by hypothetical generative relations. In discussing the relations between the *sentences* used to describe the model and the sentences used to describe the phenomenal effects of these hypothetical states I shall talk, for this case, of a *causal transform*. There is, in fact, a partial sentential modelling between sentences describing the model and sentences describing the phenomena, but of course the mere existence of a formal modelling of the sets of sentences is neither here nor there for scientific purposes. The model, as we have seen, has to satisfy the requirements of plausibility control before it has scientific value.

2. Considered purely formally, causal transforms may not, at first sight, seem to differ from any other kind of transform, and to be able to be exhaustively expressed as simple conditional statements. But from a scientific point of view, the distinction is crucial. Where there is a causal relation the cause and the effect are independent existents, and the mechanism by which they are related can fruitfully be asked for, opening up a new dimension of explanation. In the other kind of transform, which I shall call a *modal transform,* the relation between the states of the model, considered as a hypothetical mechanism, and the phenomena, is not such as would give them independent existence. The state of the model is existentially identical with the phenomena. For instance, from an existential point of view, reflecting light of a certain wave-length and the being coloured a certain hue of a surface are identical states of the world. (This needs refinement, but the necessary conceptual machinery to handle this sort of case properly will be developed in a later chapter. The point is here illustrative only.) Nevertheless, the wave model for light is a considerable scientific advance. It does explain a great many otherwise inexplicable phenomena.

A brief sketch of some of the interesting philosophical issues raised by modal transforms will be in point here. They are connected both with the role of theory in taxonomy and the metaphysics of taxonomic theory, and with the problems raised by decisions to carry out wholesale recategorization of things in the world. Consider the discovery that crystals of common salt are cubical lattices of sodium and chloride ions. Here surely we have a modal transform between the shape of the crystal and the structure of the lattice. But to describe a substance as crystals of common salt invites one, in domestic situations, to classify the substance along with peppercorns, bay leaves and parsley, and not with other things. To describe it as sodium chloride invites one to classify it along with the electrovalent compounds, a class which quite excludes peppercorns and bay leaves. In this example the problem as to where the weight lies in taxonomy is pretty easily solved simply by referring to the context and purpose of classifying. And to the question 'What is common salt really?' The answer 'Sodium chloride' seems right. But if we apply the notion of modal transform, say in the province of the science of psychology, the taxonomic weight cannot so easily or so plausibly be placed upon the technically most advanced concepts. In many cases, the identification of a certain state of the nervous system as being that in which the organism hears something, say, depends upon that organism identifying its state as 'hearing something', and this applies as much to animals making auditory discriminations as to men listening to symphonies. So that the taxonomy, or general system for classifying states of the nervous system for taxonomic purposes seems to depend upon a mode of description which is the least technical of the modes involved.

The total theory can be thought of metaphorically as a statement-picture complex. The iconic model is a 'picture' of a possible mechanism for producing the phenomena. It has already been explained in detail, above, how we come to know what the iconic model is like and how it works, deriving our knowledge from the parent model or models: the source of the model. By considering the sentences of the theory, abstracted from the total external expression of the statement-picture complex, that is by ignoring the model, picture or diagram of the state of the world, the following logical relations emerge:

(a) If the relation between the description of the model and the description of the phenomena looked at from the point of view of the complete theory, is a causal transform, then the sentences which express this relation will appear as conditionals, asserting that if the model-state obtains then the phenomenal effect of that state will come to be.

(b) If the relation between the description of the model and the description of the phenomena is a modal transform then the sentences expressing this relation will be bi-

conditionals, asserting that if and only if the model state obtains will the phenomenal aspect of that state come to be.

In the case of the modal transform, there is no separate question as to the existence of the hypothetical mechanism and its states which the model represents, for they are the same states of the world looked at from a different point of view. But a causal transform links sentences describing quite different states of the world. Since the iconic model is introduced just because we do not know the mechanism of nature at that point, it can come to be looked upon as a hypothetical mechanism. Then the question arises, does this mechanism exist? Or one something like it? Or is what is really responsible for the phenomena something so different from the mechanism represented by the model as not to allow any modification of the original model as a version of reality? For an example consider the distinction between our being permitted to go on saying that chemical atoms exist, while continually, and drastically, modifying our beliefs about their natures, and our not being permitted to go on saying caloric exists, now that heat transfer is understood as a transfer of energy. William Odling gave a lecture towards the end of the Nineteenth Century, in which he made out, partly as a *jeu d'esprit,* that phlogiston did exist, only people had come to call it 'energy'. We shall look later into the conditions under which we make this distinction in judgement.

5. THEORY AS HYPOTHESIS GENERATOR

A theory then should prompt the consideration of four kinds of hypotheses, which are separated off by both logical and epistemological distinctions.

1. *Existential Hypotheses;* such as 'There are molecules', 'There are no pores in the septum of the heart', 'There is a planet between Mercury and the sun', 'There are complexes'. Each one of these hypotheses prompts its appropriate question, like 'Are there molecules?' the attempt to answer which generates a particular kind of experimental research. I shall emphasize later that not all questions in the form of existential hypotheses are properly answered by an experimental technique, since some questions of this form are rather invitations to change our categories than to look for new things, or properties, or processes amongst those already acknowledged to exist.

2. *Descriptions of the model or hypothetical mechanism;* such as 'Molecules are moving', 'Complexes are formed by the association of traces of repressed experiences'. The empirical pursuit of answers to the corresponding questions cannot be undertaken until questions as to the existence of the *entities* in question have been settled. Since, as will emerge, settling existence questions cannot be done in isolation from the satisfying by some entity of a certain description, some hypotheses in this class are decided along with the existential questions.

3. *Causal Hypotheses;* in such questions as, 'Is gas pressure caused by the impact of molecules?', 'Is hysteria caused by repressed experiences?' the power of a hypothetical mechanism to produce the phenomena is queried. The attempt to answer such questions gives rise to some interesting questions to which we shall return again and again, from different points of view.

4. Finally, in questions like, 'Is gas temperature really only another way of looking at mean kinetic energy of the molecules?', 'Is a slip of the tongue really an admission of guilt?' we are invited to consider *modal transforms.* Very complex issues are raised in their epistemology. Since they have the form of biconditionals, descriptions of a situation in terms drawn from the vocabulary of observation or from the vocabulary used to describe the model, must be true together. Yet the model seems to introduce a new range of things and happenings, which, at least in the first instance, we cannot study independently. If all the

relations between hypothetical mechanisms and phenomena were modal transforms we should be as sure of the truth of theoretical statements as of those describing the phenomena. But then explanation by adversion to hypothetical causal mechanisms would be no more than a redescription of the phenomena. We shall see that any paradoxical air which might derive from the presence of both modal transforms and causal transforms in the same theory, as in the kinetic theory of gases for instance, can be resolved in those cases where the elementary components of the hypothetical mechanism are also the elementary parts of the entity whose behaviour is being explained. Gas molecules are the elementary parts of gases as well as the elementary components of the hypothetical mechanism responsible for the holistic behaviour of gas samples. . . .

5.1. The Analysis of Some Theories

The model of an unknown mechanism, whose invention is the core of the creation of a theory, is paramorphic with respect to that unknown mechanism, since in the absence of knowledge of that mechanism we can hope only for some kind of parallelism of behaviour. But the central model is homoeomorphic with respect to each of the sources upon which it is modelled. This must be shown in detail in some examples.

The phenomena which Darwin attempted to explain can be called generally 'natural variations'. This includes, but is not exhausted by, both the differences between the character of populations now living and plant populations of the past, as revealed by geologically preserved remains: and the differences contemporarily observed between similar populations inhabiting different environments. From this static picture, Darwin drew the dynamic conclusion that there was variation in the character of populations of plants and animals in nature. The mechanisms by which this happened were unknown. So in the first five chapters of the *Origin of Species* (1859), Darwin built up a picture of a hypothetical mechanism which would explain the changes he believed occurred. In short Darwin invented an iconic model, paramorphic to the real but unknown processes by which biological change occurred. Quickly moving from model to hypothetical mechanism, he called it 'the process of natural selection'. Its sources were the known process of domestic selection by which domestic variation was brought about, and the bellicose metaphor of the struggle for existence.

Natural selection is a homoeomorph of domestic selection. It is a selection operation but without the deliberate acts of the breeder. In short, it is the kind of homoeomorph I have called a teleiomorphic abstraction. But a selection operation without some agent which diminishes the chances of reproduction of those particular plants and animals whose structure, or constitution, or patterns of behaviour is unsuited to their environment, is empty.

The deficiency was made good by Darwin by developing a second source of concepts for the natural selection model. As a paramorph, it is multiply connected. The new source was the theory of Malthus on the effect of population pressure on the division of resources, and the law that in human societies a geometrical population increase is accompanied by only an arithmetical increase in available material resources for that population. In general, therefore, there must be competition, which supplies the source for the agent conception in the

Phenomena	Transforms	Model of Unknown Mechanism	Sources
Natural Variation	The cause of natural variation is natural selection.	Natural selection by the struggle for existence. M	1. The causal structure of domestic variation as selection. 2. Malthus theory.

Table 1

Darwinian theory. And, of course, the 'struggle' conception is an abstraction of Malthus' picture of human life, since, for example, the differential growth of two species of grass on a particular soil is only competitive in an abstract and metaphorical sense. These two sources combine then in the creation of Darwin's model of the actual causal mechanism responsible for organic change. Schematically the analysis of the theory [is laid out in Table 1]. *M* is a paramorphic model (a partial analogue multiply connected) with respect to the unknown mechanism of change. M is a homoeomorphic model . . . with respect to domestic variation and selection, and with respect to Malthus' theory.

Notice that Darwin is postulating a novel kind of process in nature, never before thought of, and up until very recently, never observed. Natural selection was as much a theoretical entity and as unobservable as the electron, for while the latter is too small to be observed, if indeed it is thing-like at all, natural selection is too slow.

Another example is the theory devised by social psychologists in which a model of rewards, costs and investments is offered as an account of various alleged social processes, such as the conferment of, or recognition of, status among people. The phenomena are the ranking of people by others as to their status, and for the purposes of the example, we will not pursue the question as to what that rather vague concept is supposed to cover. The model is constructed for the unknown mechanism which determines (if there be such) the according of status. Its source is the simple economics of costs, rewards and investments. For example, it is said by Secord and Backman (1964, p. 297) that '*distributive justice* is obtained when the outcomes or profits of each person — his rewards minus his costs — are directly proportional to his investments'. This model is then used with respect to the phenomenon of two people according each other equal status according to the 'formula'

'My investments — my rewards minus costs
His investments — his rewards minus costs'.

Here we have a case of a pure modal transform. There is no existential question raisable as to the realities of any quasi-economic process lying at the back of the according of status. The source of the model is economics. The model then is paramorphic with respect to its subject, and homoeomorphic with respect to its source. But in this case, it is not a teleiomorphic abstraction, but an idealization, since though investments, rewards and costs are genuine economic concepts, they are being used here in an idealized way. This example will serve to introduce an important new concept into the discussion. There can be no question of the model employed in this theory becoming a candidate for reality, and so a hypothetical mechanism. Instead, since *all* its connections with phenomena are mediated by modal causal transforms with phenomena. The function of such a model cannot then be to generate existential hypotheses, so it does not form part of an explanation in the usual sense. Its function, as is clearly evident in the example I have quoted, is to illuminate the facts, to throw them into a new light, to make them more readily memorable. Its function, in short, is *literary*. This is not to denigrate the theory, but to place it in its exact logical role within the structure of science. For any theory with only modal transforms, we can look only for a function as metaphor and not anything else. In more formal sciences, other heuristic advantages may derive from such theories, say ease of mathematical handling.

We have noticed that the statements describing each of the components of theory bear sentential modelling relations with respect to each other, and that the transforms which give structure to the whole are each an open set of sentences. What transforms are admitted depends partly on the degree of likeness and unlikeness between homoeomorph and its 'source-subject' which relates the model in the theory to its parent, or parents, and partly on the plausibility of the model as a putative hypothetical mechanism, that is partly on the degree to which we think it likely that the unknown mechanism for which it is an analogue is like it. *If all the transforms are modal,* then the sentences describing it become metaphorical redescriptions of the phenomena, and we commit ourselves to an effective total *unlikeness* between mechanism and model. The logical structure of the sentences describing the various components of a theory like that of Secord and Backman is now very complex, since it contains one inductive (analogical) relation between sets of sentences, namely that between the parent and the model, and one metaphorical, which comes not under logical scrutiny, but under quasi-literary scrutiny.

In theories which incorporate causal transforms, and in which the hypothetical mechanism is a possible and plausible causal mechanism, there are two inductive steps, whose validity depends upon our judgements of the likeness and unlikeness of the model and its parent, and our judgement of the plausibility of the model as a hypothetical mechanism. In neither case can the whole structure of the sentences in the expression of the theory be collapsed into a deductive system. At best it can consist of three deductive systems: that of the laws of phenomena, that of the 'laws' of the model, and that of the laws of the phenomena upon which the model is based. The sentential modelling relation between the three systems will be through a set of rules of transformation (though that itself is a modification of the real role of transform statements as causal hypotheses and modal transforms) which cannot be fully completed.

5.2. Models of Models

The account of theory given so far has assumed that the development of a model from a parent derives from laws of phenomena, known behaviour of known things. But other origins for models are possible. Sometimes the source of a model may be itself a model. Drude's theory of electrical and heat conduction is an example of this. In that theory the behaviour of metals as conductors of heat or electricity is explained by the help of a model which is itself based upon the molecular model of gases. A piece of metal is imagined to be a container in which electrons are in random motion in a manner like the molecules, in a container, which simulate the behaviour of gases. Conceiving of electrons in this way, and devising laws for their behaviour in such circumstances by analogy from the laws of gases, the model is found to simulate the actual behaviour of metals with respect to heat and electrical conductivity. In this case, the electron model is paramorphic with respect to the conductivity behaviour, and paramorphic with respect to the gas molecule parent, since there is no sense in which the electrons trapped in a metal are a gas. Considered with respect to the gas laws and the gas molecule, the electron model differs in source and subject. However, the gas molecules are supposed to be small particles in motion, and considered with respect to particles in motion (their source and their subject is the same) they are modelled on particles and models of particles, supposing as we now do, that a gas does consist of a swarm of particles. As a model for the electron 'gas', the swarm of molecules is paramorphic, as a model of a swarm of particles, it is homoeomorphic.

CHAPTER TWO

FORMAL MODELS

1. THE USE OF THE CONCEPT OF 'MODEL' IN LOGIC AND MATHEMATICS

1.1. The Standard use: models and the interpretation of calculi

The concept of 'model' is used in logic in the following way: the formulae of any abstract calculus are in need of interpretation into meaningful sentences. This can be done by choosing a semantic basis which will consist of some suitable 'universe' of objects with their characteristic properties and relations. This 'universe' can be made to serve as a source of meaning for the variables, connectives and constants of the calculus. I shall call such a set of objects and relations a 'semantic basis'.

In logical studies the meaning-creating relation through which a semantic basis becomes the source of meaning for the symbols of a calculus seems usually to be taken to be simple denotation. This step in the use of a semantic basis is rarely highlighted and even more rarely discussed explicitly. I have been unable to find any commentaries on the consequences for the interpretation of logic of building meaning in this way. Logicians have no settled terminology to describe this initial step.

A semantic basis is called a model for the calculus if and only if the sentences of the interpretation, when used to describe that semantic basis, are all true. There is, so to say, an 'internal' relation between the model and the statements made about it with the interpreted sentences of the calculus. The logician's usage is nicely exemplified by Bridge (1977, p. 34) who says, 'When O [a set of formulae interpreted by reference to U] is valid in U we say U is a *model* for O'. According to Bridge the letter 'U' denotes a 'relational structure' which consists of a set of elements, some of which can be individuated, and various relations that obtain among the elements. These relational structures are also sometimes called 'realisations'. It is important to emphasize that 'U' is entitative, not sentential. The word 'valid' in Bridge's definition comes as a surprise. One would expect that 'valid' would be reserved for correctly performed logical operations, with 'true' as the appropriate term for interpreted formulae used to make statements about the semantic basis. Bridge offers 'true' as a synonym for 'valid' (1977, pp. 33-34). I shall use 'true' in this context.

There are other unfortunate and confusing choices of terminology appearing in the writings of logicians. By flagging a particularly troublesome one I hope that confusion can be avoided. It concerns the use of the word 'interpretation'. For example, Mendelsohn (1987) says, 'an interpretation M is said to be a *model* for a set (A) of wffs if and only if every wff in (A) is true of M'. In English the word 'interpretation' is reserved for the result of interpreting something, not as that which facilitates the interpretation. In the next paragraph Mendelsohn goes on to say that an interpretation of M consists of a non-empty set D, called the domain of the interpretation, with relations, operators and fixed elements. This seems to mean that an interpretation in his sense, is entitative, not sentential. In short it is what I have called a semantic basis.

Logicians, as far as I can discern, have no interest in the external liaisons of any of the semantic bases which they use in their studies. Adequacy for the job in hand, which is more often than not testing for the consistency of the formal system in question, seems to be the sole criterion for choice of semantic basis. There is no attention to ontological considerations with respect to the viability of a semantic basis that will serve as a model.

1.2. Models in physics, taken formally

Models in physics can be seen as a species of the same genus of models as those that are used in logic. In physics the statements that can be made by the use of the sentences of an interpreted mathematical formalism are true of the semantic basis by the use of which the formulae of the mathematics were interpreted. The semantic basis must therefore count as a model, in the logician's sense, for the abstract mathematical formalism. For instance, the statement

> P (pressure) ≅ V (volume) = 1/3n (molecular number) ≅ m (molecular mass) ≅ c^2 (root mean square molecular velocity)

is true of a swarm of point molecules. But the interpretation of the formula 'pv = 1/3nmc^2' as a sentence capable of being used to make the above statement requires the variables and constants of that formula to have been given meaning in terms of a 'world' of imagined material entities, the model. No gap could open up between the model and the statements of the theory, taken in its usual sense as a discourse, for which, as the relevant semantic basis, it provides the interpretation. If one or more of the statements made with an interpretation about a semantic basis should turn out to be false, it shows that that semantic basis ought not have been used to interpret the abstract formulae of the mathematics of the theory. False statements can appear in physics because some of the statements of the theory, the interpreted calculus, can be subject to experimental trial. In these circumstances physicists change the semantic basis in just such a way that a revised version of the original statement, seemingly empirically true, is also true of the new basis, which then becomes the current model. The basic principles governing the relation between abstract calculus, interpreted sentences and semantic basis as model are the same in physics and in model-theoretic studies in logic.

There are, however, some important differences. In the construction of a physical theory the model is almost always picked before the mathematical version of the theory is created. More often than not the abstract formalism of the mathematical version is developed just by describing the pre-existing model in formal terms. From a logical point of view the molecular model is the semantic basis for interpreting formulae such as 'pv = 1/3mnc^2'. But from the point of view of scientific methodology and the history of the kinetic theory the model was proposed first, probably most explicitly by Francis Bacon. It was progressively more carefully formulated and fully described in mathematical terms by Maxwell.

I have already remarked on the fact that logicians show little interest in the ontological or 'external' aspects of their models. A closer examination of how semantic bases are actually used in physics reveals that they are subject to external constraints. Physicists require that a theory should be physically meaningful: that is, the model for a theory should be physically plausible. Of all the range of entitative systems any one of which could be the semantic basis for the interpretation for the abstract mathematics of a theory only those models are selected which are similar in physically significant ways to entitative systems which are known to exist and about which physicists already know a great deal. The current model for electric circuitry is a special case of a hydraulic system. About such systems a good deal is known. This has been called the 'plausibility constraint' on the choice of models for physical theory. Hesse refers to it as the 'material analogy'. From a philosopher's point of view the plausibility constraint can be further analysed as I shall demonstrate in the next chapter in which I shall show how the ontology of specific models is constrained by their places in type-hierarchies. From my point of view the only useful borrowing of the concept of 'model' from the formal sciences is from model-theory in logic. However, there is another use for the term 'model' in mathematics, which has been taken up by some philosophers of physics. Mathematicians sometimes call one formal system a model of another if there is a mapping which correlates the formulae of one system with those of another. There are 'models' of this sort to be found in the discourses of physicists (Redhead 1980). However, it is difficult to see that they have any significance with respect to substantive issues in the philosophy of science. The fact that one can find rules to read one set of formulae into another parallel set tells us nothing about the possible meanings of sentences which could be derived from either of these sets of formulae. It tells us nothing about the physical plausibility of either. However, if the parallel set already has a model in the sense of a semantic basis it seems clear that the mapping between the sets of formulae would, more often than not, permit transitivity of modelling, *ceteris paribus.* Whatever is a model (semantic basis) for one system could also serve as a model (semantic basis) for the other. For instance, the formal mapping of the laws of hydraulic networks on to the laws of electric circuits permits the use of the 'current' model as a semantic basis for the latter.

The standard use of the word 'model' is exemplified by Del Rey (1974). 'A simplified treatment based on a model involves replacement of the actual physical system (say a molecule) by a simpler one (the model) which is treated in a quite rigorous way.' However, Redhead's account of the use of a 'model' to achieve the virtues of simplification (Redhead 1980, p. 147) runs as follows:

> In such a case [a theory too complicated for easy development] a simplified model M may be employed ... M plays the role of an *impoverished* theory, the important ingredient being that M and T contradict one another.

Only propositions can contradict one another. Redhead's use of the term 'model' is clearly not the standard use. It seems to belong with the specialised use in mathematics. Redhead uses the term in its common use just once. On pp. 149-50 (n. 5) he writes of the 'billiard ball model of a gas', though he does not continue in this vein. One can focus one's attention on a closed cycle of mathematical forms to the exclusion of the investigation of what gives them meaning as physical theories and what determines their plausibility as statements about the world. But this is to opt out of the philosophy of physics altogether.

2. SUMMARY OF ARGUMENT SO FAR

Scientific discourses, fragments of which are picked out as theories, are about models and get their meaning from models. Models, in this sense, are a special case of the kind of models that are used by logicians. There are many similarities between the use of models in physics and their use in logic, but there are also many differences.

There are at least the following similarities:

(1) The interpretative role: both provide a 'vocabulary' with meaning, that is transform a formal or mathematical system into a theory: that is, into an intelligible discourse.
(2) The internal coherence role: in both cases statements made with the interpreted sentences of the theory are true of the model. Semantic bases leading to interpretations for which this condition is not met are just not models for that theory.

There are at least the following differences:

(1) Any relational structure which satisfies 1 and 2 above is a model for a logical formalism, but

models in physics are more tightly constrained.
(2) A semantic basis is only a model in physics if it is physically plausible relative to the common ontology of a certain epoch.

(a) The model must be abstracted from a determinate source.

(b) The source must be of a type with known entitative systems. . . .

3. THE 'SET-THEORETIC' OR 'NON-STATEMENT' ACCOUNT OF MODELS AND THEORIES

. . . [There] is no fundamental distinction between observational and theoretical concepts. Which category a given concept will fall into at any moment in the history of a science depends on historical accident. In my account of science the distinction 'observation/theory' plays no fundamental part. However, the distinction played a very large role in the attempts by the logicists of the 'Vienna' school to give a formal account of science. The positivism which infected their thought meant that the meanings of expressions which were not instantiated observationally posed some sort of problem. The problem had a logical core since the logicist account of theories was based on the ideal of an axiomatic structure, and theoretical terms would surely be characteristic of the axiomatic level of the deductive hierarchy of law statements that constituted the theory, The 'non-statement' account of theories, which, when shorn of its pretentious symbolic garb, is in most respects more or less identical with the naturalistic account, developed out of an attempt by Sneed (1971) to resolve the issue of the status of theoretical concepts in physics. Since I have no place for the distinction I shall merely note Sneed's version of it in passing, and make only the most minimal comments. My interest is in the question of whether the formal treatment of theories as sets of models has any advantage over the naturalistic treatment from which our semi-formal account is derived.

As formulated by in Stegmüller (1976, p. 41), Sneed's criterion for theoreticity runs as follows: 'quantities or functions whose values cannot be calculated without recourse to a successfully applied theory are theoretical in relation to that theory.' If we need the theory of chemical ions to calculate the value of a transport number in physical chemistry, then the concept 'transport number' is theoretical. Unfortunately the example in Stegmüller uses to illustrate the point is badly chosen. He supposes that one would need 'classical mechanics' (though he does not say which version, and that matters!) to give a numerical value to 'mass' and 'force' but would not require it to give a numerical value to 'position'. How 'position' would be measured without recourse to an inertial frame escapes us. So the example is unhappy, but one gets the general idea. However, this old debate is irrelevant for my discussion since we simply do not recognise the observation/theory distinction as having any semantic, ontological or epistemological significance.

The basic idea behind the 'non-statement' account of science is that we could imagine the world to be made up of an indefinite array of systems, each of which could be idealised as a set-theoretic structure. Each system could be presented as a set of elements and relations. Each such idealisation could serve as a model, in the logician's sense, of some abstract mathematical structure. To each abstract structure will correspond a set of idealised systems, the intended applications. A theory, on the non-statement view, consists of just such a structure and its intended applications. It does not consist of a deductive hierarchy of sentences. As Sneed remarks (1971, p. 258), '... we must begin to look at the phenomena "as if" they were models (in the sense of mathematical logic) for abstract structures' and the steps to achieve this 'already entail a significant amount of idealization.' Thus far the basic theme of the non-statement account is identical with the naturalistic treatment, but for one important detail. What I have called the 'cognitive object' or 'content of a theory or theory-family' — that is the set of models for the theory — is called the 'theory' by the non-statementists. Nothing but confusion can arise from a terminology which tempts a reader to confuse a discourse with its subject matter. I shall retain the ordinary usage of the word 'theory': that it is to be used as a word for a description of some one of the set of models—that is, taken to be the most verisimilitudinous at some moment in its evolution.

Consider a formulation of the non-statement view as it is to be found in Stegmüller (1976). I start with a reminder of the new use for the word 'theory'. To avoid misunderstandings I shall henceforth use the expression Sneed/Stegmüller or SS-theory for a mathematical structure (to be referred as MS) and its intended applications (IAs). The non-statement idea can be expressed in terms of three main notions:

(1) The 'matrix' of an SS-theory is the set of entities which may have the structure MS.
(2) The 'frame' of an SS-theory includes the set M of all actual models of MS (in the logician's sense of 'model') together with Mp, the set of all possible models of MS. In addition the frame includes the set of all conceivable applications, Mpp, and a function 'r' which serves to differentiate the set Mpp into two parts, one theoretical and one non-theoretical in Sneed's sense of that highly polymorphous term.
(3) The 'core' of an SS-theory is simply the frame of the theory plus a 'constraint'. A constraint is a function which eliminates certain intended applications from the set of possible models, essentially those picked out in the frame as theoretical.

If the non-statement view of science had been confined to principles (i) and (ii) above, and care taken not to be misled by its advocates' eccentric use of the word 'theory', there would be little to differentiate it from the naturalistic analysis, even though it is presented at a very high level of generality.

I shall now turn to examine the non-statement view in its full form. There is one obvious advantage, illustrated in a paper by Diedrich (1989). Since it is presented at a very high level of generality it should represent the root ideas in an adequate account of any scientific enterprise. But a high level of generality has certain disadvantages, particularly an inevitable vagueness when one tries to apply it in concrete cases. This is a general problem with the use of formal representations of cognitive practices, for instance the impossibility of representing the various modes of generality expressible in ordinary language in terms of the standard quantifiers of first-order predicate calculus.

A more particular disadvantage of the SS treatment *vis à vis* the naturalistic analysis of model use is the inversion of the actual order of genesis of physical theories. In the SS treatment the mathematical structure is among the givens and the problem, if I may put it this way, is to find the best model. But in real science the model or models are in hand and the problem is to find the best mathematical structure to capture their main structural features. Connected with this weakness is the inability of the SS treatment to express the way that sets of models for theories (in the traditional sense) are ordered. There is nothing in the SS view which would tell us that van der Waal's model for the gas laws is more sophisticated than that of Amagat. Connected again to both of these shortcomings is the failure of the treatment to express the dialectic interaction between models and mathematical structures which is the heart of the process of theory development. Usually, though not always, theoretical innovation and/or experimental discoveries pre-exist transformations in the mathematical structure which is altered to accommodate them. In my opinion these disparities and lacunae are serious disadvantages in a program which purports to capture the essence of theorising in the physical sciences.

But there is a much more serious problem with the SS-treatment. It seems quite clear that it presupposes a covert positivism. It purports to give a formal account of the empirical content of theories in physics. So does my treatment in terms of type-hierarchies. But there the resemblance ends. I can bring out the difficulty by focusing on the ontological implications of Stegmüller's actual formulation of the SS point of view. He refers in many places to 'classical mechanics'. There is, as anyone who has taught physics knows, no such thing, at the level of physical theory. There are at least the Newtonian, the Hamiltonian and the Hertzian formulations of a genus of theories that could be loosely called 'classical mechanics'. These formulations are not notational variants of some common theory. They are ontologically quite distinct. The Newtonian version is based on an ontology of forces, atoms and a (notorious) container version of the absolutist view of space and time. The Hamiltonian version is based on an energeticist ontology, while the Hertzian way of formulating mechanics makes no use of energy or force, and is a pure mechanics of mass-points. And there are other versions too. For us this means that the three versions are different theories and in particular they must have different empirical content. How could they be elided as one theory (in the SS sense) with the same empirical content, as Stegmüller and Sneed seem to imply in their use of the expression 'classical mechanics' as if it denoted one theory? Seemingly only under positivism. Each theory generates the same sets of numbers representing the results of actual or *Gedanken* experiments in mechanics, though by very different routes. Sneed's remark about 'phenomena', which I have already quoted, leads us to think that there is indeed a covert positivism embedded in the SS point of view. I would like to take the greatest possible distance from the idea that the empirical content of a theory is to be confined to what is non-theoretical either in Sneed's sense or in the observation limit sense of the older positivists. In my view Sneed's distinction between theoretical and non-theoretical qualities and functions (to use Stegmüller's phrase) is defective. It is no more sustainable in actual applications than was the logical positivist distinction based upon their meaning criterion. If the 'ontologically reductive' aspect I have detected is indeed a feature of this point of view, it accounts for the attraction the non-statement view has had for such quasi-positivists as van Fraassen. In the end all that the SS view amounts to is a high-tech version of the old point that a physical theory and the world meet in isomorphisms between sets of numbers, those generated by the relevant apparatus and those generated by the theory. In this banal observation, note the crucial importance of mappings between mathematical structures and their models, as intended applications.

4. *ICON* AND *BILD*: HERTZ'S ACCOUNT OF MECHANICS

The double interpretation of certain cognitive objects used by physicists, as a formal treatment of the content of physical theory, is by no means original to the present generation of philosophers of science. It was the central insight of Hertz's philosophy of physics and descended directly to the 'picture theory' of Wittgenstein's *Tractatus.* 'We form for ourselves images *(Bilder)* or symbols of external objects; and the form which we give them is such that the necessary consequents of the images in thought are always necessary consequents in nature of the things pictured' (Hertz 1894, p. 1). According to Hertz such 'images' must be permissible, appropriate and correct. That an image is permissible 'is given by the nature of our mind'; that image is the more appropriate which 'pictures more of the essential relations of external things'; an image is incorrect if 'its essential relations contradict the relations of external things' (Hertz 1894, p. 2).

What exactly does Hertz mean by an 'image'? Clearly an important constraint on a scientific image is the set of laws governing the behaviour of phenomena, and the hypotheses introducing the unobservable entities necessary to complete the image given empirically, together with what

Hertz calls the 'normal connections' of things (Hertz 1894, p. 28). What could a row of symbols, a formula, and the motion of a body relative to some frame of reference have in common so that the former could be an image of the latter? Hertz seems to have had in mind a formal correspondence, some kind of isomorphism of structure. Such an isomorphism would need to be mediated somehow. One possibility is that a formula is a *Bild* of a physical process if there is a one-to-one correspondence between the set of consequents of applying the formula in some specified conditions, an ordered sequence of numbers, for instance, and the set of consequents of running the process in some apparatus yielding another set of numbers, the numerical measures of the successive states of that physical system while the process is running. In this way the formula '$s = ut + 1/2gt^2$' could be an image of the falling of a weight in an Atwood's machine.

I shall retain the German word *'Bild'* for this very abstract sense of picture in which the picture-world relation is mediated by a formal mapping between ordered sets of numbers. We could use the Greek word *'icon'* for the more concrete pictures that are created by the kind of modelling [associated with the construction of theories]. By extending the concept of 'picture' in this way I too can offer a general account of theorising without abandoning the insights into detail that were derived from the naturalistic approach. Just as the ionic theory of chemical reactions maps phenomena onto states of affairs specified in a common ontology (irrespective of any contingent constraints on observability), so theories like Galilean kinematics, special relativity and Newtonian dynamics also function as just such maps. The sparse conceptual structure of the various forms of mechanics is the result of the collapse of some of the distinctions drawn in my general account of theorising into one another. In the naturalistic account I identified two 'regions' or submodels of the basic iconic model, on which a theory is dependent. There was a descriptive model abstracted from observed phenomena. And there was the explanatory model by which a physicist could anticipate some form of human contact with an as yet untouched realm of physical beings. In classical kinematics and in special relativity nothing is hidden. Descriptive model and explanatory model collapse into one another, or to put it another way there are no differentiable regions in the general models or *Bilder* of kinematic phenomena constituted by the formulae of classical kinematics or special relativity. In the descriptive account I distinguished between statements describing iconic models and those models. In the Hertzian treatment even that distinction collapses, since the formulae which serve as the sentences with which kinematic statements can be made are also constitutive of the model. The same collapse is to be seen in the various versions of classical dynamics. Though each draws on a different common ontology, in the Hertzian interpretation the set of formulae is the model, and the theory-to-World mediation by isomorphic sets of numbers is just the same.

There is a danger for philosophy in using mechanics as the prime example of a physical theory. The textbook presentation of a system of mechanical laws as a deductively organised ordered set of formulae looks very much like an example of the Cartesian or deductive-nomological account of physical theory. Hertz's treatment of theories in physics as abstract *Bilder* permits a view of the structure of many important physical theories as collapsed versions of the structure revealed in the general naturalistic analysis of theorising.

5. SUMMARY

(1) In general a model is related to its description 'internally.' An entitative interpreting domain which is to serve as a semantic basis for a formal structure is a model for a theory only if the sentences created by interpreting the formulae of the theory by that semantic basis can be used to make true statements about the domain. The dynamics of theory development largely

controlled by this imperative. Theory and model are continuously mutually adjusted, in so far as that is possible. (2) However, in physics models are subject to an ontological constraint. They are related to the states of affairs they purport to represent 'externally.' Issues of faithfulness and degree of resemblance to their external correlates can be raised. (3) In the naturalistic analysis of model use, we see that models are created for practical purposes and then described in a theoretical discourse, which comes, so to say, already interpreted. The main constraint is physical plausibility. (4) In the logical analysis of model use we see that models are introduced to interpret a pre-existing abstract formalism. Models in this context are constrained only by the internal requirement of a common formal structure. In real science models are types in a type-hierarchy.

CHAPTER THREE

PROPERTIES AND IMAGES

The problems of scientific progress, of conceptual development, of 'external' influences on science, of the relationship between science and the social order, cannot be properly enquired into unless an adequate theory of scientific thinking and its relation to its subject matter, as well as an adequate theory of society, has been developed. I hope to show in this paper how theories of science and theories of society which might be adequate to each, must develop together. The inter-linked theories I shall be proposing involve a number of new and unfamiliar ideas, both about the nature of scientific thinking and about the world to the understanding of which that thinking is directed, as well as the re-emphasis of some well known ones.

Many of the difficulties that beset the philosophies of science and society derive from the continued acceptance of unthinking assumptions about the nature of science and the nature of the social world. It is still often assumed that the world to which science is directed is a flux of atomistic events independent of the conceptual structures which we bring to its understanding. At the worst, this point of view generates a corresponding atomism in language and concepts in that the atoms of the world are thought to be matched by linguistic or conceptual atoms, the notorious independent empirical predicates of logical empiricism. There is a sibling theory of the social world supposed to exist in some sense 'out there' beyond the experience of individual human beings whose task in subjugating that world to concepts is conceived of on precisely the same model as the positivistic theory of a science of nature. I have sufficiently examined and criticized logical empiricism (positivism) as a philosophy of science in other places, and I shall not be examining it further. I proceed, then, to the positive construction of a more adequate theory of science linking it to the most reasonable theory of the social world I know of, namely that of ethnomethodology. By this means I hope to provide a schema for handling the problem sketched above.

1. A THEORY OF SCIENCE

At the back of any theory of science lies a theory of concepts — the means of scientific thinking — and a theory of individuals and properties —the subject-matter of that thinking. Recent theories of science in the empiricist tradition have assumed that a theory of concepts adequate to science can be elaborated in terms of a language model derived from elementary predicate logic and that a theory of properties need proceed no further than the elements of sensory experience. Neither theory can stand close examination and I shall proceed to construct new theories for each domain.

1.1. A Theory of Concepts

Not only is it not at all clear that all the concepts essential to a scientific theory can be carried by simple, independent predicates and logical constructions out of them, I shall show that, at least, it is essential to invoke concepts in scientific thinking that are not naturally expressed verbally. These will prove to be particularly prominent when thinking is directed towards structures. I shall be concerned, unashamedly, with a theory in which the concepts and properties I shall be suggesting invoke a world of active agents, though I do not wish to be understood to be promoting any sort of Pan-psychism or *anima mundi* theory.

Our problem, then, is to identify in scientific thinking some vehicles for thought, bearers of concepts, which are capable of forming the ground of thinking in structural and dynamic terms. For the bearers of such concepts I shall introduce the term 'icon'. I prefer this, now, to the much-abused notion of model, which has recently fallen into dissuetude since it has been used carelessly both for sentential structures and for the objects which are described in such structures. Furthermore, the use of models in the latter sense has been seriously misunderstood in that some critics have supposed that the introduction of real or imagined model objects was in contrast to a conceptual theory of thinking. Nothing could be further from the intention of those who have directed their attention to such models since they are intended to express the manner of real scientific thinking when it is concerned with structures and dynamic or power concepts (Harré 1970). They are, thus, no less the bearers of concepts than are words. However, not wishing to embroil myself in what is largely a resolution of a mere misunderstanding, I shall speak in this paper only of icons.

An icon is a non-verbal image (often a pictorial or quasi-pictorial symbol) which is a vehicle for thought or a bearer of a concept, but, unlike words and such constructions, is projectively related to reality. The concept of an icon will become clear only when I have explained what it is for an entity to be projectively related to reality.

When a projective relation exists, two objects are related by structural isomorphism in that the elements of each correspond according to some rule and the invariants by which the integrity of the structures of each are maintained also correspond according to rule. An icon is a member of a pair of objects related by such an isomorphism. In scientific thinking the object to which the icon is related need not be a concrete entity. It may be an abstract entity, or entities equally related by systematic relationship. This can be made clear by comparing the iconic character of a wiring diagram for, say, a television receiver, which is an object related to another object of the same ontological character, with the system of military insignia, with which for example in the British Army, three chevrons are related to two chevrons as the rank of sergeant is related to the rank of corporal. The relation of the ranks is an abstract relationship exemplified in a wide range of relations between those individuals who have that rank. The icons which are sewn on the sleeves of the appropriate uniforms show in a non-verbal projective relation the structure of the abstract relation.

An adequate theory of concepts must, at least initially, assume that conceptual thinking is carried by both verbal and iconic means. I shall not assume lightly that there is necessarily any verbal equivalent to a specific icon. We shall see instead that the central role that the conceiving of icons performs in scientific thinking, suggests rather that the icon is an indispensable element in that thinking.[1]

1.2. A Theory of Properties

The theory of properties which I shall sketch considers a property to be assigned to an individual. From the scientific point of view a property is treated as the cause of a certain manifestation as a quality or pattern in experience. In the language of philosophical logic, such a property will be a power or a grounded disposition (Fisk 1974). For example, considered naively, the property of redness as assigned to an object, is nothing but the quality that object shows when

[1] A theory on this view must describe an icon of reality to be truly scientific. Thus purely formal theories, consisting only of signs or other free linguistic constructions, must be regarded as primitive. Thus constraints on, or a *priori* universal invariants in, icon conception constrain theories.

viewed by a human being, but considered scientifically, the property of redness is the power or grounded disposition of an object to cause such an experience. Thus, when the property of redness enters a scientific context, it is no longer a simple quality but a complex property. It is clear that, for a scientist, redness is a grounded dispositional property or power of many, but not all, red things. I think it is indisputable that this is how properties are understood in science. Are we obliged to accept this way of understanding? Philosophers have sometimes disputed the propriety of the scientific understanding of properties in favour of a theory of properties which identifies them with the elements of experience, particularly qualitative elements. This is the theory of phenomenalism. Despite tremendous efforts, it has not proved possible to construct a satisfactory phenomenalism, one which does not collapse into unconnected, individual, experiential worlds of atomistic elementary experiences. I have no wish to add to the mountain of objections which has been built up to that view. I shall assume that the only theory of properties worth advancing is that which is current in the sciences.

A property, then, is to be considered as a power or grounded disposition: that is, to say that an object is red, speaking scientifically, is to say that it has such an internal structure, and, if the quality is used to identify the object, such an intrinsic nature, that it looks red to well-placed observers. What is the nature of that internal property in which its disposition to look red, or power to produce such experience, is grounded? In scientific thinking there is a remarkable uniformity in the answer to that question. The grounds of dispositions or the natures of things upon which their powers depend are almost always thought of as structures, usually spatial, though not necessarily,[2] of elementary agents. Such elementary agents are themselves the possessors of properties which can again be treated as grounded dispositions or powers leading to further and more refined structural hypotheses. It is my general contention that in actual scientific thinking the property of an object which is thought of as a structure of elementary agents is thought of by means of an icon. So, the concept of that property is carried by the appropriate icon. Sometimes that icon will be a physically realizable diagram or picture; sometimes, when the structure and elementary agents are abstract, it will not. And iconic representation may extend to the power of the elementary agents, as well (c.f. Maxwell's fluid images for thinking about potentials).

I can now bring together the theory of concepts and the theory of properties in a brief description of the nature of the predicates required for the assignment or attribution of such properties and can say a word or two about their truth conditions. Such predicates as 'red', conceived as incorporated in a science, will have identical logical form to such predicates as 'charge' or 'potential'; that is, they will ascribe to an individual, either concrete such as a thing, or abstract such as a field, a certain nature which is the cause of the manifestation of the power or disposition as the given quality in experience, so that the logical form of a scientific predicate will be:

> If certain unspecified conditions obtain, then a certain effect will be produced by virtue of the nature of the individual involved.

Thus, the disposition to present a certain quality or pattern to experience is grounded in an unspecific hypothesis about the nature of the things and substances involved. This logical structure allows us to link the predicates by which properties are ascribed to real things with a

---

[2] A non-spatial structure would be exemplified by the old theory that substances were made up of different proportions of the Four Elements.

sensory language, since, of course, it is required that the properties of real things appear either directly in personal experience as qualities and patterns, or indirectly in the responses of instruments. But this linkage is loose, since of course, the qualities may appear for other than their usual causes. Something may appear red, or hot, or a galvanometer may give a twitch, for reasons other than the standard reasons which are captured in the structure of scientific predicates.

It is possible now to see why one wishes to say that science is the attempt to conceive and describe (or depict) icons of reality under certain constraints, in terms of which dispositions to produce experience can be grounded, and thus the powers of things to act on other things can be explained. It is clear, then, that there is a logical reason for the necessity for thinking with such icons. Since experiental predicates *must* be transformed into grounded dispositional predicates, or power predicates, to have a place in science, thought is forced on to the task of conceiving an unexperienced world in terms of which those dispositions can be grounded and those powers explained.

There is, then, no substance whatever in the idea that science must be in some way or other inductive, growing by accumulation of little bits of knowledge. Generality is achieved in science in the manner conceived by William Whewell—that is, by the addition of what he called 'an idea' to a field of experience. Whewell's ideas are not concepts corresponding to elements of that experience, nor are they merely logical orderings of it. They are such ideas as that of chemical atom. Clearly, they enable us to identify genuinely similar experiences — that is, similar beneath their skin of qualities — and to identify within experience those patterns which are genuinely general. For example, Darwin's theory of evolution is a great scientific theory because it describes an icon of reality, an icon of biological history and its dynamic process. Empirical evidence enters Darwin's work, not as particulars from which generalizations are drawn, but as accompanying anecdotes which illustrate the capacity of the icon to represent, and thus its description to explain, diverse matters. The world of animals and plants is made intelligible by Darwin's theory, because Darwin's theory describes an icon of the mechanism by which recognised patterns are generated, and previously inchoate experiences come to be seen as intelligible and orderly. It is to this notion of making intelligible that the iconic theory of science is directed. Notice that in special circumstances a mark or sign of such intelligibility is the fact that a description of the pattern generated can be deduced from a description of the generating mechanism, together with a description of the conditions under which it is active. But, according to this view, this is a consequence of the satisfaction of the more fundamental condition rather than being itself the necessary and sufficient condition for the existence of an explanation. On this view, the paradoxes of the deductive conception of explanation are very easily resolved in that there may well be logical forms which allow predictions to be made which our intuition tells us could not count as explanations, and the iconic theory makes it perfectly clear what the ground of such an intuition might be; namely, the absence, in the content of the propositions satisfying that form, of any reference to the structure of elementary agents which generates the pattern upon whose generality the prediction depends.

2. CONSEQUENCES FOR SCIENCE

2.1. The Form of a Scientific Problem and Its Solution

The specific form that a problem and its solution takes, depends upon very broad, general, almost ideological differences in conceptions of knowledge and experience. I have hinted at such differences already, but I would like at this point to make them a little more precise. In a great many fields of scientific investigation one seems to come across a dichotomy of approaches, which I shall call for convenience, atomism and structurism. (One version of structurism being structuralism).

It is typical of the atomistic ideology to see the world of experience as a flux of independent, non-overlapping, unconnected and passive elementary parts and the dynamics of such a world to be structured in terms of an efficient cause externally imposed. Thus, the explanation of the appearance of a particular elementary unit of experience is to be achieved by reference to a specific antecedent element which is concomitant with it. The only relation, it is assumed, which we can legitimately suppose to obtain between those elements, is one whose ultimate explanation lies in the psychology of the people who experience such concomitances. Such an approach is exemplified in psychology, for example, by the theory of language which is based upon the ideas of Skinner. In that theory, the sentence produced by a person is a linguistic atom, which is under the control of features of the environment which call it forth.

The Skinnerian theory is in strong contrast to the theory of language based upon the ideas of Chomsky and others, in which the sentence which is produced by a speaker is thought of as a structure which has evolved from another, more fundamental structure— the underlying form of semantic representation. Such an idea, when generalized, forms the basis of the competing ideology—structurism. The world process is conceived in terms of the operation of formal causes, that is, of the imposition of structure or pattern on an unstructured, unpatterned base. Such imposition is conceived by structurists as the evolving of an antecedently given structure, under the constraint of certain invariants into the structure or pattern which is manifested, sometimes under a very different appearance, in experience. The nature of this evolution and the manner in which the evolved structure imposes its pattern on the unpatterned base, varies from science to science.

It is within the structurist ideology that the theory of grounded dispositions or powers is most naturally fitted, since the pattern of experience in which the disposition or power of an object is manifested, is derived from structural properties of that object, those very structural properties which are the grounds of the disposition.

It should now be clear that the fundamental form of a scientific problem is 'What is the source of this or that pattern in experience?' And the fundamental form of the answer to any such question will be that the icon of the real world, by which the grounds of the disposition or power to produce those patterns are conceived, contains a plausible template for the production of that pattern in experience. That, of course, is a very abstract way of putting the matter. I can dispel uncertainties best by citing some examples. The pattern of a crystal of diamond is to be explained in terms of the structure of carbon atoms. Indeed, the pattern of crystal of diamond can be shown to be a derivative of the pre-existing structure of the carbon atom. A synchronic pattern such as the anatomy of an organism is the product of a diachronic process by which the elementary parts of that organism are assembled in the pattern by reference to a pre-existing template. This is the form that hereditary explanations now take, so that one synchronic pattern is the source of another. Such patterns are the patterns of things, but there may equally be patterns amongst events, diachronic patterns, the structure of processes. And symmetrically with the explanation of synchronic patterns, so diachronic patterns will be explained by the existence of synchronic templates such, for example, as is exemplified in a punched tape which determines and controls the production of a process. This is how, currently, we understand the patterns to be observed in physiology or in chemical interactions. It seems, then, as if the structural properties of our icons of the real world are central to the form that specific scientific explanations take.

. . . [We] must ask whether there is any general feature of icons of reality which mark off those which are likely to prove satisfactory as the bases of scientific thinking. We might look then for the sources of these features in certain constraints upon conceptions of reality. It has been a commonplace since the later middle ages that the construction of scientific theories, whether conceived of logico-verbally as instruments of prediction, or realistically as icons of reality, must come under some constraint since there are certainly infinitely many instrumental theories possible in any field of the phenomena and a finite though large number of icons of reality are certainly possible. One of the most serious objections to the use of purely logical concepts in analyzing scientific thinking is that by deliberately eliminating attention to content from the philosophy of science, this point of view leaves theories constrained only by logical possibility, which is hardly a constraint at all, or by some formal criterion such as simplicity, whose application is uncertain. The shift to the iconic theory makes the problem of the identification of constraints more tractable, though it does not, by itself, solve it.

If we regard theories as descriptions of icons of reality produced by the human imagination, it is clear that there must be some account of the constraints upon that imagination, for the human imaginative faculty is well-known for its capacity to generate mere fantasy: and yet, it is plain that the conceptions of reality which scientists have drawn upon from time to time, are not fantasies, though in the end some have been abandoned as unrealistic.

To understand the complex intellectual process involved in the constraints on products of the imagination, I will proceed by first looking at the structure of the thinking that is involved in the creation of particular icons of reality for the solution of particular problems. One must remember that the reality represented is to be causally responsible, usually in the formal cause sense, for the structure of the behaviour or anatomy of some thing or class of things. So the reality conceived must come under the primary constraint of behaving in such a way as to match the behaviour of the real world. We know the way the real world behaves since it is those patterns we have observed. What we do not know is how the real world generates those patterns. Thus, the first constraint which we could call 'behaviour-matching', can be put like this:

> The icon of reality must represent a structure which could produce behaviour similar to that we know the world produce.

But in order for the form and elementary parts of the icon to be satisfactory as a source of content for the theory there must be thought to be a possible reality. We might describe this as 'content-matching'. This leads us to a statement of the second and more constrictive constraint:

> The structure of elementary agents which constitutes the specific content of an icon must be a possible existent.

Thus, the behaviour of the icon is matched to experience and its content is matched to existing conceptions of reality. The constraints under which the content structure and elements of the icons must come are much more rigorous constraints than the instrumentalists demand for matching behaviour. The constraint can be considered under two headings:

1. The behaviour of the structure and the elements of the icon should, in general (though there are exceptions) be according to the principles of already accepted science, for example, gas molecules must obey Newtonian laws (Bunge 1974).
2. The most pervasive but the least easily examined source of constraint is what is usually

referred to dismissively by philosophers as the 'background'.

The background consists of what Toulmin called the Principles of Natural Order (Toulmin 1961) or what others have called the General Conception of Reality, etc. Since this background is so vitally important to the kind of icons of reality that the minds of a given generation will generate, we must not simply dismiss it in these rather general terms and return to logic-chopping.

Before proceeding to look into the structure of the background, I would like to draw attention to the correspondence between the pair of constraints, behaviour and background, and the elements in our theory of properties, namely the disposition and its ground. The behavioural component corresponds to the disposition; the constraint upon the content of the icon corresponds to the ground of that disposition. In this way, a theory constructed in accordance with the iconic paradigm serves as a scientific explanation, in that it satisfies both parts of the scientific attribution of a property to a logical subject. It has long been my contention that no other theory of theories has this capacity and no other theory of properties is adequate to an understanding of science.

2.2. Sources of Sources

The background to icon-conceiving, which serves to place rather sharp constraints upon what is possible, can be thought of in terms of choices among certain options with respect to the main Aristotelian categories. A background will then consist of a certain path chosen amongst alternative views as to the nature of substance, individual, quality, relation and so on. For example, substances may be regarded as continuous or granulary, bounded or unbounded, complex or simple. Individuals may be regarded as atomic or inter-related, qualities may be regarded as the manifestation of powers or as the units of sensory experience, and so on. Only certain choices amongst the background options form consistent wholes. It is the traditional task of metaphysics to examine the consistency of various choice patterns amongst background options.

Recently, a quite new feature has entered our discussions of such matters. Namely, the idea that the icon of reality generated in accordance with a certain set of choices amongst background options, has certain structural features, that is requires that certain invariants be preserved during the history of the universal entity conceived in accordance with the choices. This approach has become particularly prominent in physics, in linguistic studies, and in anthropology. Indeed, in anthropology we have had a movement, the main burden of which has been to draw attention to the value of the search for structural invariants in a wide range of what, in this paper, I am calling Icons of Reality, e.g. Levi-Strauss's alleged demonstration of the structural isomorphism between the totemisms of various Australian Aboriginal tribes (Levi-Strauss, 1962, p. 127). I do not myself attach much importance to the fact that this approach has come under strong criticism since the critics have, it seems to me, contented themselves merely with a critique of details. I do not think any successful general attack has been made upon the approach. However, the undoubted rightness of many detailed criticisms has led me to use the word 'structurism' rather than 'structuralism' for the approach I wish to advocate; thus standing firm with the structuralists only on the most general issue, namely that the world and the thought matching it are both to be conceived of as structures which evolve in time, preserving certain invariants.

The problem then, of sources of sources, resolves itself into two distinct areas:

(1) What are the invariants, and what are the elements of the various structured icons which form the most general background to thought? This problem is a matter for an informed historical investigation, informed that is, by the iconic structurist theory.

(2) But . . . the dynamics of change in such structures is a much more interesting since much more obscure area to investigate. I shall, in the end, in this paper, devote some attention to the sketching of some preliminary points about it.

At present there is little more that we can do but remark upon the enormous influence that something we might call 'fashion' has upon the background. One can perhaps put the matter in a preliminary way in a quotation from Henry Fielding. 'Fashion is the governor of this world. It presides not only in matters of dress and amusement, but in law, physics, politics and religion, and in all other things of the gravest kind. Indeed, the wisest of men would be puzzled to give any better reason why particular forms in all these have been at times universally received and at other times universally rejected, than that they were in or out of fashion'. I am sure that we can do better than Fielding's Wisest of men. But our problem is part philosophical (what are the invariants?) and part social-psychological (what are the processes by which the evolving structures are passed from mind to mind?) Our business will be with a general account of the nature of such structures and of the possible dynamics of their evolution. But we are not likely to get far in speculative psychology until the psychologists can offer us at least the glimmerings of a theory of fashion, even in that most obvious world fashions in clothes. However, there is, even at this early stage, something that can be said and I shall return to the matter in the last section.

3. THE DIRECTION OF SCIENCE

. . . I think it abundantly established by the difficulties that Popper and others have run into in attempting to formulate their intuitions that science does develop towards something that might be called 'truth', that any attempt to discuss this matter in terms of a logic coextensive with the logic of deductive reasoning runs into enormous difficulties. I think there can be no doubt that Popper has established once and for all the limitations of the deductive approach to rational thinking, in his distinction between the logical structure of science and the methodological rules, where it seems to me, all we want to say about science as a unique activity lies in the latter (Popper, 1959). I do not propose, therefore, to consider progress in any quasi-inductive fashion in terms of universality or anything of that sort. However, I do believe that we can make considerable progress in terms of the iconic theory.

3.1. Stratification

The idea that science is pursued according to the grounded disposition or powers theory of properties and predicates, leads us to a stratified theory of knowledge which is reinforced by a further, very pervasive feature of scientific thinking.

(1) The theory of powers or grounded dispositions (Madden and Harré 1971) leads us to say that the behaviour of some class of substances and their anatomical structure derives from an inner and, in general, other set of properties, which we have found frequently to be expressible in terms of a structure of elementary agents; thus the valency of a chemical element is explained in terms of the structure of the elementary charges in terms of which we conceive its atomic constitution. So the dispositions of things and substances are grounded in hypotheses as to their natures. In expressing the results of a power-natures investigation, we are naturally led to lay out our knowledge in strata, each stratum consisting of two levels. Thus in one stratum there would be, in level one, a description of the pattern of behaviour of a chemical element, and in level two of that stratum, a description of the structure of the elementary agents that constituted the nature of the atoms of that element.

Now this is a *stratum* because the expression of our hypotheses in this form leads immediately to the question of the explanation of the behaviour of the elementary agents referred to in level two. This explanation will take exactly the same form, that is the behaviour of electrons will be conceived as the manifestation of certain dispositions to behave and, in theory, those dispositions will be grounded in structural analyses of the nature of electrons, which must on this paradigm, be conceived of as some structural form of yet more elementary entities which themselves have powers. At present physicists are unable to make this step, but it is not clear to me that there is any prohibition in principle upon it being made. In other sciences, the stratified form is even more clearly exemplified. It has been one of my contentions in recent years that this stratificatory form should also be aimed at in the social and psychological sciences, in preference to the one-dimensional instrumentalist positivism by which they have been recently plagued (Harré 1971).

(2) I have taken for granted in talking of the setting up of a grounded disposition account of behaviour that the powers of things and substances are grounded in structural hypotheses as to the natures of those things and substances. It has certainly been a very pervasive feature of scientific thinking to prefer structure to quality. I can perhaps illustrate this preference in an example. There are at least several hundred thousand different substances to be found on the surface of the earth. Let us say that there is a million-fold diversity among the substances which we can find or synthesize. Chemical thinking involves the substitution of the million-fold diversity by a structural hypothesis that there are a million different structures involving, roughly, one hundred different kinds of atoms. Thus a million-fold qualitative differentiation is reduced to a hundred-fold difference of substance and the disparity is taken up by structural theories of chemical compounding. Chemists have not rested at that. The one-hundred-fold diversity of chemical elements has itself been transformed by a structural theory in which the one-hundred-fold qualitative diversity of chemical elements is explained in terms of a hundred different structures of three fundamental substances — protons, neutrons and electrons. Unfortunately, nature has not exactly sustained the original beauty of this conception, but nuclear and particle physics is at a very early stage of development as yet and I am prepared to hazard the guess that any further development will consist in the reduction of diversity amongst so-called elementary particles to a structural diversification of some yet more and yet fewer elementary substances.

It is clear that the structural explanation of qualitative diversity matches almost exactly the powers-natures explanation of the behaviour-dispositions of natural substances, in that the same structure is responsible both for their disposition to manifest certain qualities and their disposition to behave in other chemically relevant ways. So these factors are in essence the same and they define the strata of a science, each stratum overlapping with the next and consisting, as always, of two levels; in one a substance or individual is described in terms of dispositions or powers, and in the other those dispositions are grounded in structural hypotheses about its nature.

3.2. A Tentative Definition of Progress

Assuming a fairly stable background to iconic thinking (an assumption which we will examine in the last section of this chapter), a limited conception of progress can be defined in terms of the stratified theory of knowledge. Thus progress would consist in supplying an icon of reality such that three conditions are satisfied:

(1) The icon represents a reality that exists in a stratum further from experience than all previous icons, so that, for example, the invention and promulgation of the double helix theory of genetic material is an example of scientific progress in that in the strata of knowledge we have

about heredity, the double helix is the structure of an icon of a reality which is further from experience than the simple atomistic gene of the Mendelian theory: that is, it is further from simple experience than the count of the proportions of green and yellow peas, and further still than the chromosomes which were the individuals at the second level of the first stratum of hereditary theory.

(2) An icon of reality represents progress if it maximises structural diversity at the expense of qualitative difference; thus the electron theory of the chemical atom is progressive over the Dalton theory which differentiated chemical atoms wholly in terms of their qualities and powers.

(3) The appearance of an icon of reality is progressive if the same icon reappears at the deeper levels of many stratified fields of knowledge. Something like this was proffered many years ago by Ernest Nagel as a criterion for the reality of a substance, process or individual, postulated in a theory; thus, the physico-chemical atom could be taken to be real insofar as it appeared in the explanation of a very wide range of very different phenomena (Nagel 1961, pp. 147-149). Nagel, I believe, had perceived something important, but had misunderstood it. The idea of there being a criterion of reality which does not involve any kind of demonstrative reference, nor in strata remote from experience, transcendental arguments, seems to me a serious error and yet there is something important about the reappearance of a certain icon and its associated concepts all over the place. I believe that this marks not the reality of the structures and entities imagined, for that question is preempted by the metaphysical choices amongst background options long before (logically speaking) any specific icon is imagined. Rather the pervasiveness of a certain conception of nature is a measure of the progress which has been achieved by its introduction.

4. SOCIETY AS A PRODUCT OF THEORIZING

In recent years a number of trends in the social sciences have led to a critique of traditional views about the nature of society and the mode of its being as society, that has been at its sharpest among ethnomethodologists (Filmer, *et al.* 1972) and at its most articulate in the work of Scott and Lyman (1970). I must confess myself wholly convinced by their line of argument and their criticism of positivist sociologists, who suppose that society has a nontheoretical mode of being and that truths about it constitute social facts. I shall call this the 'naive' theory.[3] It is in accordance with the naive theory that much contemporary empirical work in sociology is still carried on as if there were a reality 'out there' which, by various techniques, we could sample and investigate and ultimately describe. I am now very much of the opinion that we should begin by insisting that there is no such reality and regard the very existence of patterns of conformative

[3] Durkheim's difficulties with this crux are worth noticing. In the *Rules of Sociological Method,* and *Suicide,* his general view that society is a system of ideas, the animating theory of the *Elementary Forms of the Religious Life* seem to be negated by a reified conception of 'social fact', forced upon him by his acceptance of three criteria of objectivity: (i) our social ideas are not self-created, (ii) an individual can come to know more about his social situation, (iii) shared social conceptions are coercive over an individual's desires. While proposing that social facts consist of 'representations and actions', an opinion from which I would not dissent, he nevertheless argues that 'their substratum can be no other than society'; the same kind of error as the Cartesian reification of minds on the grounds of the incompatibility of physical and mental properties, so that he, Durkheim, proposes that we should look for the sources of social facts in other social facts.

behaviour as the problematic element in human life. It follows then, that in order to give an adequate account of the relation between social factors and the sciences we must attempt a new departure, taking the ethnomethodological conception of society as the groundwork of our argument. Thus, the social sciences will be in a very different position from the natural sciences with respect to their ontology and metaphysics, though not, I believe, with respect to the intellectual skills and theoretical methods which are brought to bear upon their problems.

According to the view which I wish to advocate, society and the institutions within a society are not to be conceived as independent existents, *of* which we conceive icons. Rather, they *are* icons which are described in explanations of certain problematic situations. Thus, the *concept* of a Trade Union, or a University, is to be treated as a theoretical concept judged by its explanatory power, rather than a descriptive concept judged by its conformity to an independent reality. The form of a social and a scientific explanation are identical, but their ontological commitments and metaphysical structures are quite different. Beyond the icons of reality which are conceived for purposes of explanation in the natural sciences, lies a real world of active things; but beyond the icons conceived for the explanation of social interactions by social actors lies nothing but those very actors, their conformative behaviour and their ideas.

For what purpose, then, are social explanations attempted and icons conceived? Social thought, whether expressed in ordinary language or in one of the official rhetorics of sociology, is aimed at the injection of meaning into a flurry of individual, face-to-face, encounters, in the course of which messages are sent from one person to another, which we *assume* reach their intended recipients and are understood. Without the active intervention of sense-giving explanatory activity, this flurry is without meaning though, as we shall see, not wholly without form. I can illustrate the point with a simple and restricted example. Physical actions enter social reality when they become embedded in semantic fields and syntactic structures. For example, a gripping of hands comes to be meaningful when it is embedded within an introductory ceremony, and acquires within that fundamentally syntactical structure a certain distinct meaning in relation to other actions and words, and certain structural relations with other parts of the ceremony. Thereby a hand-grasp becomes part of social reality, but only as the bearer of a recognised meaning, conformative to its intended meaning. The gripping of hands has no social meaning taken outside all possible embodiments. Within a karate encounter, of course, its meaning is quite different from that which it has within an introduction ritual.

So an analysis of the social world must begin with the fundamental thought that the components, or elements, of that world, have no meaning in themselves and therefore do not exist as social objects. Once we pass from the simple and restricted, and immediately apprehensible, structures, such as are exemplified in introduction ceremonials, the joint procedure by which meaning is given to an element or set of elements, and structure is imposed upon them, has to be achieved by reference to some icon, for example, an introduction ceremony makes sense only if people are perceived according to the social dichotomy, intimates and strangers.

I will illustrate the role of such icons in two examples of increasing relative generality.

(1) *A microsocial icon.* Sometimes, perhaps even within a single building, the flurry of personal encounters is made intelligible for those within it by theoretical explanations, based upon descriptions of a shared icon in terms of which some of the flurry can be treated as consisting of activities within an institution, for instance, a police station. Such an icon frequently appears in pictorial form as a chart showing, for example, the hierarchical organization of the management structure. Of what is that diagram a picture? Certainly not any structure existing independently of the icon itself which the icon mirrors. Rather it is a

conception of the institution in terms of which actions are performed and made meaningful, and in terms of which the flurry of such actions has structure.

(2) *Macrosocial icons.* It happens from time to time that we become aware of differences in the dress, speech, habits and tastes, jobs, etc. of other people. Along with this perception may go certain feelings of respect, resentment or contempt, and we may provide an explanation of such feelings by reference to the differences we perceive between us and them. In real life this phenomenon occurs in an unordered flurry of momentary, personal interactions, some few of which may persist. Sociologists provided us some time in the nineteenth century with a societal icon in terms of which this flurry of isolated experiences could be made intelligible. I refer to the class theory. It describes a well known image or icon. It is not a theory which describes an independent reality, but it is the basis of a wide variety of explanations in terms of which various differences between people can be made intelligible. It replaced an icon of reality which paid attention to other features of the pragmatic boundaries of concepts like respect, namely, the situational or place icon which allowed us to conceive of society rather as a network of transformable relations than as a layered structure, within each layer there being more or less random movement. Now it is my contention that classes exist only insofar as they are thought to exist and the function of such a notion is to provide a standard, ready-made, easily acceptable explanation for understanding what has happened on some occasion.[4]

The point of view which I am sketching here is certainly not new but has rarely been defended in a pure form, since the temptation to suppose that there are 'social facts' and structures in the social world independent of thought, is perennial, a temptation based upon the correct perception that there is a problem of how an individual's forms of thought are related to the shared thought of his community, which community is created by the sharing of the thought. I should like to describe the pure form of the theory as the Borodino theory, in acknowledgement of the exquisite exposition of the theory by Tolstoy. Tolstoy uses the Battle of Borodino both as an example and as a model, for understanding society. If I may remind you of the account he gives, the battle is seen as a flurry of relatively isolated, inter-personal encounters occurring both on the battlefield and amongst the officers of the High Command. Messages are sent from Napoleon to the various parts of the battlefield but of course they rarely arrive, and if they do are either garbled or not understood. And if rarely understood, are not acted upon. At various levels in the army hierarchy an icon physically exemplified in the badges of rank, generates the illusions of command that exist throughout the day, though smoke and the geographical structure of the battlefield prevent any empirical verification of individual hypotheses of command. Yet, Tolstoy reminds us, the battle was written up by historians in a rhetoric which conceives of that flurry of personal interactions as a structured entity controlled from one side by Napoleon and from the other by Kutuzov. Historians' accounts include such items as 'brilliant tactical decisions' and so on, and *assume* a structure of command with reliable interrelations. All of this accounting, Tolstoy makes clear to us, is essential since it makes sense of an affair which though concentrated in space and time was without overall structure, but being human we persist in pursuing the theoretical activity of endowing it with one. When reading *War and Peace,* were we momentarily in the presence of the greatest of all social analysts? I am inclined to think so.

I see myself then, as accepting and advocating a Tolstoyan or Borodino theory of the social. That is, that we cannot take for granted that without an implicit social theory describing assumed structural icons we are left with a mess or flurry of individual encounters which are then given

[4]E.g. 'He did so and so *because* he was middle class' which would be interpreted on the iconic theory as 'In doing so and so he represented an aspect of sociologist S's layered icon of reality.'

form and meaning by accounting procedures which draw upon various rhetorics. For example, we cannot assume that any people in society are endowed with *actual* power, since messages are rarely understood, seldom reach their targets, and are usually ignored, but in accordance with some picture of the way things are, people are endowed with the property of 'having power' and so occupy a certain niche in the icon, generated by and at the back of the accounting rhetoric. Such a person can then become the focus of certain important social feelings, not in himself, in which he is frequently a pathetic figure, but rather in terms of the imaginary capacities which, by our act of supposing him to have them, serves to make our experience intelligible. It is in terms of a Borodino theory, then, that I propose to examine the relation between science and society.

But before I proceed to the final section of my argument, there are two qualifications to the pure Borodino theory that must be entered.

(1) There are patterns in social action which are not represented in any individual consciousness — for example, the patterns of unintended consequences — which we can *come to know,* and which must, therefore, exist prior to our knowledge. It is the existence of these patterns which has encouraged the reification of 'social facts'. I contend that though there are such patterns as, for example, the socially divisive migration of people inside cities which has been associated with changes in the structure of the school system, the activity of representing such patterns in thought is a bad model for the activity of representing institutions, ceremonies, role-presentations and the like; but when we come to attempt to explain the patterns of unintended consequences, we must return to the individual consciousness and the theories which individual citizens form concerning their situation....[5]

Thus the patterns of unintended consequences are not produced by independent societal causal processes, but are either,

(a) the accidental products of myriads of stages modulated through individual consciousnesses, as, for example, messages are passed and habits are acquired; or

(b) the intended products of myriads of individual understandings of messages and acquisitions of habits according to some promulgated scheme, for example, the successful promulgation of the idea that we should drive on one or other side of the road in any particular country.

In the case of (a) the patterns are imposed by us in the search for intelligibility and appear in our accounts. Their objective status is like that of the castles in the embers or the cherubs in the spring leaves.

Even in the case of (b) the promulgation of a pattern through society is an uncertain matter and dependent upon matters of which we are quite ignorant. Compare our understanding of someone poisoning everybody through the water supply and our lack of understanding of someone ordering everybody through the media.

(2) The icons which make our actions intelligible as social actions are shared conceptions, and the conception of social order held by any individual may fit it more or less well. I am not the first to suggest the analogy of this problem with Saussure's distinction between language and speech. But is an individual who only gradually comes to understand the structure of a university like someone who only gradually comes to understand the structure of a protein? Or like

[5] There is a corresponding error in transferring the causal mode of the pattern of unintended consequences to the arena of political action. One should notice here the folly of the socialist radical in attacking 'institutions' while leaving untouched the system of meanings that sustain them, and of which they are part representations. The hyper-radical would aim first at semantic change.

someone who only gradually comes to understand the current theory of electrical conductivity? It is my contention that the person adapting himself to society is like the latter as far as the form of his intellectual activity is concerned. But whereas for a scientist there is a three-fold system of structures — that of his own icon, that of a common icon in the shared theory of the scientific community, and that of the bit of the real world represented, for the social actor there are in many crucial cases only two.

In terms of *his* icon, the individual constitutes himself *a* mayor, and in terms of the *shared* icon he is constituted by his co-societals as *the* mayor; and they can verify their attribution in his way of going on because his icon sufficiently matches each of theirs. Only if the other recognise him as such does his constitution of himself leave the realm of fantasy. And it is well worth noticing that his fantasy does have a measure of latitude even with respect to the shared icons of those who recognise him as mayor in that he has considerable latitude in the style in which he performs those duties which devolve upon him to his place in the picture of the village that the villagers have.

5. TRUE RELATION OF SCIENCE AND SOCIETY

Taking the theory of science together with the theory of society, we reach the inevitable conclusion that the relation between science and society is an internal one. The icons of reality conceived by scientists, and society itself, since it is just another icon — though of enormous importance for human life — are generated by the self-same process and according to the self-same constraints. However, one should notice that while it is necessary that society be generated as an icon by the same process as scientists conceive of the structure of the world, it is contingent that scientists fall under the very same set of constraints at any given time as do those who conceive the social world. There may well be historical times in which science and society are out of joint, but in general we would expect, in the absence of a solution to the problem of the generation and promulgation of fashion, that contingently the icons of science will be constrained by something of the same set of background options as are their contemporary conceptions of society.

It follows from this that there are no causal connections *from* society (or even patterns of conformative behaviour existing pre-theoretically to any particular consciousness) *to* thought patterns. Even so sensitive a student of symbolic forms as Mary Douglas slips into an unexamined assumption that such connections exist. "The body", she says, "is a symbolic medium which is used to express particular patterns of social relations" (Douglas, 1970, pp. 13-14). Not so. Patterns of social relations and bodily symbolism could equally be regarded, as far as the ethnographic evidence goes, as both representations of a conceptual pattern, syntactically structured and intersecting with semantic fields. Again this assumption shows up in her remark that each set of metaphysical assumptions is *derived* from a type of society. Let us say instead, each type of society is a representation of a set of metaphysical assumptions. By taking this turn one can quite readily understand Douglas's true perception "that the ideas about the human body, its potential and its weaknesses, which are found in particular social types, correspond uncannily well with ideas current in the same social types about the potentials and weaknesses of society."

In Bernstein's well known terminology (Bernstein 1965), I would claim there is no elaborated code, there are only different restricted codes, since there is no code which is independent of a social form, in that some social patterns may develop to be another representation of its syntactic and semantic structure.

Thus the communality of 'backgrounds' and structural principles in 'social reality' *construction* and natural reality *matching,* explains how social factors are actually involved in scientific thinking and its products. That is, societies are conceived according to the same set of options that are present in natural science in terms of which icons of physical reality are conceived. For example, two well-known and well-articulated sets of background options are:

(1) Numerous passive atoms in random motion interacting by collision.

(2) A structure of agents transforming itself under the constraint of some invariant.

According to the hypothesis I am proposing in this paper, it should be possible to show that when nature is conceived according to one or other of these sets of options, so society is conceived and thus created in a similar mould.[6] Now as I have pointed out, it may be historically the case that there are timelags and disconnections in thought. For example, it may be that in some epochs the societal icons in terms of which social actions are made meaningful remain fairly fixed, while the ways of conceiving of the structure and nature of the real world through iconic representation are changing. And the reverse may also be the case in other epochs. There are, I believe, two very striking cases where the background and structural invariants across the whole range of human thinking had a marked universality. These are the Romantic movement, culminating in the science of Michael Faraday and the social forms of Victorian Europe, and the Positivist movement culminating in the psychology of Skinner and the conception of society as bureaucracy. Skinner has proposed an explicit theory of society which is intended to promote an icon of the social world which is precisely adjusted to the set of background options that appear in classical positivism — namely a theory of human beings as passive automata and a conception of moral virtue centred, one might say, from outside the theory, around the notion of perfect obedience, i.e. all actions as produced by the environment. The internal relation between the scientific and societal icons is very clearly evident in Skinner's recent work, *Beyond Freedom and Dignity* (1971).

I offer, therefore, as a tentative hypothesis, yet to be fully developed in the context of certain specific examples, the idea that societal factors are not causally efficacious in the thinking of scientists. The occasions on which they seem to be so are, according to this theory, an illusion. There is no mysterious causal process which Mannheim and others looked for in vain.[7] Students of the sociology of knowledge have found no plausible causal relations, since there are none to find. However, it is my firm belief that the connection between science and society is very strong and in the theory I have proposed, the connection has certain features which are no less than necessary in that the icons of the inner forms of things which form the subject matter of science, are generated by the very same intellectual processes by which all of us, scientists and laymen alike, generate societal icons in terms of which our experience can be made meaningful, and are likely to share the same set of metaphysical options and be subject to the same constraints. Thus the social order is one set of realizations of the basic structure of our icons, scientific theories are

---

[6] So far as I can understand, this is the meaning of Foucault's proposal for an archeology of knowledge, identifying the structures common to forms of thought, modes of practice, and explanations of conformative behaviour, which each, in its own way expresses, and then seeks a theory of change in their transformation (Foucault 1972).

[7] Mannheim's analogy between artistic style and forms of thought as socially conditioned, and indeed any of the examples he cites (Mannheim 1936, pp. 264-290), is susceptible of explanation in structurist theory, without *assuming* any direction of influence *from* social position (situatedness) to intellectual product, since in the structurist view

another.

6. THE DYNAMICS OF ICON CHANGE

While the facts of social and conceptual change are treated as wholly separate problem fields, it is unlikely that any progress will be made in understanding either, but the sociology of knowledge, as conceived by Mannheim, was a false start, since it assumed an implausible causal theory. Equally, conceptual change in science has been, in my view, wrongly studied by paying exclusive attention to its products, verbal structures and the meaning of words, as opposed to its subject matter—Conceptions of Reality. The dynamics of change in the logico-verbal part of science is, according to the theory of this paper, derivative from a more fundamental dynamics—that in which the background options are revised, and in accordance with certain, as yet not clearly identified invariants in the structure of icons of reality, control the transformation of one icon into another plausible picture of the world. Meaning change is then derivative from this underlying process. On the other hand, if society is conceived of as an 'out there' real entity, we may be tempted to look for an underlying dynamics—for a process which simply does not exist. If societies are theoretical constructs for the explanations of our situations, then they will be subject to the same dynamics as are the conceptions of nature which lie behind the sciences, but as I have emphasized, whereas there is an independent reality to the natural world, society has no being except as an icon. Thus, I can offer no answers to the problems of change, but instead what I hope is a well-articulated theory as to the problem-fields in which such answers may be sought. Only when we direct our attention to the right field of study are we likely to come up with some useful answers.

I would, however, like to offer two more specific ideas, which could be applied both to science and its social analogue, political controversy—say that surrounding the behaviourist-Skinnerian paradigm—discussed in Chomsky's political works (Chomsky 1973). If I am right in a general way, then, if for the sake of reducing the problem to manageable dimensions, we ignore, though of course only temporarily, the whole moral dimension, then we seem to have two quite distinct methods by which we justify a background. And, of course, it may well be that in the means of justification, particularly in the failure and success of justification, that we may get the first hint of a dynamics of the structure of icons.

The two methods which seem in fact to be involved in the justification of backgrounds, are as might be expected, one theoretical and the other empirical. Theoretically, one can take up the hint that the study of backgrounds is a form of metaphysics and look for transcendental arguments by means of which items and structures present in the background and conceived as the products of choice among various category options, might be justified. By a transcendental argument I mean a demonstration that, preferably, the only conditions under which our experience is possible, are realised by a particular set of options. Of course, in any transcendental argument, a very great deal depends upon the prior analysis of experience. The particular form the necessary conditions for experience take will be dependent upon how that experience is understood. Thus, if experience is held to be inherently atomistic, analyzable into independent elementary parts, then this will influence rather strongly the form of demonstration by which the set of background options are shown to make it so. But at this point the ethnomethodology of everyday life, our common-sense sociology and psychology, begins to diverge from the ordinary natural sciences, since there is no external criteria by which societal thinking can be brought under some kind of existential constraint. But it is clear that background options and structural properties assigned to the physical world do generate existential hypotheses which can, under certain restrictive conditions, be subject to test. Thus, if the background is 'randomly moving atoms', then there are certain situations in which it may be possible to ask whether such entities exist, and even to find an answer empirically. Or if a more sophisticated theory of the background is in play, it may be possible, for instance, to provide an existential test for the existence of a potential or a charge, or something of that sort, at a place for a time. Now such testing proceeds by a progressive passage through the hierarchy of strata that I have identified earlier as an essential feature of an adequate theory of scientific knowledge. For example, we pass from diseases to bacteria, to viruses. We pass from the anatomical features which differentiate species and individuals one from another, to genes and chromosomes through to molecules. Existential tests are relative to strata and what is shown to exist in one stratum may be further analysed and indeed re-categorized from the point of view of a deeper stratum. Thus the products of the background, even if established in being, are liable to continuous categorical transformation as that background changes in essential ways.

But for institutions and other societal entities, to believe in them is to bring them into being. So the existential test for an element or a structure in a societal icon is liable to be self-fulfilling; though not, therefore, invalid.

There remains the problem of transformation. Here again I can only propose some problems. Transformation of the structure of icons occurs in two dimensions in this theory. We can ask how does the underlying synchronic structure, say, a conception of a system of elementary agents, transform into an icon of the world, natural, social or even artistic, in each of which the mode of representation may be very different. This is the problem of the mechanisms of contemporary corepresentation. But that underlying structure may itself suffer diachronic change, and we can pose the problem of the manner and means of such transformations too.[8]

---

[8] I am grateful to Dr. I. McGeochie for some useful criticism of an earlier draft.

SECTION TWO

POLICY REALISM

SECTION TWO

POLICY REALISM

INTRODUCTION

Models are used for two main purposes in scientific practice. Some are constructed to be used to display salient features of things and processes which we have already observed. Different ways of constructing models will highlight different features of the subject of the model as salient. However, some models are constructed to represent that which we have not observed. There many ways in which the target of a model making procedure may be hidden from observation.

This distinction in role parallels a distinction that can be drawn between realms of beings based on the way that they are related to the unassisted and extended senses. Let us call the realm of beings that have been observed, 'Realm 1'. This realm includes everyday objects as well as those kinds of beings that the sciences have made visible, audible and so on, such as bacteria and the asteroids. The realm of beings that have not been observed at any moment in the history of a science splits fuzzily into two subrealms. There are those which could be observed if we had the technical means. Let us say that beings of this kind are in Realm 2. Before the advent of electron microscopy, viruses and molecular lattices were then and there unobservable. They were Realm 2 beings. The development of new technologies of instrumentation changed the status of such beings so that are taken now to be included in Realm 1. However, the developments of modeling in physics and chemistry, cosmology and even psychology, have forced scientists to include in the scope of science a realm of beings which could never be observed. Let us call the domain of such beings Realm 3. Beings typical of Realm 3 include charges, fields of potential, magnetic poles, background radiation, and many others. Creating models to represent the beings of Realm 3 is a very different cognitive process from modeling beings in Realm 2.

In the course of developing the theory of thinking with models the three realms will be more carefully characterized, and the modeling procedures typical of each will be displayed in detail. This brief section is meant to serve only to introduce the 'Realms' terminology. It will be used throughout the chapters in this book.

Realms 1, 2 and 3 overlap. This will turn out to be a fact of great importance in coming to understand how scientific research projects can be planned and executed. The broad divisions among possible targets of modeling can be laid out in a simple tree diagram:

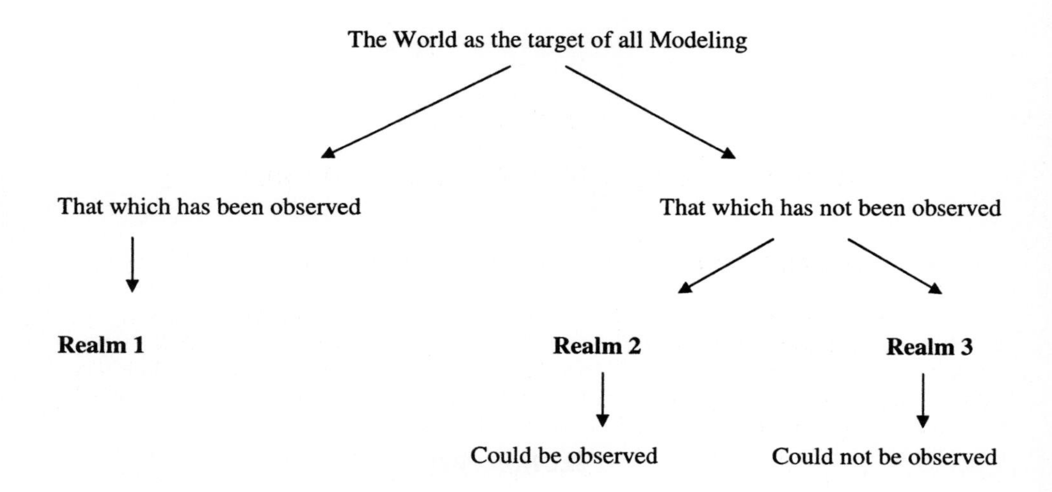

Figure 1
Realms of Reality

CHAPTER FOUR

MODEL MAKING AND THE DEFENCE OF REALISM

1. IDENTIFYING THE TARGET: FICTIONISM OR SCEPTICISM?

In the recent literature we can discern two main anti-realist positions. There is anti-realism as relativism and there is anti-realism as scepticism. Realists and relativists differ over a metaphysical thesis, that there exists a world independent of human beings which is identical with the subject matter of the physical sciences. Realists and sceptics share the first clause of this thesis but differ over the degree to which the second clause can be realised.

That the Realism/Scepticism contrast was at base a debate about the proper interpretation of theories was clearly discerned in the sixteenth century. For instance Ursus (1588), in discussing the range of interpretations of geometrical theories of the solar system distinguished between Realism, that geometrical models of the heavens were to be judged by how accurately they depicted the actual cosmos, and a Fictionalist version of Scepticism, that geometrical models were merely pictures of possible heavenly realities. Each was no more than a likely story to be used as a convenience in reasoning. The question of the relative verisimilitude of one picture rather than another could not arise, since the matters of fact which could distinguish them were strictly unobservable.

The arguments that follow are intended only as a defence of Realism against Scepticism, for instance as Scepticism about the Realist interpretation of physical theories has been exemplified in the writings of Laudan (1984) and van Fraassen (1980). Realism as anti-Fictionalism will be defended by bringing together two recent insights to extend the range of existing inductive arguments. These insights are as follows:

a. The manipulative efficacy of a technique is not reducible to the empirical adequacy of the theory which suggested it (Hacking 1983; Harré 1986).

b. The verisimilitude of a theory is not to be assessed directly by how near it is to the truth, but indirectly by how well the model or models upon which it is based, depict some relevant aspect of reality (Giere 1988; Aronson 1991).

Quite different arguments are needed to defend Realism against Relativism.

2. THEORIES AND THEIR MODELS

The inductive arguments to be developed below are concerned with justifying the projectability of a certain property of theories, namely their degree of verisimilitude, from cases in which the match between model and world has been assessed to cases where it has not. The verisimilitude of a theory will be defined as the degree to which the most ontologically plausible model for the theory resembles some feature of the real world in relevant respects. Neither the concept of 'theory' nor of 'model' are univocal, so an account of what is I shall mean by them is needed.

It might be objected at the outset that the very idea of assessing the match between a model and the aspect of the world that it represents is misconceived. It has been argued that there are no theory-independent 'aspects of the world'. It is enough for the purposes of this paper that it be allowed that there is a hierarchy of theories, the most comprehensive of which engages with the world in a way relatively independent of more local theories. A thorough philosophical exploration of this objection would centre on the Realism/Relativism debate to which the arguments in this paper are not addressed. Sceptics and Realists share a metaphysical thesis, that there is an independent world which the physical sciences purport to take as their proper domain of enquiry, but differ in the degree to which they think this world can be known.

A model for a physical theory is to be understood as a set of entities and relations which fulfils the following functions:

a. It provides an interpretation of the terms of the theory, such that the theoretical propositions thereby created are true of the model. In this respect models in the physical sciences are like models in logic.

b. It serves as an idealised representation of those entities, structures, processes, properties and so on in the real world which are implicated in the production and origin of the phenomena in the field of the theory. In this respect a model serves as a virtual or surrogate world (Giere 1988; Aronson 1991).

For the purposes of this paper a theory will be taken to be a structured set of propositions, a certain fragment of the discourse of science. The function of a theory is to map phenomena onto those real world states, processes, entities or structures that are thought to be implicated in their production. For example the theory of continental drift maps the shape and location of the continents, the lay out of mountain ranges, the occurrence of earth tremors and so on onto the locations and relative motions of tectonic plates.

The mapping of phenomena onto (generally) unobservable but real states, structures, processes and so is brought about by establishing a discursive relation (logical, semantic, analogical and so on), between a description of an abstract or idealized version of the phenomena of the field of the theory and a description of a model of the states and processes thought to be implicated in the production of the phenomena. For example the abstract and idealized empirical law describing the relation between conductivity and temperature can be derived, with suitable assumptions and boundary conditions, from theoretical laws describing the behaviour of electrons in atomic lattices. To take a more familiar example, '$PV = RT$' describes the behaviour of an idealized gas sample, while '$pv = 1/3nmc^2$' describes the behaviour of a model of the unobserved process which produces the gas phenomena (Cartwright 1983).

A theory is internally related both to the idealised phenomena it purports to explain and to the virtual world of its model. If the idealised version of phenomena or the model of the states and processes implicated in the production of the phenomena change so do the propositions describing them. The fit of theory to idealisation and model is always perfect. For example when Amagat's experiments led him to revise the idealisation of the relation between pressure and volume of an enclosed sample of gas, he revised Boyle's Law to fit the new version of the phenomena, while his revision of the point-molecule model to include molecular volumes ('b'), led him to revise the law of molecular behaviour to '$p(v-b) = 1/3nmc^2$'.

Schematically the total 'cognitive object' of which a theory forms the discursive component has the following structure:

1. Phenomena represented by Idealisations described by Empirical Laws

2. Unobserved (-able) states of the world represented by a Model described by Theoretical Laws.

Models and what they represent are related by relevant (weighted) similarities and differences. Ideally the Empirical and Theoretical Laws are linked together into a coherent structure by deductive chains and semantic rules.

The relation of theories to real states of affairs, structures, processes and so on, is always indirect, mediated by the match between their models and the relevant aspects of reality, a thing to thing relation.

3. FOUR RELEVANT PROPERTIES OF THEORIES

I shall assume that it would be widely agreed that a 'good' theory should display empirical adequacy, ontological plausibility and manipulative efficacy.

a. Empirical adequacy: that the theory (as a set of statements) coupled logically with another set of statements which describe some conditions of application of the theory, yield, deductively predictions and retrodictions that turn out predominantly true.

b. Ontological plausibility: that the model for the theory be a subtype of a well-established supertype, for example that molecules be of the same type as Newtonian particles.

c. Manipulative efficacy: that operations performed on some material system, guided by the assumption that observed features of reality do resemble the properties of the relevant model are more or less successful, for example that attempts to manipulate the orientation of the nuclei of silver atoms by means of a magnetic field (the Stern-Gerlach experiment) succeeded.

d. An advocate of Realism would require a fourth property for a 'good' theory, namely that it should display some degree of verisimilitude, that is that the model on which the theory is based and the reality it purports to represent are well matched.

Of the above four properties of theories ontological plausibility and verisimilitude are directly related to the ontological status of models. How should we present this? A useful procedure has recently been proposed by Way (1991). Her treatment depends on borrowing a theory of knowledge representation from the field of Artificial Intelligence. According to that theory the basic structure of a knowledge representation system is a type-hierarchy. The properties of type-hierarchies can be employed to make sense of the organization of the cognitive structure of the kind of theories we find in the physical sciences, in such a way as to present clearly the role of ontological constraints on the construction of models and the assessment of their verisimilitude.

The basic principle of Way's analysis is that a type-hierarchy represents the structure of a model world as a hierarchy of subtypes under a common supertype. The supertype condenses the working ontology of some field of natural science. In the case of the molecular model of gases the relevant supertype expresses the Newtonian mechanical, world picture. . . .

A subtype 'inherits' the properties of the supertype(s) under which it falls, but not vice versa. Subtypes resemble or differ from one another in numerous respects, which can be ranked, relative to the task in hand. Degree of resemblance between types in a type-hierarchy can be measured by the following metric: the fewer the nodes between subtypes the more similar they are (Aronson, Harré and Way 1994).

The relative ontological plausibility of a model can be expressed as the degree of weighted similarity in relevant respects between the type instantiated in the model and the types known to be instantiated in some real world, entities, events, structures, processes and so on.

The relative verisimilitude of a theory can be expressed as the degree of weighted similarity in relevant respects between the type instantiated in the model and the type or types instantiated in the reality it represents.

A theory, the discourse describing the model or virtual world in point has, vicariously, that degree of verisimilitude that expresses the degree of resemblance in relevant respects of the type of the model world it describes to the type of the aspects of the real world it purports to be about. The doctrine of Realism, from this point of view, can be expressed as the thesis that *models* of the world can be assessed for verisimilitude (Aronson 1991).

It is important to bear in mind that both the ontological plausibility and the verisimilitude of a theory, when that can be determined, are always relative to the type-hierarchy currently underpinning the physical sciences.

4. BASIC REQUIREMENTS FOR A PRO-REALISM ARGUMENT USING ARONSONIAN VERISIMILITUDE

To cut any ice a versimilitude-based argument for realism must lead to two conclusions

(i) that models do more or less resemble the relevant aspect of reality;

(ii) that there are criteria by which the degree of relevant resemblance can be assessed prior to any independent comparison between model and reality.

These propositions express the thesis of Convergent Realism in terms of the concept of a model as I have introduced it above. However, as an essential preliminary to arguing for Convergent Realism, it is necessary to defend the practice of reading theories 'realistically', that is to defend Policy Realism: that it is rational to read theories fulfilling certain desiderata as if the models which they described resembled aspects of the real world in relevant ways. This will lead to model control of research projects, since it would make sense to set about trying to test the degree of resemblance between the model controlling the research and the aspect of the world it purports to represent. The rationality of Policy Realism is equally compelling when research fails to disclose any resemblance as when it succeeds.

Only if Policy Realism can be vindicated does Convergent Realism make sense, that is only if it makes sense to read theories as possibly verisimilitudinous does it make sense to ask which models, if any, of some set of theories do resemble aspects of the world, and how we could know this. The programme for the defence of realism must then proceed by first showing that Policy Realism is well supported inductively, and then go on to show that there is evidence from the history and practice of science which supports Convergent Realism.

5. THE FORM OF THE INDUCTIVE ARGUMENTS FOR REALISM

The evidential premise of an inductive argument is a report that, as a matter of fact, in cases which have been examined, that is in which the fit of the model and what it represents has been assessed, all or some of the three basic properties of a 'good' theory are correlated with a reasonable degree of verisimilitude. The inductive conclusion is that the same correlation holds for theories which have not been examined.

The evidential properties are empirical adequacy, ontological plausibility and manipulative efficacy, and the projectible property is the verisimilitude of the model the theory describes. The argument must, in the end, be inductive, since it is logically possible that a theory should be empirically adequate, ontologically plausible, manipulatively efficacious but not verisimilitudinous.

The sort of historical evidence which would be needed to support the induction would consist of a catalogue of cases in which the majority of theories which exhibit the three evidential properties also exhibit the projectible property.

The degree to which any theory meets criteria of empirical adequacy, ontological plausibility and manipulative efficacy can always be ascertained, though sometimes with great practical difficulty and in some cases, such as geology and astrophysics, only indirectly. It is the alleged theoretical difficulties involved in ascertaining the degree to which the model related to a theory is verisimilitudinous in the above sense when the relevant states of the world are unobserved or worse, unobservable, that create the philosophical difficulties for a thorough-going realism. If all we can have are the above three criteria to assess the value of a theory then the question of its match to any reality other than the field of phenomena to which it is addressed, is empty. The progress of science can be described only in pragmatic terms. The inductive arguments will be constructed out of these concepts. They involve reasoning from known cases in which theories which were empirically adequate, ontologically plausible and manipulatively efficacious could be shown to have incorporated verisimilitudinous models because the states of the world the models purported to represent were, or became observable, to cases of theories which satisfied the three 'working' criteria but for which the verisimilitude test of their models was impossible. I shall be offering arguments to support the thesis that it is legitimate to conclude by induction over the population of theories meeting the three working criteria that all should be deemed to be built on models which are verisimilitudinous. Whether a head count of theories and their fate favours the induction is a question I will not address here. My aim is to defend the legitimacy of the inductive form of argument.

6. DEFINING THE INDUCTIVE DOMAIN

The domain over which the 'Realist' inductions will range is a set of theories ordered by reference to the epistemic status of their core models. Models, in the sense of this paper, fall into three classes:

R1: those which are readily assessed for their resemblance to perceivable entities, properties or processes.

R2: those which could be assessed with respect to their resemblance to possibly perceivable entities, properties or processes with a technology by means of which they might be observed to become available.

R3: those which are thought to resemble entities, properties or processes which, in the present state of our knowledge we believe never could be observed.

Evidence for a 'Realist' induction accumulates unproblematically by establishing correlations between one or more of the three properties of 'good' theories and the verisimilitude of their models, for theories whose models are of the R1 type. The cognitive/practical successes of plumbers, surgeons and car mechanics provide a rich basis of evidence for a strong correlation between epistemic adequacy, ontological plausibility, manipulative efficacy and a highly 'resemblant' model at the core of their working theories. The problem is to justify the induction from cases of this sort to theories the models of which are of the R2 type and finally of the R3 type. I will show first that it is possible to justify an induction from evidence for 'Realism' accumulated in the R1 domain to theories whose models are of the R2 type. It will emerge that the evidence adduced by philosophers (for instance Lipton 1991) for this induction is not strong enough to support a 'Realist' induction theories of the R3 type. Without that induction Convergent Realism is atrophied.

7. THE 'MOVING BOUNDARY ARGUMENT'

This argument has been presented in several different forms (Harré 1961; Lipton 1991). It is based on the principle which Aronson (1988) has called 'epistemic invariance'. It is important to be clear about what it can and cannot be used to establish. In order to set out the argument we must call on the distinction made above between the three possible epistemic standings that a model of a theory may have.

The history of science shows that the boundary between virtual worlds, that is models which belong to the R1 category and models which belong to the R2 category, is not fixed. It has turned out that some aspects of the world or some class of entities which were not observable have become observable and can now be compared with the model which was offered as a surrogate for reality in the construction of the relevant theory. For example the improvement in microscopy, the development of telescopes of increasing power and sophistication, the invention of tunneling microscopes, the possibility of sending probes into regions previously inaccessible and so on, have allowed a vast range of models belonging to the R2 category to be compared for their resemblance to the entities, properties, structures and processes that they stood in for.

The fact that the R1/R2 boundary is not fixed shows that the policy of making a realist reading of models drawn from a type-hierarchy of virtual worlds belonging the R2 category is entirely reasonable. Technological advances have allowed for their direct empirical assessment. Both positive and negative outcomes of research aimed at verifying the verisimilitude of R2 type models support Policy Realism. Policy Realism can be extended by induction to theories whose models represent causal processes etc. which are assumed to be forever and in principle unobservable. It would generally be conceded, I believe, that it is reasonable to make a realist reading of such theories, proved they meet the two criteria of empirical adequacy, and ontological plausibility.

What about Convergent Realism? Convergent Realism could be established for theories whose models were of the R1 or R2 type, and met the above two criteria, the evidence favoured the induction from the possession of the evidential properties, empirical adequacy and ontological plausibility, to verisimilitude, since for such models their verisimilitude could be tested empirically. The induction to Convergent Realism runs as follows: The models of theories which are empirically adequate and ontologically plausible have been shown to be pictorially resemblant for RI models. Many cases of models of empirically adequate and ontologically plausible theories being pictorially resemblant for R2 models have been recorded. So, in general, empirically adequate and ontologically plausible theories whose models belong in the R2 category, prior to the shift in the boundary of what is observable and what is unobservable can be expected to be verisimilitudinous, that is to have models of a high degree of pictorial resemblance to the reality they represent.

Allowing for a difference in terminology we find Lipton presenting the 'moving boundary' argument but then extending it, without further augmentation of the evidential grounding, to theories whose models are of the R3 type, (Lipton 1991, p. 179) as follows:

> [the realist's] reasons for trusting his method in the case of observables also supply him with reasons for trusting his method in the case of unobservables ... just as his success in sometimes observing the causes he initially inferred supports his confidence in his method when he infers unobserved but observable causes, *so it supports his method when he infers unobservable causes* (my italics).

But since the boundary between what is observable and what can never be observable can never shift so as to reveal the real entities for which R3 models are built to resemble, the inductive argument from empirically established verisimilitude to empirically establishable verisimilitude cannot be extended to them. A determined anti-realist can always point to the possibility of alternative models for some formal structure (such as quantum mechanics) or for fully alternative theories (such as the one and two fluid theories of electricity). We cannot extend the induction to theories with R3 type models by reason of Clavius' Paradox, the underdetermination of theories for truth (or models for verisimilitude) by any observational data.

The purpose of this paper is to argue that the addition of the third evidential property, manipulative efficacy as a criterion, does permit an induction to the verisimilitude of theories with models of R3 type, sufficiently strong to give some support for Convergent Realism, but not as strong as the argument that supports an overall Policy Realism. The importance of the 'moving boundary' argument is not only that it supports Policy Realism, but it also identifies a segment of the set of all scientific theories for what empirical adequacy, ontological plausibility *and manipulative efficacy* are correlated with verisimilitude, in the model-as-picture sense. This segment will provide part of the evidence for the projectibility of the property of verisimilitude to the 'untested' cases for which the degree of relevant resemblance between model and the aspect of reality it purports to represent cannot be empirically ascertained.

8. USING THE HAND TO REACH BEYOND THE RANGE OF THE EYE: ADDING THE THIRD CRITERION

In his *Origins of forms and qualities* Robert Boyle (1666) offers an interesting argument for the reality of the mechanical corpuscles which he took be the unobservable basis of the physical world. The argument makes essential use of the idea of indirect manipulation. Boyle was well aware that the doctrine of primary and secondary qualities that he shared with John Locke meant that the 'textures' (molecular level structures), responsible for the observed qualities of bodies were beyond all possible perception, at least with the equipment available at that time. From this Locke drew his pessimistic 'nescience' conclusion, that though we knew there must be real essences we would be unable to incorporate them into natural science. Boyle argued that it was possible to manipulate material corpuscles at the molecular level to bring about observable changes in another sense modality than that in which the manipulation had been performed.

Manipulations which could only lead to changes in shape could nevertheless be used to bring about changes in colour. The philosophical question to be addressed is of course how we would know that a certain manipulation of a material set up was indeed effectively a manipulation of unobserved constituents of a well-defined type.

Boyle argues that we know from observations that 'mechanical' procedures bring about observable changes in 'mechanical' properties, in many cases. 'Mechanical procedures' are manipulations which either change the internal structure ('texture') or the state of motion of the parts of a material thing or both, that is they bring about changes in the quantitative or primary qualities of bodies, their bulk, figure, texture or motion. We know, for example, that by chopping a log we can produce chips, that by the impact of a moving body on a stationary body we can put it in motion, and so on. We also know that there are cases in which quantitative or mechanical procedures acting upon the primary 'mechanical' qualities of observable material stuff bring about changes in its secondary or 'non-mechanical' qualities. Crushing a green emerald yields a white powder. Crushing is a mechanical procedure, an action upon the primary qualities of the stone, breaking it up into parts, and thus changing the texture. The key move is an induction over observed cases of 'mechanical' procedures which have 'mechanical' effects, to the general principle 'Mechanical procedures have only mechanical effects'. I shall call this the 'Boyle Principle'. The corresponding generalization for electromagnetic manipulations I shall call the 'Faraday Principle'. If a mechanical procedure can have only effects on the primary qualities of a body, and yet sometimes such a procedure can be seen to bring about changes in a secondary quality, such as its power to induce ideas of colour or warmth in human observers, either the principle is false, or the procedure must be a manipulation of unobserved primary qualities, such as the texture or molecular structure or motion of parts, resulting in a change in its secondary qualities. By citing many instances of changes in ideas of secondary qualities (taste, colour, felt temperature, medical powers etc.) brought about by the manipulations that should change only primary qualities, Boyle sets up an inductive argument for the corpuscularian metaphysics. In our terminology he has tried to show that the virtual world of Newtonian mechanics provides through its local models, the best, most verisimilitudinous picture of the physical universe.

Boyle's argument can be reconstructed using a great many modern instances (Hacking 1983) ['if you can spray them they are real!]; Harré 1986). The Stern-Gerlach apparatus and its use of magnetic manipulations to separate the constituents of atomic and molecular beams according to the magnetic orientation of the ions is striking example of a manipulation which produces observable changes via the effect upon some unobservable property of the real system guided by the model which depicts unobservable aspects of the world. This example is important in the generalization of Boyle's argument, since it shows that the force of the reasoning does not depend on the highly contestable distinction between primary and secondary qualities. It requires only that the effect be of a different type from that which the manipulation is usually observed to produce, a change in orientation of an image rather than a change in magnetization. So we seem to have an argument which permits the extension of the induction from an 'observed' correlation between empirical adequacy, ontological plausibility, manipulative efficacy and representational quality of models of the R1 category to the conclusion that the fourfold correlation of properties can be ascribed to theories based on models of the R3 category.

There is an obvious Humean objection to Boyle's claim that the empirical observations of cases of mechanical procedures bring about changes in non-mechanical qualities. If the Humean regularity theory of causality is correct we are not obliged to go beyond the observed correlation of mechanical cause and non-mechanical effect. Why should we, so to say, dip into unobservable background processes, hidden changes, when we can simply correlate the crushing of an emerald with a change in its colour? Thus, however ingenious and original Boyle's argument may be, it also seems to need upgrading to meeting the Humean objection.

9. 'BOUNDARY' WITH 'BOYLE': STRENGTHENING BY MUTUAL SUPPORT

We have seen that neither the moving boundary argument nor the Boyle manipulation argument is free from reservations. But if these arguments are taken together they are complementary in that each makes up for the weaknesses of the other. 'Boyle' takes care of the objection to 'Boundary', that physical science has always made use of models purporting to represent real world entities, properties, processes etc. which at best have the status of fictions since they could never be observed. According to 'Boyle' we can make reality claims for the unobservable *since we can manipulate what we cannot perceive.* Our ground for claiming this, as we saw with Boyle's presentation of the argument, is a good induction from the observation of the effects of this or that type of manipulation. 'Boundary' takes care of the Humean objection to 'Boyle'. The justification for the claim that, say, magnetic operations modify unobserved structures of matter, is strengthened by cases where changes which, by reason of technical problems, were unobservable became observable through technical advances. For instance the fact that we can now observe the reorientation of elementary magnets as an effect of the magnetization of iron strengthens our faith in the general principle that magnetic manipulations always have direct magnetic effects. So in cases in which we observe a non-magnetic effect to be correlated with a magnetic manipulation we are licensed to infer that it is brought about indirectly.

10. THE INDUCTION ARGUMENT IN FULL

Let us now develop the inductive argument in full, that we have good inductive grounds for holding that if a theory achieves a certain degree of empirical adequacy, is ontologically plausible and has a certain degree of success as a guide for effective manipulations, then the model which it describes is of greater verisimilitude than that of a theory fulfilling the three criteria less adequately, that is, it resembles some entity, property or process in the real world more accurately than does the model of the rival theory.

For models (and the virtual worlds that sets of models subsumed under a common supertype make up) that can be compared with relevant aspects of the real world by observation we have, as a matter of fact, found that empirical adequacy, successful predictions and retrodictions, ontological plausibility and manipulative success, that is bringing about the results we would expect from such manipulations guided by the model taken as a picture of the real world, is usually followed by observational proof that the model resembles the world in the relevant respects. Digging out foxes' earths, fixing cars, unstopping the drains and a host of everyday practices testify to the criterial force of the three 'evidential' attributes of theory.

For models (and virtual worlds) which could be compared observationally with the relevant aspects of the real world if certain technical advances could be made, it has also turned out, as a matter of fact, that when the boundary between the observed and unobserved but observable has been moved, the models in theories which have met the demands of the three criteria have, for the most part resembled the world, as it is now revealed, in relevant respects, and those which have not, do not. There are many examples of such cases to be found in chemistry and biology. Ionic and surface chemistry provide a rich source of examples. The ionic model of the imperceptible processes and ephemeral structures of the chemistry of solutions offer all sorts of hints and guides as to the development of manipulative techniques for operations to be performed that would affect ions, which for the most part, had the results that were to be expected. With modern microscopy ionic dissociations, transport and so on can be observed.

Similarly there is scarcely an issue of *Science news* without a false colour photograph of the atomic structures that form the surfaces of all sorts of solids.

For models (and virtual worlds) which, we believe, never could be compared observationally with relevant aspects of the real world, we have, in the considerations just advanced, the basis for extending the inductive argument just sketched to theories whose models are of the R3 type. I claim that the inductive arguments already set out for the first two cases give inductive support to the philosophical thesis that models (and virtual worlds) from theories describing which are empirically adequate (that is enable the successful prediction and retrodiction of experimental and observational results), and ontologically plausible (that is the models are instantiations of subtypes under a supertype which also comprehends subtypes instantiated in entities etc. which we have already observed) and which best facilitate successful manipulation of the unobservable substructures of the world, are the most verisimilitudinous, in the sense of thing-to-thing resemblance in relevant respects. In extending the induction from R1 type models to R2 we were able to carry through both observational and manipulative criteria. But in extending it from R2 type models to R3 type the observational criterion necessarily drops out. This is where Boyle's great insight bears upon the issue. The manipulation criterion is common to models of all three epistemic standings!

In summary the sequence of arguments runs as follows:

a. For theories whose models are of the RI category, empirical (in)adequacy, ontological (im)plausibility and manipulative (in)effectiveness and observational (dis)confirmation of the relevant similarities and differences between models and the world are regularly found to go together.

b. For theories whose models are of the R2 category, empirical (in)adequacy, ontological (im)plausibility and manipulative (in)effectiveness are good predictors of observational (dis)confirmation of the relevant similarities and differences between models and the world when the boundary between the observable and the unobservable is moved. The extended inductive argument can now be stated quite shortly.

c. For models (and virtual worlds) which could never be compared observationally with relevant aspects of the real world successful manipulation in accordance with their picture of reality allows us to claim that we have inductive evidence that the entities etc. that we believe we have been manipulating resemble the entities which etc. constitutive of the model we have been using to guide the manipulation. That inductive evidence is, of course, the successful research programmes for models and virtual worlds that have the R1 or R2 epistemic status. Boyle's argument supports the generalization of the realist thesis to theories whose models are in the R3 category.

The induction from the successful methods of motor mechanics and pest controllers to a realist interpretation of the Stern-Gerlach experiment involves a move from direct to indirect manipulation, the central step of Boyle's argument. Why should we accept it? The answer is very simple and in accordance with the spirit of the way of tackling metaphysical problems exemplified in this paper. *The Boyle Principle (together with the Faraday Principle) and their generalization in the above argument should be taken to be a part of physics.* There is no conceptual, a priori argument which could be given for it. So far as I can see it would not be self-contradictory to deny it. Indeed we have examined the possibility that a determined Humean might do just that. The Humean was defeated only by an empirical induction. Taking the moving boundary argument with the Boyle manipulation argument we can find inductive support for the claim of the ontological significance of indirect manipulations. In case they succeed we have inductive support for a belief that the relevant ontology is verisimilitudinous, and in case they fail we have inductive disconfirmation of the verisimilitude of that ontology.

11. THE FINAL STEP: THE INDUCTION OVER TYPES

It has been a feature of the line of argument laid out in this paper that all comparisons between models, one with another and with aspects and features of the real world, have been carried out through their mutual relations on a type-hierarchy, dominated by one supertype. But is it not the case that our supertypes are underdetermined even by the joint use of the three criteria brought together in the realist inductions? Could not we reapply the Paradox of Clavius, and raise the objection that there are indefinitely many systems from which a picture of our system, as we observe, imagine and manipulate it, could be derived? In order to meet this objection we need another induction, an induction over types. Suppose that, as a matter of fact, the historical record shows that a certain supertype has engendered a type-hierarchy, instantiations of the subtypes of which have been observationally assessable for relevant similarities one with another, and that certain aspects and structures of the world have also instantiated subtypes of this type-hierarchy. Should it not be adopted, *ceteris paribus,* for the construction of models for unobservables of the third epistemic standing? The Newtonian type-hierarchy is ontologically plausible since it is instantiated in many familiar objects and the models used in hosts of theories have met the two remaining criteria, empirical adequacy and manipulative efficacy. If the Newtonian supertype has been successfully employed for so many research programmes should it not be extensible to those yet to come? History is not so favourable to this induction. The Newtonian type-hierarchy has not been so successful when generalized as the source for all physical models. For a century the 'advances of science' have required the invention of type-hierarchies resting on exotic supertypes.

At this point in the argument the first intimations of a general limit to scientific progress appear. The argument so far has tended to show that placing the limit of our knowledge of what there is at the bounds of human observation, as it has been placed by van Fraassen (1980), is not rationally sustainable. The argument with which this conclusion was established does not, however, license the conclusion that there are no bounds to the metaphysical advances of science as it can be practised by human beings. The Boyle manipulation argument can tolerate a wide variety of ontologies, that is type-hierarchies based on a variety of different highest supertypes. The essential point in Boyle's argument is the induction on manipulative procedures. Mechanical procedures are observed to have mechanical effects, magnetic procedures magnetic effects and so on. Just so long as an exotic type-hierarchy includes subtypes that are susceptible to manipulations of the kind certified by a Boylean induction then the realist meta-induction of the last section can be applied to them. Just so long as the models instantiating subtypes of the exotic hierarchy are taken to be sensitive to changes in the gravitational, electrical and magnetic fields, the fields with which physical manipulations are commonly accomplished, the beginnings of an argument to the existence of a class of exotica can be constructed.

But what of the other two criteria, empirical adequacy and ontological plausibility? The former presents no problems other than a challenge to the ingenuity and skill of the theory constructors no different in kind from that presented by any theory if the first and third criteria are met. It seems to me that what is taken to be ontologically plausible must be allowed to change. That means that science must countenance inductions over type-hierarchies. The Newtonian type-hierarchy has failed the inductive test, while the extended electromagnetic type-hierarchy has passed it. For example the empirical adequacy and manipulative success of the photonic type-hierarchy, subtypes of which include not only the luminiferous photon-type but *W* and *Z* particle-types, in providing models for the virtual particles of quantum field theory, must speak in favour of the ontological plausibility of the generic photonic supertype. This suggests that it is not ontological plausibility that limits model construction in accordance with Way's type-hierarchy treatment, but the manipulability criterion.

12. A FINAL OBJECTION REBUTTED

The induction to Convergent Realism would fail if it could be shown that the 'pragmatic' property of manipulative efficacy of a theory/model is simply a special case of the logical property of empirical adequacy. It might seem that it would be enough to point out that manipulative efficacy is revealed in a theory-guided material practice while empirical adequacy is revealed in discursive procedure the criteria for the correctness of which are logical. The practice and the procedure are of radically distinct categories, and therefore irreducible.

But it might be objected that this short way with the problem overlooks the fact that the claim that a practice is efficacious is supported only if the outcome, the phenomenon produced to order, so to say, fulfils certain criteria, that is it is a phenomenon of the type to be expected. These criteria are defined discursively, by the same hypothetico-deductive procedure that would have established empirical adequacy.

The reply to this objection takes us to the heart of the argument. There are two possible hypothetico-deductive procedures which would permit an efficacious manipulation to count as a test of empirical adequacy of the theory which guided it. One case is when the empirical adequacy of a mere Humean correlation between implementation of the procedure and its outcome provides the hypothetico-deductive structure, on the basis of a 'covering law'. Thus the Stern-Gerlach experiment merely demonstrates the empirical adequacy of the 'law' that activating a certain kind of circuit is correlated with a characteristic change in the pattern of light on a screen. But this requires physics to include an indefinite number of ad hoc 'laws', involving standard physical concepts, which are 'out of step' with such well-established principles as those I have called Boyle's and Faraday's Principles, each of which is a summary of a well-founded branch of physics. If indeed one were to accept the 'covering law' reduction of the Stern-Gerlach experiment it would be a disconfirmation of the Faraday Principle, and so effectively an abandonment of electromagnetism. But this point might be conceded while the determined reductionist could argue that manipulative efficacy should be taken as test of the empirical adequacy of the full-scale theory/model/type-hierarchy, and no more than that. But in that case the reductionist must invoke either the Boyle or the Faraday Principle (or something like it) as part of the theory. But this is tacitly to accept the ontological account of the efficacy of the manipulation with respect to the Humean phenomenological correlation revealed in the Stern-Gerlach experiment between activated circuits and reoriented images. The empirically well supported Boyle and Faraday Principles block the use of Clavius' Paradox (underdetermination of theory by data) to undercut the proof of the ontological assumptions involved in the experimental procedures by reference to the efficacy of the manipulation.

So either the putative reduction tends to privilege a weak phenomenological correlation over well-established laws and principles, or it involves tacit acceptance of an irreducible ontological assumption. This assumption amounts to the principle that what is being manipulated by the overt procedure is a covert structure, process, entity or property. The reduction is either ad hoc or ontologically concessive. So it fails to convince in either case.

A different kind of defence of the autonomy of the manipulative efficacy of procedure as an independent criterion for the verisimilitude of the model that guides it, invokes the general incompleteness of the discursive presentation of many theories. There are many procedures that can be demonstrated to be efficacious by 'rule of thumb' in the absence of a discursive presentation of the theory sufficiently articulated to permit a water-tight hypothetico-deductive demonstration that the phenomena produced to order by the manipulation are proofs of the empirical adequacy of the theory. Medical science is full of examples of the efficacious but indirect manipulation of unobservable structures, processes, entities and properties in the admitted absence of a well-articulated hypothetico-deductive derivation of propositions describing the results.

We can now rank the three criteria on which the piggy-back sequence of inductive arguments was constructed. Ontological plausibility is parasitic upon empirical adequacy and manipulative efficacy, in that an exotic type-hierarchy which is ontologically innovative, is defensible for verisimilitude in the sense of Aronson, the sense made use of throughout this paper, just in so far as instantiations of its subtypes meets the primary criteria, as we have just identified them. This step provides a philosophical ground for the defence of the important thesis that ontologies have an empirical basis, even when none of their subtypes, in a Way-style presentation are instantiated in entities that can that can be commonly be observed. Without this step the inductive argument for Convergent Realism, in the version defended here, would be open to the objection that it is implausibly ontologically conservative.

I have tried to show that induction from 'observed' cases of correlation between the satisfaction of the three criteria based on the 'evidential' properties of theories and the 'projectible' property, that is between the satisfaction of the criteria for 'good theory' and the demonstrable verisimilitude of the model of the theory, to unobserved cases, that is cases in which the verisimilitude of the model cannot be assessed empirically, is reasonable. Theories which meet the three criteria are more likely to be verisimilitudinous than those which do not, and the more a theory meets them relative to the performance of a rival, weaker by the same criteria, the more likely is the model of the former than the latter to resemble the relevant aspect of reality. Since this is an inductive argument, the future of science may throw up a countervailing multitude of cases that defeat the induction. I hope to have shown the inductive form of argument in this context is rational. It is another matter to accumulate sufficient evidence for Convergent Realism.

CHAPTER FIVE

MODELS AND THE REALISM DEBATES

In recent years we have all come to see that 'realism' is not the name of one philosophical position, even within the philosophy of science. Nor is there just one range of arguments for and against a realist position. Nevertheless, 'realism' and 'anti-realism' are generic terms for clusters or families of doctrines for each of which there are loosely related ensembles of arguments. Part of my purpose in writing *Varieties of Realism* (1986) was to demonstrate that there was a significant polarity in the cluster of realisms, Realist doctrines differed not only in their metaphysical underpinnings, but also in their relative vulnerability to fairly traditional anti-realist arguments. At the time I wrote that book I thought that no one realist doctrine could be successfully defended for the physical sciences at every stage of their development. I now see that there are ways in which the two main varieties of realism I want to defend—viz, a strong 'policy realism' and a weak 'convergent realism' — are inter-related, though they are not, I believe, either mutually reducible or simultaneously applicable.

The first step in clearing one's mind on the nature of realism should be an attempt to catalogue the most obvious varieties. For the purposes of this commentary I want to highlight a major division into epistemic approaches and pragmatic approaches.

1. The epistemic approach uses concepts like 'truth', 'falsity' 'verisimilitude' and so on to characterize its variety of realism. The most conspicuous modern form of the epistemic approach is 'convergent realism'. According to this approach the greater predictive success (empirical adequacy) of a theory, the more truly it depicts the world (the greater its verisimilitude). By varying one's conception of empirical adequacy, say by taking it as persistent survival in the face of vigorous attempts at falsification, one can arrive at a cluster of 'realisms' that includes the doctrines of both Newton-Smith and Popper.

In common to all versions of the epistemic approach there is the *Principle of Bivalence.* According to this principle the statements of a theory are true or false by virtue of the way the world is whether we know it or not. To apply the principle of bivalence to scientific research we need another principle to support claims for verisimilitude for as yet untested statements. Of equal prominence then in recent discussions of realism has been the *argument to the best explanation.* This argument has a mundane use in which we say that the best explanation of the predictive success of this or that theory is that it is true. And it has a transcendental use in which we say that the best explanation of the long-run success of the physical sciences is that science, as a whole, is getting nearer the truth. There are, it is not hard to show, alternative explanations of this success, that are not so good — for instance, that the success is the result of a vast, long-running coincidence. Of course, much of the force of the argument hinges on the criteria for 'best'. Lipton (1985) has shown that there is an essential ambiguity in the criteria for 'best' which greatly weakens the argument.

2. The other pole is occupied by a family of positions based on pragmatic notions like 'intervention', 'manipulation', 'material practice', and so on. The concepts of truth and falsity give way to notions like reference and denotation. Science is seen as a practical rather than as a cognitive activity and its products as material things rather than propositions. . . . .

I have presented the two main families of realisms as polarities rather than antitheses. They interpenetrate one another to some extent. By treating the establishment of reference in terms of the satisfaction of certain propositional functions, the relation of reference is transformed from a physical link between an embodied scientist and a material being into a semantic indication determined by a set of true and false propositions. On the other hand, polar oppositions germane to practice, such as success or failure, seem to mimic the polarity between truth and falsity.

Realists of the epistemic persuasion seem to take for granted that the aim of science is the enunciation and testing of laws. If the truth of laws eludes us by virtue of traditional objections to inductive universalizations, then perhaps we can be sure of the falsity of some conjectured laws. The tendency to focus on laws in this way is closely correlated, not surprisingly, with a tendency to take theories simply in their discursive form, as sets of propositions ordered by the deducibility relation. However different their epistemologies may have been, this conception of theory is shared by both Hempel and Popper, and it accounts in part for the disparities between their philosophies of science and scientific practice.

For those of the 'pragmatic' persuasion the role of laws is secondary to that of interlocking structures of analogies, models and metaphors in the organization of scientific knowledge. The Hempel-Popper deductivist conception of theory is seen as the realization of a rhetorical convention for scientific writing rather than as the necessary basis for an account of scientific cognition and of the genesis and development of scientific theories in the thought collectives of real science. Theories, as philosophers have tended to analyse them, are momentary abstractions from evolving theory-families. The theory-family idea has appeared from time to time in different guises. Whewell wrote of 'the development of an idea'; Ludwig Fleck used the expression 'thought-collectives'; Thomas Kuhn called such entities 'paradigms' and Lakatos described them as 'research programmes'.

Given the vulnerability of the epistemic variety of realism to quite simple sceptical arguments can we find a better line of defence for the pragmatic variety? If we can, how can at least some of the valuable aspects of the epistemic variety of realism be reconstituted? I have in mind such concepts as 'scientific progress', 'increasing verisimilitude', and so on.

The historical pageant of science has been presented as if it were a linear progress from the verifying or corroborating of the laws of the observed to those of the unobserved to those of the unobservable. At each stage greater hazards to fortune are offered in that subsequent work stands a greater chance of revealing inadequacies in what has gone before. But there were richly elaborated accounts of the transcendental realm as integral parts of physics long before the present era. A case can be made for saying that the overall pattern of thought in the physical sciences has changed very little, despite huge changes in content and in the sophistication of the experimental equipment and of the mathematical tools. A methodological insight of Archimedes or of Robert Boyle ought to hold good for today's physics. The physics of the past ought to be as good a test object for efforts to make judgements about it intelligible under our alternative realisms as contemporary physics.

Be that as it may, any variety of realism needs a platform in perception relative to which the status of beings described by theory, but which are currently unobserved or even unobservable, can be assessed. If the apparatus I see before me is a subjective phenomenon, hazardously projected into interpersonal 'space', all further discussion about the rights and wrongs of realism is pre-empted.

To escape the traditional shackles of phenomenalism, I turn first to the psychological work of J. J. Gibson, not only for its technical sophistication and its remarkable experimental programme, but also for the depth of the new conceptual system he proposed. In sharp contrast to the traditional picture of the perceiver as the passive recipient of stimuli-producing sensations, integrated thereafter into perceptual structures, Gibson presented his vision of the human perceiver as an active being exploring his or her environment (1979).

The longest-running threat to a realist reading of natural science has been the idea that perception is, at bottom, subjective. Locke's ideas and Hume's impressions are modes of the consciousness of individual people, strictly incomparable with the ideas and expressions experienced by anyone else. It is an easy step forward into Machian sensationalism or back into Plato's cave. How can it be known that there is a common world which scientists collectively study by experimentally exploring it and theorizing about it in a way that is indirectly disciplined by the results of these explorations? At best the existence of this world is a hazardous assumption. In response to this, it could be claimed that we have little idea what it would be like to discover that our hypothesis of a common world and of a community of scientists like ourselves was false. But this kind of move, however nicely elaborated, is not as good as a positive demonstration of the psychological plausibility of the common-sense view that there is a common world and that we jointly explore it.

The traditional view took it for granted that perception is the result of a synthesis of atomistic sensations which are presented subjectively. Gibson held that sensations had little importance as such in the perceptual process. He believed that our senses were integral parts of perceptual systems, which had evolved to explore the ambient flux of energy in which we ourselves cast a shadow, so to speak, and to detect certain higher-order invariants in that flux. These invariants were the effects of physical objects. Gibson called this 'pick up' and the whole process 'direct perception'. There was no synthesis and nothing mediated in any sensory way between things and people. This robust account was supported by a huge range of ingenious experimental evidence. . . .

General ecology is in debt to von Uexkull (1909) for a number of refinements of the idea of an environment. He introduced his distinctions to try to differentiate the various ways that animals and plants are integrated into the physical world. For the purposes of this discussion his most important contribution was the concept of Umwelt. The Umwelt of a species of organism is that part of the material world that is available as a living space to the members of the species by virtue of their specific modes of adaptation, such as distinctive perceptual and manipulative capacities. The same 'total' world contains any number of possible Umwelten. Gibsonian ecological psychology encourages us to think of the physical world we share as human beings, as an Umwelt. It is the living space made available to people through their perceptual and manipulative capacities. If Gibson is right it is the human Umwelt which is the object of study of the physical sciences. I propose to treat experimental apparatus and the advancing techniques of observation as prosthetic extensions of or as 'organs' added to our perceptual systems e.g. telescopes, as Gibson saw them. It would follow that the human Umwelt is changing historically. It would be in the spirit of Gibson's psychology of perception to say that scientists are enlarging or diminishing the human Umwelt, rather than that they are revealing more of a universe which was, neutrally, there. Of course, the universe is richer than the current Umwelt, and I am the last to deny that there is scientific progress.

According to Gibson the most important properties observable in the Umwelt, that is, available to us by the use of our perceptual systems actively to explore the ambient flux, are such attributes as durability, solidity and so on. These are properties of the Umwelt but affordances of the 'total' world, the universe. Why 'affordances'? Because they are material dispositions relative to human activities and practices. 'Using' is the activity correlative to 'durability'. A paved patio affords walking and a particle accelerator affords the photographing of tracks. I do not think it is doing too much violence to Gibsonian ideas to go a step further and take the tracks so photographed as affordances of the apparatus. In the advanced sciences the apparatus with which we manipulate the material world also delimits it as an Umwelt.

According to convergent realism, successive theories are better and better approximations to a perfect representation of a fixed and given but partially unknown world. But in the Gibson-von Uexkull framework I have been putting together there is an enlarging human Umwelt which is that aspect of the 'total' world that our perceptual systems and the apparatus by which we have extended and enlarged them, makes available to us. In one version or another this is an old idea.

For Vico and for Kant it was the root of the intelligibility of society and of empirical experience respectively.

The world will not always afford what we expect (the thin ice that does not, after all, afford walking; the gold that was to be the outcome of alchemical manipulations). This seems a natural way of expressing an insight but ultimately it is an unsatisfactory way of talking. We should not say that the world does not afford these activities or products. There would be no walkability to afford or not to afford were the vertebrates to be wholly avian or aquatic. We now know that the world does afford gold under another manipulative procedure. This is the kind of relativism which I believe Niels Bohr was trying to express in his 'correspondence' principle. The occurrent properties of the world, the 'total' world, which ground the dispositions we ascribe as affordances, can never become available to us independently of the apparatus that we have the ingenuity and technical skill to construct. 'The limitations of my equipment are the limits of my world!' . . .

A main thrust of my own point of view is against the assumption that the prime product of science and the main instrument of its creation of knowledge is the statement (or proposition). Before the proposition is the 'scientific act' (cf. Bachelard 1934, p. 11), a purposeful intervention into a natural system, guided by theory and assessed by reference to criteria of practical success or failure. Of course, theories are cognitive entities. My point is that it is not their truth or falsity that is of importance but their role as guides for action. The significance of such acts was seen clearly by Robert Boyle and exploited by him in his attempts to provide a firm empirical grounding for the 'corpuscularian philosophy'. A 'scientific act' has the following structures: a person acts on something, *X*, by the use of a certain manipulative technique, *T*. The point of the action is to manipulate something else, *Y*, through the medium of *X*. (A person heats a gas-filled tube so as to increase the mean kinetic energy of the molecules of the gas.) Our actor conceives this aim by virtue of tentatively holding a theory about *X* in which the concept '*Y*' figures. Changes in *Y*, that which has been indirectly manipulated, have effects *Z*, which a suitably alert and well-equipped person can observe or detect. (The increase in mean kinetic energy of the molecules of the gas results in an increase in pressure which changes the observable state of a manometer.)

What can now be said about the gas? The pragmatic claim is that we now know that *X* has a certain disposition, a Gibsonian affordance, that is, it affords Z on condition of manipulation *T*. As an affordance this disposition cannot be reduced to *Z*, its overt display, since the human act, *T*, and perhaps a humanly devised detector, is required for the display. By performing scientific acts we can explore the boundaries of the human Umwelt. But we must bear in mind that the boundaries are jointly determined by the nature of the material world, whatever it is, and the range of manipulative techniques we have invented with which to explore it. For me scientific knowledge is not a collection of true beliefs (or if you prefer so far unfalsified conjectures) *about* *X*, *Y* and *Z*. It is the totality of scientific acts we know how to perform in an environment bounded by *X*, *Y* and *Z*.

The remarks above already foreshadow the ontological claim that a policy realist will make on the basis of successful scientific acts. The indirect target of the human manipulations, Y, must also be included in the human Umwelt. This needs argument and I propose to provide it in what follows. It was plainly assumed by Boyle and I hope to show that this was not without some reason. The theoretical sciences gain credit with us only in totalities of such acts, because our concepts of what we are manipulating occur intelligibly only within evolving theory-families. The view I have called 'policy realism' requires that we read theories not as sets of true or false statements but as guides to possible scientific acts. Manipulative practices can be successful or unsuccessful. Theoretical concept denote states, structures, individuals, properties and processes which tentatively enter the Umwelt as manipulables. The metaphysical categories just listed are conservative. This is partly a consequence of the origins of most theoretical concepts in displacements of existing concepts, and partly a consequence of the role of apparatus in fixing affordances. The states of apparatus must, as Bohr pointed out, be perceptible. Once all this is in place programmes for developing techniques for extending the human Umwelt, through its manipulable contents, can be set in motion. The Boylian idea that I have labelled 'scientific act' lets us shift from the epistemological dichotomy between the observable and unobservable to the pragmatic dichotomy between the manipulable and the nonmanipulable, with a consequential shift of criteria of adequacy for theories and, of the demarcation of the boundaries of the knowable.

. . . On the classical account by Richard Gregory, every perception is a kind of judgement, a hypothesis subject to all the troubles of the underdetermination of theory by data, and other forms of inductive scepticism. But for Gibsonians, perception is a practice, an exploration of the Umwelt not a conjecture about it. On the conjectures (Gregory) account, the gap between the observable point of application of my efforts and the ultimate target of my interventions (to borrow Hacking's (1983) term) is a yawning chasm. On the neo-pragmatist view this gap does not exist. Both the point of application and the target of our manipulations are material beings. ('If you can spray them they are real': Hacking.) Basing ourselves on Gibsonian 'direct perception' and 'affordances' we have a shared platform from which ontologically secure inductions can extend the compass of reality, the human Umwelt, in all sorts of directions.

But this cannot be the end of the matter. What of 'galactic jets', 'continental drift', and so on, to which the idea of a scientific act as direct or indirect manipulation can hardly apply?

Of course, a jet of matter several lights years in length, moving under the influence of powerful fields, is not a manipulable as such. But I would argue that a study of how it is accommodated within our systematic physics shows that the concept is such that it has its root in manipulability. From whence comes the conceptual cluster which I used above to characterize the phenomenon? Clearly the origin is in some laboratory manipulation in which plasmas are guided, constrained and so on. Just as the familiar explanation of the auroras is rooted in the experimental manipulations by Ramsay and Rayleigh, so too is the sketch above of galactic jets.

. . . Whatever kind of perceptual judgement we are considering, be it scientific or even of the states of our own bodies, there are historical, cultural and conventional aspects of the total repertoire of reasons for making the judgement. Philosophers have tended to take for granted that that aspect of scientific judgements can be cauterized to leave a crystalline statement of fact. For instance, in an otherwise excellent study of Wittgenstein's philosophy, Robert Ackermann (1988) seems to rest content with the account of language in the *Tractatus,* provided it is taken as the topography of only one of the suburbs of Wittgenstein's city, that district we could call 'Science'.

However, if we take a closer look at the way scientific communities use even simple, singular declarative statements, the *Tractatus* model begins to lose its plausibility. Sociologists of science have collected a corpus of material that illustrates how far the situated discourse of scientists is from that picture. We cannot assume that their utterances are adequately understood if we treat them as attempts to describe the natural world; nor is their linguistic significance exhausted by displaying their logical form say in terms of the predicate calculus. We should look at the larger context of scientific discourse that spreads far beyond the confines of the written scientific paper. In generalizing Boyle's argument, I have portrayed the statements of a theory as guides to action, rather than as hypotheses to be tested for truth or falsity. As such, they appear as parts of a network of fiduciary acts fully intelligible only in the light of our knowledge of the structure and history of those scientific communities within the moral orders of which they have their place. This is a very far cry from the picture theory of the *Tractatus* or any other account of truth as correspondence.

How much of the traditional conception of science can be salvaged once we start to study science as a human phenomenon? . . . There is something of great importance to be learned from studies of scientists at work, but it is not wholly clear just what it is. Recourse to logic by scientists themselves seems to be best seen as a socially motivated strategy of defence of one's own competence or as an attack on the competence of one's rivals, rather than as a standard technique of knowledge production or criticism. It is very difficult to find unambiguous cases from the history of science in which the demonstration of an internal contradiction has led to a theory being dropped or dismissed. Provisional remedies are attempted and case studies show that these proceed only so far as to end controversy. The mid-eighteenth-century demonstrations that Newton's version of 'Newtonian physics' was internally contradictory led to a proliferation of remedial innovations, motivated, it seems, more by metaphysical predilections than by the aim of restoring any purely logical coherence. The history of a theory-family is a story that perhaps would best be told in the framework of dialectics. I know of no serious attempts at such analyses. Most philosophers of science are heirs of Russell and share his distaste for the German tradition. I believe that, in abstracting the logical form of finished scientific writings, philosophers of the logicist persuasion have not revealed, the true skeleton which endows the discourse with its epistemological, ontological and pragmatic qualities as science. Rather, they have picked out a relatively superficial stylistic convention whose investigation belongs to study of scientific rhetoric. . . .

The position I have called 'policy realism' is epistemologically modest. Under certain conditions, it is reasonable to read the terms of a theory as denoting as yet unobserved phenomena. By 'phenomena' I mean Niels Bohr's sense of that notion, that is, the products of the interaction between a noumenal reality and the apparatus and techniques of observation devised by human beings. In such a reading a tentative extension of the human Umwelt is proposed. The connotations of theoretical terms are then taken as the basis of a programme of exploration designed to bring to light the right kind of Bohrian phenomena, if any such are to be found. Taken as moments in the development of a theory-family, individual theories can be shown to differ in respect of a cluster of properties I have summarily called their relative 'plausibility'. An assessment of plausibility is based, so it seems from historical studies, on a history of growing empirical adequacy, that is growing power to predict and retrodict the results of experiments and observations accurately, while metaphysical propriety is maintained. Giving the molecules 'volume' improves the empirical adequacy of the kinetic theory, while maintaining the metaphysical status of molecules as material things. According to policy realism it is reasonable to read a theory that meets these conditions as if its terms denoted real things, and to use the sense of those terms as guides to setting up practical procedures for attempts to manipulate or perhaps actually to disclose their putative referents. Plausibility assessments do not license prior inferences to the correctness or incorrectness of the picture the theory offers of a part of the human Umwelt, before we know the outcome of the use of the exploratory procedures. The view that plausible theories are *a priori* more likely to be found to depict reality accurately is a much stronger form of realism than the policy realism I have been advocating. The stronger view is sometimes called 'convergent realism'. . . . [The] advocate of convergent realism hopes to show that, by and large, the scientists of one generation are not only better at manipulating the material things of the world than their predecessors, but also are able to tell more true stories about it.

Before turning to examine the suggestion in detail I shall briefly review the original arguments for policy realism. The argument proceeds by two steps.

1. There are exemplary cases which demonstrate the wisdom of taking a theory in the policy realist manner. In these cases both the original field of phenomena in which the empirical adequacy of the theory has been tested and the metaphysical propriety of the kind terms in the

theory have instantiations which are objects of ordinary unaided perception. In these cases it is clear that the theory used in this way is plausible in the sense referred to above. Plausibility is not a sign of the verisimilitude of the theory but of the rationality of testing it for referential adequacy. Verisimilitude on this account of realism is always a matter to be assessed *a posteriori*. Scientific progress is conceived ontologically, not epistemically—not an accumulation of truths but of things disclosed in practices.

2. Many scientific theories denote (that is, are used by scientists to refer to) beings that are currently unobservable. With suitable technical developments, beings of such kinds could appear as part of the human Umwelt. There are huge numbers of cases like this: extra-solar planets, bacteria and viruses, geographical features of the sea bed, 'subterranean' Martian water, and so on. The distinction between the extensions of denoting expressions in theories whose domain encompasses observables from the extensions of those whose domain encompasses possible observables is historically contingent because it is relative to the state of technology which sets bounds to the human Umwelt.

The conclusion of this argument is the rationality of making policy realist readings of plausible theories. The objects to which scientists can refer using these theories are of kinds familiar from the domain of the observables. So the procedures for searching for instances and what counts as their finding are also familiar. This is achieved automatically in the model making process. . . . Thus, natural selection is the displaced version of domestic selection, and both denote processes in which only certain plants and animals of a given generation breed.

3. A final step to incorporate theories the extensions of whose domain of beings is doomed to lie beyond the bounds of possible experience cannot be made without severe qualifications. The point is this: the role of theory in our first two domains is, among other things, to sketch the properties of underlying objects and processes that ground the dispositions we assign to things on the basis of the Bohrian phenomena which they afford to a human observer or manipulator. The rationale for taking such theories in this common-sense way is that by virtue of the arguments employed in steps 1 and 2 above we stand some chance of checking out claims about those groundings. But there is no chance whatever of undertaking some programme for a theory whose domain of reference in principle unobservable. The properties we can confidently ascribe to that domain are never more than affordances.

Of course, there is an element of contingency even in the boundary we currently think separates the domains of the unobserved and the unobservable. But the general argument that no extension of the human Umwelt could encompass the whole of that which cannot be observed remains unaffected by that consideration. Yet the structure of theorizing, even given the restriction to affordances, is more or less the same for all domains. So there is sense in the idea that we should read theories in all three domains alike. The difference lies in what can be revealed by theory-guided experimental programmes. . . .

The possibly observable and the in principle unobservable can be tied together more strongly by the manipulation argument of Boyle and Hacking. The first step in enhancing the plausibility of a theory abstracted from a theory-family denoting possibly observable being would surely be successful manipulations. The theory would seem more implausible if the manipulations for which it is being used as a guide persistently failed. Davy became more enamoured of the 'ionic' hypothesis the more success he had in manipulating unobservable electrically-charged particles by electrical techniques, for instance his success in the decomposition of the alkaline earths.

In many case theories that passed the 'manipulation test' became available for a 'perceptual test' when technical advances promoted their denotata from possibly observables and observables. This was what happened when Pasteur's 'manipulative' success with anthrax bacilli was perceptually confirmed by Toissant's microscopic examination of the contents of the gut of earthworms living near buried victims of the disease.

But we have clear cases of manipulative success and failure for theories whose denotata are in the domain of the unobservable in principle. Sub-atomic physics provides a very rich trawl of examples. For instance the Stern-Garlach apparatus is a device for manipulating particles distinguished by their catalogue of quantum numbers to yield an observable effect. It fits the manipulation schema perfectly.

Scientist performs operation *A* which manipulates an unobservable of type *B*. (The *A—B* manipulation has an observable counterpart from which the concept for it was displaced in the construction of the relevant theory.) Changes in *B* have observable consequences *C*.

The inductive argument now runs as follows: For possible observables, manipulative success has been correlated with ontological success, after a technical advance has moved the contingently located boundary between observable and possibly observable. The location of the boundary between possibly observable and unobservable in is also contingent (but resting on a different contingency—the limits of the human Umwelt). Therefore (inductively) manipulative success with beings in the domain of the unobservable principle is a good ground for a (revisable) ontological claim on their behalf.

. . . The policy realist thinks that scientists progress in their projects by achieving a better sample of what there is in the world. The convergent realist thinks they progress by achieving a better description of the world. The policy realist stocks a museum. The convergent realist stocks a library. Can we convene a dialogue between 'curators' and 'librarians'?

Notions like 'a better picture of the world' and other perceptual metaphors tend to figure in attempts to make out a useful sense for the vague notion of verisimilitude, as well as familiar epistemic notions like 'truth', 'correspondence', etc. Popper had no luck at all with his idea of reducing relative verisimilitude to a relational property of two theories based on a measure of the balance of their respective true and false consequences. By examining the most recent version of this idea, Oddie (1986) has shown it to be *generally* misconceived. A closer look at what experimental programmes actually accomplish may help to sharpen ideas about what a theory-led development in a scientific field actually accomplishes. It is a commonplace of scientific research that things very rarely turn out to be just as one expected. Sociologists of science (Knorr-Cetina 1981) have shown how research programmes are rewritten to present the actual results as programme-relevant. A great many research programmes involve technical developments that enable scientists to scrutinize beings that exemplify the theoretical categories in question. Success for the theory that controls the search for exemplars usually consists in the successful demonstration that this or that kind of being exists. Idiosyncratic properties of exemplars are discounted. A theory is counted as a failure if it appears that no such exemplar can be found and the putative referent kind must be dismissed. In the course of the successful procedure reference has been established. For me, reference is a physical relation between an embodied scientist as a user of the theory and an exemplar of the right sort of entity within an extended human Umwelt. Manipulative success is one of the signs that one has established that kind of relation. Observational success (disclosure) is another. . . .

Can we use demonstration of the existence of a being of the right sort to be a referent for our theories to claim epistemic success, the production of knowledge? At best we can say we now have epistemic access to a researchable domain. A bridging principle is needed because it is quite possible to establish a referential relation with the help of descriptive statements that are or that turn out to be false. It is *characteristic* of scientific development that there are three stages to the enlargement of our knowledge. In the first a new vocabulary is devised by displacement of concepts. With the help of this vocabulary the community of scientists can discuss their next step. Embedded in the 'grammar' of that vocabulary is sufficient content to abstract a set of instructions for engaging in a search for exemplars of the kind of beings in question, and for establishing physical contact with one or more of them. The third stage lacks the glamour of the initial discovery of such a being, but should be of the greatest interest to philosophers. For during this stage further, often tedious, research leads to massive changes in the content of the original speculative theory, as more and more is learned about the beings now disclosed. In the course of these changes most of what we first claimed about the being of the kind in question turns out to have been false. There is no such thing as an unrevisable discovery.

But the question is—unrevisable to what level? 'So the dinosaurs were not cold-blooded, but they were organisms!' (and not strange petrous deposits). 'So the craters on the moon were not volcanoes but they are geological features!' (and not visual illusions like the canals on Mars). The second phase of research usually, if not always, involves 'backing up a revisability hierarchy', through species to genera and ultimately to onto-logical kind. Is that where we stick? If so the librarian can permanently catalogue some of his materials, and the curator need not fear too drastic a reorganization of his exhibits. In short, the librarian can now at least name the exhibition halls of the museum in some permanent way that will satisfy the curator.

Whether we can 'stick', that is expect no further revisions, depends on two matters.

1. How good is the double-inductive argument sketched above? It links Realm 1 to Realm 2 theorizing by the citation of historical examples of a preponderance of successful perceptual investigations guided by plausible theories. The argument links Realm 2 to Realm 3 theorizing by citation of historical examples of preponderance of successful manipulative investigations guided by plausible theories. The whole scheme is welded together by the citation of historical examples of successful manipulative investigations guided by Realm 2 theories which have also been perceptually successful. If this argument is good, then ontological attributions are empirically defensible for Realm 3 beings.

2. But historical examples also show that claims to scientific truth and knowledge are apparently indefinitely revisable. However, even a cursory examination of these revisions shows that they are hierarchically ordered. Accidental attributes are revised first, natural kind classifications next, and finally, *in extremis,* ontological classifications may be revised. For instance, the question of whether neutrinos perhaps have a very small rest mass has been raised — a revision of the first order. The suggestion that light propagators are not distributed wave fronts but localized photons is a revision of the third order, a revision of category or ontological kind. This revision occurred despite the enormous manipulative success of the wave theory.

Our conclusion must be that while global scepticism (at the lower level of the two levels of inductive scepticism) is rebutted by the double-inductive argument (1) above, the evidence of revisability, right up to ontological kind, demonstrated in argument (2) above, shows that there is no support for global antiscepticism either.

To the question 'Is it really true that light is propagated by photons and really false that it is propagated by waves?' we still have to say, 'Wait and see.'

In the discussion so far I have been presenting what I take to be the strongest inductive *argument* for a variety of realism since it recruits both the successes and the failures of past scientific research projects to its evidential basis. But a price has to be paid. The variety of realism which this argument supports is weaker than one would hope ideally to establish. To see this one can juxtapose the varieties of anti-realism to the argument.

We can classify anti-realisms by reference to the verificationist principle of meaning. Logical positivism and Machian sensationalism are anti-realisms of the first kind. Policy realism, as defended so far in these comments, is strong enough to defeat anti-realisms of this variety. But there are weaker anti-realisms which may escape. The oddly named 'critical empiricism' of van Fraassen is a case in point. In many ways it seems to be a revival of sixteenth-century fictionalism, say as advocated by Ursus, together with the late Berkeley of *Sins.* According to this view theories mean what realists take them to mean. But a rigid distinction between observables and unobservables is imported to control the interpretation of the kind of success now customarily called 'empirical adequacy'. Only observables denote the empirically real. The fact that a policy realist must believe that the entities denoted by theory probably exist for the setting up of a project to hunt for them to be rational, could be sloughed off by adherents of this view as a mere psychological condition. *Until* the unobservable becomes an observable no claims for truth or falsity, denotation, etc. can legitimately be made.

This position is fraught with internal difficulties. The most serious concerns the mismatch between prospective and retrospective inferences from the success or failure of research projects. A theory cannot be said to denote anything the nominal term for which is currently an unobservable. So in 1600 the term 'magnetic desmesne' which was used to refer to the elementary magnets that were supposed to have been oriented in the magnetization of a magnetic material, does not denote anything in the world. Nor do the capillaries required by Harvey's circulation theory. Improvements in microscopy disclose both elementary magnets and capillary vessels. What can we say retrospectively? If van Fraassen concedes that we now know that the terms denoted real entities all along, realism is also conceded retrospectively. Those who took plausibility to be a good ground for a realist reading were right and those who did not were wrong. Prospectively, it must now be the case that some of us are right and others are wrong about our readings of theories, though we will not *know* which are which until some future time. Nevertheless, we do have pretty good inductive grounds for judging who are right and who are wrong, namely the relative plausibility of the theories. Those who have refused to give a realist reading to a relatively plausible theory in the past have usually turned out to be wrong. Since this is an inductive argument we can accommodate those few cases when they turned out to be right. One might also press the critical empiricist on how he thinks existence is related to observability. If he denies that there can be retrospective *validation* of ontological hypotheses (and there will be retrospective invalidations too) it looks as if he has to hold the absurd view that whole categories of beings are brought into existence by the validation procedure itself.

We can now assess our two inductive arguments in the light of this discussion. Realm 1 and Realm 2 were linked through the 'moving boundary argument'. We were inductively justified in believing that terms used to refer to Realm 2 beings did indeed denote something real provided that the theory in which they were embedded conserved natural kinds from Realm 1 was empirically adequate. The second argument was used to link Realm 2 and Realm 3 theories through the manipulation argument. A technique of manipulation in which manipulative act, target of manipulation and subsequent qualitative change were all in Realms 1/2 can be applied in the context of a Realm 3 theory. But in that context the target is necessarily unobservable. Successful manipulation is a good inductive ground for making a claim about the terms in the Realm 3 theory which are used to refer to the unobserved targets as actually denoting beings of that natural kind, say the charged ions in a Stern-Gerlach apparatus.

Policy realism is obviously strongly supported by these arguments. It is clearly rational to read plausible theories as if they denoted real beings, and tailor our research efforts in the light of that reading. The question is how much further can we go? Is there support for convergent realism too? Of course, the support will be weaker. It is one thing to expect a determinate answer to a question about whether a certain class of beings exists or is possessed of this or that property, another to expect the answer 'yes, it is!' Nevertheless there are two further arguments that point in that direction.

Let us argue it out in terms of beliefs about existence. The point to be established is that it is not only rational to give a realist reading to plausible theories in the expectation of some determinate answer to the question as to whether one was right in one's reading, but that it is rational to believe prior to the determinate answer that the terms in the plausible theory denote what they seem to denote.

The first argument depends on a generalization of Strawson's presupposition theory. A statement can be assessed for truth and falsity only if we are willing to accept the existential presupposition of its terms. Likewise the implementation of a policy realist reading is only rational if we are willing to believe in the existence of the beings apparently denoted. Of course, if we are thinking in terms of there being a Nay answer as well as a Yea answer to the *existential* question this argument cannot be exactly Strawson's argument for existential presuppositions. But our argument is for the rationality of belief, not certainty. One would surely be irrational to spend huge sums of money and invest a great deal of time in looking for something one did not belief probably did exist. So to proceed with a policy realist project if one had in mind only the even-handed possibility of getting either determinate answer would be irrational. Looking for something as a project has the pragmatic presupposition that one is more likely than not to find it. But one can go further. It is not just a matter of there being an even chance of being right or wrong. The realist reading is rational because there is a much greater payoff if we are right or wrong about the existence of the beings we refer to by means of a theory, than if we adopt any of the anti-realist readings and do not pursue the relevant research programme, except by accident.

The second argument comes from Aronson, a point he has urged repeatedly. Provided we understand verisimilitude aright (and that means not in the way it is taken by those who build their realism on bivalence, such as Popper or Newton-Smith) the inductive argument from greater plausibility to greater verisimilitude does go through. The key to rebutting the arguments of such as Laudan lies in how one handles the fact that all our empirical claims seem to be revisable. My own response to this argument, as used by the anti-realists, was to point out that there is a revisability hierarchy. I owe to Aronson the further observation that revisability hierarchies are not indefinitely open.

To argue that the 'realist' induction is invalid because all subsequent empirical disclosures and manipulations cannot protect a result against revision embodies a serious philosophical error. Revision of results is always hierarchical. That is, when a hypothesis is revised the revision follows the following pattern. First, the specific attributions are revised (it is not a sheep, it's a goat). Then the more generic (it is not an animal, it is a bush). Finally the metaphysical category or ontological kind may go (it is not a material thing, but an optical effect). We have seen just this kind of hierarchical revision applied to the tales of UFOs for instance. Now Aronson's point is that revision does not *and cannot* extend beyond a supertype or ontological kind, because that is the Realm 1 footing on which the hierarchy rests. I proposed a similar argument in *Varieties of Realism* in using Gibsonian psychology of perception to defend Realm 1 realism. If we admit the Realm 1 to 2 to 3 induction, via disclosure and manipulation links, then we must also admit that the boundaries of observability shift outwards from Realm 1, and this at least is shared by the anti-realists of the second van Fraassen. There is a limit to revision, and it is set at just the point that allows the inductive argument beyond strong policy realism to weak convergent realism. It must be *weak* convergent realism because it would be hopeless to try to argue that plausible theories give us unrevisable access to the world. That we do not need. All that is necessary to save a realism which preserves a measure of verisimilitude is to show that the revisability of results is not indefinitely open. A somewhat similar line of argument is pursued by Devitt (1984) in which he emphasizes the weakness of the revisability argument or 'pessimistic meta-induction' when it is spelled out in detail.

Finally, it is worth pointing out that disputes with anti-realists of the second kind do not involve the whole of Hume's version of the 'revisability' argument. Hume's scepticism comes at two levels. At the first the argument goes like this: no scientific reasonings are fully rational because our knowledge of a stable world is unproveable. That is the form of the revisability argument with which I have been dealing. Neither Aronson nor I have to deal with the argument at the second level of inductive scepticism, namely that no reasoning whatever about nature is rational because we can have no surety that the world will not so change that whatever we did know about its current state is worthless as a guide in the new conditions. Madden and I (1975) called this the 'neurotic problem of induction' and it is clearly irrelevant to our debates.

SECTION THREE

SCIENTIFIC EXPLANATION

SECTION THREE

SCIENTIFIC EXPLANATION

INTRODUCTION

We have been feeling our way, from several different directions, towards the culmination of our study, the elucidation of the ideal form for theories. Theories are the crown of science, for in them our understanding of the world is expressed. The function of theories is to explain. We have already identified some of the forms of explanation.

Two important paradigms of theory have appeared upon which we can base our ideas of what a theory should be. . . . The science of mechanics with its central concept of force is one, and the science of medicine with such concepts as the virus is another. They are opposing paradigms as we have had occasion to notice already. They present two different kinds of theory as seen from a logical, epistemological, and metaphysical point of view. Must we accept both paradigms? Does each have a particular role to play? Can the one be reduced to the other? We shall come some way to settling these questions in this chapter.

The concept of force and the concept of virus seem to play similar roles since each is used to explain observations, on the one hand concerning motion, and on the other concerning, the course and development of disease in plants and animals. In the normal course of events neither a force nor a virus is observable in the way in which the happenings which they are designed to explain are observable. Finally, both the conception of force and the conception of the virus are concepts devised by analogy. They are descriptive of entities analogous to certain things with which we are familiar. Forces are analogous to the efforts that people make in shifting things against a resistance, and viruses are analogous to the bacteria which had been found to be the causes of many diseases. We have looked already at some of the detail in the development of these analogies. But if we look a little further at the science of mechanics, and compare it with pathology, a deep difference appears. The concept 'force', and with it the analogy with human effort, is inessential to the science of mechanics, as has been shown by the several ways in which that science can be reformulated without this concept. Its function is entirely 'pragmatic'. It serves the function of an aid to understanding, a device by which intuition is engaged in the business of understanding motion. But it is perfectly possible to understand motion without the concept of force. It is possible to understand all the phenomena of motion using, say, the concept of 'energy', and its redistribution among the bodies involved in a system of moving particles according to certain laws. The analogies are quite inessential to mechanics. But compare this situation with that in pathology. Without the concept of the virus as micro-organism the whole theory of the transmission and cause of a wide range of diseases would be quite different. The theory is an essential part of the understanding of the observations. A description of a disease is one thing, its pathology is quite another. This shows up for example in the difference between bad doctoring, in which the symptoms only are treated, and good doctoring, where a diagnosis of the cause of the disease is made and that cause is treated initially and it is this which explains all the other differences between the concept of force and the concept of the virus, it makes sense to ask whether or not there are viruses, and it makes a tremendous difference to medicine which way that question is answered. But though it makes sense to *ask* whether or not there are forces, whether or not individual things exert efforts as people do in bringing about motion, it makes not the slightest difference to the science of mechanics whether there *are* or *are not* forces. The science would be differently formulated no doubt, but there would be not the slightest difference in the predictions of future states of moving systems which could be made by means of it.

Each of the paradigms marks an important ingredient of theory, at least one of which must be present for a theory to be satisfactory. The science of mechanics is organized by the use of a mathematical mode of expression into a logical system, where there are certain fundamental principles and the practical laws of motion are deduced from these in a logical and rigorous way. It is obvious that there are certain great practical advantages in being able to express a theory in a logical system. A great many particular laws and even particular facts can be comprehended in a very economical way in the principles of the theory. Systematization has considerable pragmatic value. But a theory would still be a theory and would still explain the facts it did explain if its laws did not fit easily or at all into a logical system. The laws might only hang together because they were the laws of the same subject matter, that is the laws describing the behaviour of the same kind of things or materials. We have a great deal of knowledge about human behaviour for instance. But this knowledge cannot be formulated in such a way as to fit into a deductive, logical structure. Our theory of human behaviour is a rag-bag of principles united by virtue of the fact that they all concern the same subject matter, namely, the behaviour of people. We may never find a systematic formulation of these laws. We may never achieve the pragmatic advantages of system.

A scientific explanation of happenings, whether individual happenings or sequences of events, consists in describing the mechanism which produces them. Only in the most minimal sense does the science of mechanics explain any course of motion. The laws of mechanics are descriptive laws, not explanatory laws. Apart from the tenuous and rather feeble concept of 'force' there is no attempt in that science to advance any account of the mechanisms of motion, of why the laws of impact, of momentum conservation, and so on are what they are. So far as I know the only attempt that has ever been made at this is the grotesque set of explanations offered by Descartes in Book II of his *Principles.*[1] But the virus theory has exactly what is required for a scientific explanation of the course of the disease with which it is concerned. The presence of the virus explains what is described in the syndrome or course of the disease, and the more we know about the nature and behaviour of viruses, the more we know about the disease. It is the interaction between body as host and virus as parasite that produces the symptoms of the disease and explains the course of it. The virus theory of poliomyelitis is truly a scientific explanation, where the beautifully systematized laws of mechanics are not. Of course in certain cases the mechanics of particles in motion explains other phenomena, because then the laws of mechanics serve as perfect descriptions of the causal mechanism at work. Such for example, is the often quoted example of the kinetic theory of gases, where the mechanics of the molecules of the gas sample serves as a causal mechanism which explains how samples of the gas behave under various conditions. The kinetic theory is an explanation, and a scientific explanation at that, of the behaviour of gases, but it follows the paradigm of the virus explanation of poliomyelitis, and not the paradigm of the force formulation of mechanics. The fact that mathematical means of expression are used in the kinetic theory and in mechanics, and are not used in the virus theory should not blind us, as philosophers of science, to the essential difference of the former and the essential likeness of the latter.

The generation of the concept at the heart of a theory, what Whewell called the Idea of the theory, is . . .. a matter of analogy. Building a theory is a matter of developing an appropriate concept by analogy. This is the essential heart of science, because it is the basis of explanation. Why is it that we cannot just go out and find out what the basic mechanisms are? Why can we not eliminate the need for analogy, and go directly to nature? The answer is that science proceeds

[1] R. Descartes, *Principles* of *Philosophy,* Bk. II, xxxvii, xxxix, xl; in (Descartes 1954, pp. 216—19).

by a sort of leapfrogging process of discovery. As soon as a field of phenomena is identified as worth studying and comes under scrutiny we can find all sorts of regularities and pattern's among phenomena, but we do not find among these phenomena their causes, nor do we find the mechanisms responsible for the patterns of behaviour we have found. Chemistry proceeds both in the study of the chemical behaviour of different substances and materials, and in the discovery of the mechanisms of these reactions. In studying the reactions we do not study the mechanism of reaction. In many, many cases a great many facts about a certain kind of phenomenon can be found out without it ever being possible to study the mechanisms of the phenomena directly. In such circumstances the necessarily mechanisms have to be thought out, to be imagined, and to be the subject of hypotheses. And once they have been thought out, then we know what sort of observations would lead to their independent discovery. Sometimes a wholly different line of investigation leads indirectly to the discovery of the causal mechanisms underlying some phenomena. Such, for example, was the case with the study of radioactivity and chemistry, where the examination of the disintegration of certain rather unusual materials led to discoveries of the greatest importance about the structure of the elementary parts of materials, the chemical atoms. These discoveries were turned by Lewis and Langmuir into a theory of chemical reaction, a description of the mechanism of chemical bonding and the circumstances under which chemical change took place. When we do not know what are the mechanisms underlying the processes we are studying, then we must imagine them, and they must be plausible, reasonable, and possible mechanisms. To achieve this we proceed by the method of analogy, supposing that they are like something about which we all ready know a good deal, and upon the basis of our knowledge of which we can imagine similar mechanisms at work behind the phenomena we are investigating.

What is an analogy? An analogy is a relationship between two entities, processes, or what you will, which allows inferences to be made about one of the things, usually that about which we know least, on the basis of what we know about the other. If two things are alike in some respects we can reasonably expect them to be alike in other respects, though there may be still others in which they are unlike. In general between any two things there will be some likenesses and some unlikenesses. The art of using analogy is to balance up what we know of the likenesses against the unlikenesses between two things, and then on the basis of this balance make an inference as to what is called the neutral analogy, that about which we do not know. Suppose we compare a horse with a car. There are certain likenesses in that both are used as means of transport, both cost a certain amount to buy and to maintain. There are unlikenesses in that one is wholly an artefact, and only in the choice of breeding partners does the hand of man interfere in the production of horses. Horses are organisms, cars are machines. Cars can be repaired by replacing worn-out parts from an external source, but this technique is of limited application for the horse. Suppose we learn that a certain city uses only horse transport, but we know nothing else about their system. We can make certain inferences about the traffic density from what we know about cities which use mechanical transport on the basis of the likeness between horse and car as means of transport, and we can make other inferences about the air pollution based upon what we know of the unlikenesses between them. In this way, by the use of analogy, we penetrate our area of ignorance about a city whose transport is by horse.

In many cases in science we are operating from one term of an analogy only. Molecules are analogous to particles in motion, but we cannot examine molecules directly to see how far they are analogous. Since the molecule is an entity which we imagine as being like a particle in motion, we are free to give it just such characteristics as are required for it to fulfil its function as a possible explanatory mechanism for the behaviour of gases. The neutral analogy is just that part of what we know about particles that we do not yet transpose to our imagined thing, the molecule. The molecule is analogous to the particle not because we find it so but because we make it so. And there is another analogy which completes the theory. A swarm of molecules must be analogous to the gas, otherwise we should not be able to use the molecule concept as an explanatory device. These distinctions are not really well brought out in terms of the simple notion of analogy. From the point of view of the notion of analogy, the relation of molecules to material particles, and the relation of the laws describing their behaviour to the laws of mechanics, and the relation between a swarm of molecules confined in a vessel and a gas, are analogies. But whereas a gas sample might really be a swarm of molecules and molecules might really be material particles, the relationships are, from an epistemological point of view, quite different.

The distinctions which we are looking for can best be made by introducing a new concept, which allows us to analyse analogy relationships a good deal, more carefully and finely. This is the concept of the *model.* In the technical literature of logic there are two distinct meanings to the 'model', or perhaps it might be better to say two different kinds of model. In certain formal sciences such as logic and mathematics a model for, or of a theory is a set of sentences which can be matched with the sentences in which the theory is expressed, according to some matching rule. We shall not be concerned with such formal, sentential models here. The other meaning of 'model' is that of some real or imagined thing, or process, which behaves similarly to some other thing or process, or in some other way than in its behaviour is similar to it. Such a model has been called a real or *iconic model.* It is with iconic models that we are mostly concerned in science, that is, with real or imagined things and processes which are similar to other things and processes in various ways, and whose function is to further our understanding. Toys, for example, are often iconic models, that is things which are similar to other things in some respects, and can indeed play something of their role. For example, dolls are often models of babies, that is, a doll is a thing which is like a baby, and can be treated for certain purposes as a baby. And baby-models can be used quite seriously in training mothers and midwives in baby-handling where it is inconvenient or even dangerous to employ a real baby for the purpose.... A toy car is often a model car, a toy plane a model plane.

Models are used for certain definite purposes, and in the sciences these purposes are (i) logical: they enable certain inferences, which would not otherwise be possible, to be made; and (ii) epistemological: that is they express, and enable us to extend, our knowledge of the world. To sort out these purposes rationally yet another idea is needed, that is, the difference between the source of a model and the subject of a model. A doll is a model *of* a baby, and also modelled *on* a baby. Its source is the real thing, the baby, while its subject is, in this case also the baby. Its source and its subject are the same. Such models are called *homoeomorphs.* But when one is using the idea of the molecules as the basis of a model of gas, the molecule is not modelled on gas in any way at all. The molecule is modelled on something quite different, namely the solid, material particles whose laws of motion are the science of mechanics. Such a model for which the source and subject differ is called a *paramorph.*

Science employs both homoeomorphs and paramorphs, and indeed the proper use of models is the very basis of scientific thinking. A theory is often nothing but the description and exploitation of some model. The kinetic theory of gases is nothing but the exploitation of the molecule model of gas, and that model is itself conceived by reference to the mechanics of material particles. We have seen how our lack of knowledge of the real mechanisms at work in nature is supplemented by our imagining something analogous to mechanisms we know, which could perhaps exist in nature and be responsible for the phenomena we observe. Such imagined mechanisms are models, modelled *on* the things and processes we know, and being models *of* the unknown processes and things which are responsible for the phenomena we are studying. This important fact leads to our having to acknowledge that a theory has a very complicated structure, one in which there are at least two major connections which are not strictly logical in the formal sense, but are relations of analogy. The gas molecule is analogous to the material particle, and the swarm of molecules is analogous to whatever a gas really is, and both these analogies are tested by the degree to which the model can replicate the behaviour of real gases. Gas molecules are only like material particles (in some versions of the theory they do not have volume, for instance), and a swarm of them is only like a gas, since even the most sophisticated molecular theories do not quite catch all the nuances of the behaviour of real gases.

But there is a further and final point of the utmost importance. Since we do not know the constituents of a gas independently of our model, we can scarcely be in a position to declare any negative analogy between the model and the gas of which it is a model. Any defects in the molecule concept can be made good so long as we can change its properties without contradiction. We can make and remake the molecule so that in swarms it behaves as near as we like to the gas. When we are considering a model of something we wish to understand we are presented with a neutral analogy and a positive analogy only. The fact that our model may be modelled on something with which it has, in addition to its positive and neutral analogy, a strong negative analogy is of no consequence, since it simply means that the model does express the concept of a new kind of entity or process, different from the one upon which it is modelled. Now as a model of some process or mechanism or material responsible for the phenomenon we are studying becomes more and more refined, a new question gradually presents itself. During the process of refinement we were concerned only with so adjusting our model that it behaved in a way which *would* account for the phenomena. Gradually we are brought to consider the question as to the reality of what had previously been only a model of the real mechanism of nature. Perhaps, we might say to ourselves, gas molecules are not just models of the unknown mechanism of the behaviour of gases, perhaps there really are gas molecules, and perhaps gases really are nothing but swarms of these things.

I want to present some examples now to show the different ways in which this deeply penetrating question of the reality of an iconic model can be pursued. Darwin's Theory of Natural Selection provides an excellent example of the use of iconic model building to devise a hypothetical mechanism to account for the facts which were known to naturalists. Students of nature had come to see that the populations of animals and plants that at present existed on the earth were different from those that had existed previously. They had also come to see that in nature there was a great variety of forms of many plants and animals closely similar to each other. Many people were very familiar with the possibilities of breeding, particularly gardeners, and stock-breeders of various animals. How are we to explain the variety of species that had existed, and to explain the distribution of that variety of species which now exists? What process in nature is responsible for these striking facts? Now whatever process it is works very slowly, so slowly that Aristotle, one of the greatest biologists of all time, had been deceived into thinking that the species of animals and plants were fixed, so impressed had he been by the similarity between parents and offspring. But there are also minute differences, and it was upon these that Darwin's theory was fixed. Darwin did not know what were the processes by which change in the animals and plants of nature came about, so he constructed a model. He knew very well that there is change in domestic animals and plants and he knew that that change is due to the fact that the breeder *selects* those plants and animals from which he wishes to breed, which are more suited to whatever purpose he has in mind, and that after several repetitions of selection a quite different appearing creature can be derived from appropriately chosen individuals solely by breeding. There is a variation in nature, and Darwin conceived of a process analogous to domestic selection which could be a model of whatever process was really taking place in nature. He called this process, modelled on domestic selection, ***natural selection***.

Had Darwin proceeded with the same model source, for filling out the details of his imaginary process of natural selection, he might have posited the active intervention of a breeder, who like the gardener or stock-breeder had some purpose in mind in bringing together those particular plants and animals which did breed, and which produce the subsequent generation. Now Darwin was looking for a process which was wholly natural and which did not involve divine intervention as a part of the model. He found another source which contributed to his model process. This was the theory of population pressure, originated by Malthus, and elaborated by Herbert Spencer, who had coined the term 'Survival of the Fittest', as a brief description of the outcome of the competition for space, light, food, and so on, which by analogy with the conditions which Malthus reckoned to detect in human society, could be projected on to nature. Darwin's Theory of Natural Selection became in effect the elaboration of the various ways in which different varieties of animals and plants were caused to breed at different rates, and so to explain why, in each generation certain individuals were favoured and bred more freely than others. All this was an elaboration of a basic model of whatever process was really responsible for change and development of animal and plant forms in nature. Did the things that Darwin's model suggested really happen in nature? By this question I do not mean 'Did evolution occur?' Rather I mean, 'Is it the case that some individuals are able to breed more freely because they' are more suited to their environment, and do they therefore transmit to their offspring whatever characteristics favoured them?' The reality of the mechanisms of evolution as proposed by Darwin is a separate question from the reality of the evolution process, that is the gradual change of species. Nowadays it is hardly conceived by most biologists that Darwin's theory began as a model of the real processes in nature, so much is it taken for granted that Darwin's model is real. I suppose that it is still just possible, though extremely unlikely, that it may eventually turn out that quite different mechanisms are responsible for the evolution of species.

Recently an interesting example of model building has taken place in the theory of electrical conduction. Somehow the electrons that are in metals are responsible for the conduction of electricity in metal. Drude produced a very successful model of the mechanism of conduction by supposing that there were free electrons in the metal which behaved like the swarm of molecules which we have seen as a most successful model of gas. From supposing that the electrons were like a gas confined within a container, he was able, with very few supplementary assumptions, to work out an explanation of the known laws of conduction, that is he showed that a swarm of electrons obeying the gas laws would behave analogously to a conductor. Here the model is modelled on another model, and is a model of a truly unknown mechanism, the unknown mechanism of the conduction of electricity in metals.

To explain a phenomenon, to explain some pattern of happenings, we must be able to describe the causal mechanism which is responsible for it. To explain the catalytic action of platinum we must not only know in which cases platinum does catalyse a chemical reaction, but what the mechanism of catalysis is. To explain the fact of catalysis we need to know or to be able to imagine a plausible mechanism for the action of catalysis. Ideally a theory should describe what really is responsible for whatever process we are trying to understand. But this ideal can rarely be fulfilled. In practice it becomes this: ideally a theory should contain the description of a plausible iconic model, modelled on some thing, material, or process which is already well understood, as a model of the unknown mechanism, capable of standing in for it in all situations.

Finally this ideal of explanation is complemented by another and final demand. The ideal model will be one which not only allows us to reason by the complex structure of double analogy which I have described, but is one which might be conceived to be a hypothetical mechanism which might really be responsible for the phenomena to be explained. This is what prompts that deepest of all scientific questions, 'What is there really in the world? Are those hypothetical mechanisms which we believe might exist really there?'

If knowledge is pursued according to this method it will tend to be stratified. Perhaps this can be seen best if we look at the way causes are elaborated. There are two conditions which have to be fulfilled for there to be truly said to be a causal relation among happenings or phenomena. The first condition, ensuring that there is prima facie evidence, is that there should seem to be some pattern or structure in what we observe to be happening. This might be that simple kind of pattern which we call regularity or repetition, when we find one sort of happening followed regularly by happenings of a certain, definite other kind, when for instance those who are deprived of fresh fruit and vegetables develop scurvy, and those who have plenty of the above commodities do not. We have *prima fade* evidence that there is a causal relation between the deprivation and the disease. But to eliminate all possibility that something else, some third factor, might be responsible both for the shortage of vegetables and for the scurvy, we must find out what is the mechanism involved, and that involves us in a study of the chemistry of the food materials and of the physiology and chemistry of the body. That study supplies an idea of the mechanism which explains the pattern of happenings involving presence and absence of fresh vegetables, and the onset and cure of scurvy. Satisfying this second condition, that is, describing the causal mechanisms, completes one causal study. Our knowledge falls out into two strata as it were: in one stratum the facts to be explained are set out and their pattern described; in the underlying stratum we may imagine or describe the causal mechanism.

Now, that mechanism is described in terms of chemical reactions and physiological mechanisms. These exhibit their own characteristic patterns and regularities, and these call again for causal explanation. But now a new kind of fact must be adduced. Chemical reactions are explained by the theory of atoms and molecules and chemical valency. By means of this model we can describe a causal mechanism for chemical reactions, and similar considerations apply to the explanation of the physiological and biochemical facts. We have reached another stratum. Then that stratum itself becomes the occasion for prima facie hypotheses that there too are causal relations, that there is some mechanism which explains the combining powers of chemical atoms, and some model of the chemical atom which would explain the diversity of chemical elements. Such a model is to hand in the electron— proton—neutron picture of the atom and the electronic theory of valency. This forms another stratum. Finally, and this is where we are today, if we are to be true to our scientific ideals, we must ask what is responsible for the behaviour of protons, neutrons, electrons, and the other subatomic particles, and we must try to penetrate to yet another stratum. As we have seen in this chapter, this must first be a work of the disciplined imagination, working according to the principles of model building, the method which has enabled us to proceed to such depths in uncovering the strata of the mechanisms and processes which make up the natural world.

In each era scientists find themselves at a loss, incapable of proceeding deeper into nature. And in each era scientists explain this temporary ending of scientific penetration by a metaphysical theory in which what is basic for one time and one limited scientific culture is elevated to the status of the ultimate. [The] the metaphysical theories of the past have presented forms of explanation as ideals, and those ideals, expressing the ultimately conceivable models for that culture, end with a seemingly impenetrable stratum, that closes the layers of knowledge. But we have also seen that metaphysical systems are not systems of facts. They are systems of concepts which we invent, and which we adopt if we will. Without them we could not think at all, but we must not allow any particular one to stand in the way of scientific progress. Perhaps science may come to an end for us, by reaching a stratum beyond which we have neither the

imagination nor the technical resources to penetrate. But that end will not be the end of nature, it will be the projection upon nature of our own limitations. In the meantime we have no alternative but to follow the methods of science as we know them.

CHAPTER SIX

THEORY-FAMILIES

A number of authors, including T. S. Kuhn (1962) with his 'paradigm' and I. Lakatos (1970) with his 'hard core of a research programme', have pointed out that the unit of scientific thinking is not the theory, as it might be 'the Clausius-Maxwell theory of gases' or 'the theory of cognitive dissonance'. These are static, synchronic descriptions of moments in the development of cognitive entities of higher order. For various reasons I prefer to call these higher order cognitive entities 'theory-families'. They are the bearers of the content of the successive theories that are 'taken off' them as they evolve. Their structure is the content structure of a set of theories, unified by the incorporation of a common metaphysics, or a common categorical framework (in the terminology of Korner, 1974) or a common ontology (to adopt Aronson's vocabulary).

At the heart of a theory-family is an entity I shall call 'an ideal cognitive object'. I believe that what we recognize as theories and as taxonomic systems are manifestations of states of such ideal cognitive objects. An object of this sort can belong to an individual human being, or it may be distributed among several people, or it could be the possession of a wider community. When anthropologists of science claim that the social structure of scientific communities reflects the necessary conditions for the production of scientific texts, these texts are best seen as drawn off from a more fundamental kind of cognitive being, the ideal cognitive object. Such beings are extended in time, but they have no special mode of existence. They can be represented in several different ways, iconically, linguistically or by means of abstract mathematical structures. In writing about such beings I shall be using language to describe iconic expressions of such objects as a matter of expository convenience. In discussing the role of visualizability in the formation of physical concepts, Miller (1984) has highlighted the subtle influence of iconic expressions on the thought of physicists. His quasi-psychological observations closely match a philosophical point to be emphasized in this chapter, namely the natural-kind constraints on conceptions of unobserved entities upon which the possibility of policy realism depends. A distinguishing feature of the structure of such objects is that their constitutive relations are semantic and intensional, creating an organization of content. They are not ordered, in any fundamental way, by principles drawn from logic.

1. THE STRUCTURE OF THEORY-FAMILIES

Behind explicit scientific discourses lie ideal cognitive objects formed by the union of two major components. There is an 'analytical analogue or model' through which the world of human perceptual experience is made to manifest patterns of various kinds of order. And there is the 'source analogue or model' from which theoreticians draw their concepts for building plausible explanations for the existence and evolution of such patterns.

[One] must assume that common experience is first differentiated and categorized with respect to some cluster of loosely organized common sense schemes, scarcely well integrated or simple enough to be described as theories. There are no brute facts. But further selections from common experience and more refined categorizations of phenomena require the use of supplementary schemes. Many of these take the form of analogues 'brought up to' items of common experience. They sharpen our grasp of the patterns that are implicit in the experience or that can be made to emerge from it, by the similarities and differences they force us to take account of. When an analogue is used for such a purpose I call it 'analytical'. Analytical analogues can be used in a great variety of ways. Sometimes entertaining an analogue simply helps an observer to see a pattern that is already there, so to speak, in what ordinarily can be seen. The young Darwin looks at the bewildering diversity of plants and animals, both living and extinct, with the eye of an English countryman, that is with the analogy of farming, gardening and breeding in mind. He sees lines of descent, blood ties, etc., where another observer (Captain Fitzroy, for instance) might see the manifestation of God's munificence. But sometimes the analogue transforms experience by suggesting an experimental programme. Largely I believe by reason of its theological implications Boyle had a keen interest in the nature of the vacuum and in finding an explanation for the apparent absence of vacua in nature. If the air were springy it would expand to fill any vacua that tended to form in natural processes. A natural phenomenon forbade vacua, if that were so. To study the 'spring of the air' Boyle made use of an explicit analogy between metal springs and the way they could be studied, and air springs and how they might be investigated. His famous apparatus is a gaseous analogue of a coil spring suffering progressive compression under increasing weights.

I have described these analytical analogues iconically, but they could just as well have been described as conceptual systems. I claim that it is cognitive entities of this sort that are an essential part of the ideal cognitive objects that underlie theorizing, and so must form part of evolving theory-families. Their role is to provide the classificatory categories by means of which experienced reality is given texture, both as a patterned flow of phenomena and as differentiable into kinds. There will be as many clusters of phenomena available in common experience and its experimental extensions as there are analytical analogues to engender them. Nature, as experienced, may not 'take' a particular analytical analogue. There may be no emerging facts. The theory of 'signatures' was just such an analytic analogue—that there were iconic illustrative properties by which plants, flowers, fruits, and so on, with medicinal virtues, were marked. In Boyle's researches into the spring of the air, the patterned phenomena (volume/pressure proportions) are not natural phenomena, but are properties of an artefact, the apparatus constructed on the basis of the analytic analogue. In other cases analytical analogues serve to reveal texture and pattern without the use of an intervening apparatus. Darwin's 'agricultural' point of view is a case in point, but the use of such analogues is ubiquitous in good science. Goffman (1959) asked his readers to look on the loose groupings of people that act together in everyday life as 'teams', intent on maintaining the impressions they make in the eyes of others. This famous analytical analogue brings out aspects of the behaviour of all sorts of people, including nurses and receptionists in health clinics, that would have been difficult if not impossible to discern without the potent Goffmanian image. There are no 'given' patterns in nature and human behaviour. The results of observation and experiment are the product of sometimes quite complex chains of analogical reasoning.

The second major component of a theory-family is its source analogue. It is from the source analogue or analogues that the material for building concepts or representations of unobservable processes, mechanisms and constitutions is drawn. Deep within the cognitive foundations of the kinetic theory lies the analogue relation that molecules *are like* Newtonian particles. The way the concept of 'molecule' is developed in successive theories of the behaviour of gases (within the framework of the one developing theory-family) is controlled by the possibilities inherent in the concept of the Newtonian particle. One of the most elegant and one might even say spectacular uses of an explicit source model is in Darwin's own exposition of the theory of natural selection. The steps that lead up to the introduction of the concept of natural selection are managed through an analogy with domesticity. The first part of Darwin's book is occupied with detailed descriptions of the breeding of plants and animals in domesticity, together with discussions of the variation that is found in successive generations of domestic animals and plants. The upshot could be expressed in a kind of formula:

> Domestic variation acted on by domestic selection leads to domestic novelty (e.g. new breeds).

As the second chapter unfolds Darwin takes his readers through a great many examples of natural variation and natural novelty, the appearance of new species. We are carried along by the narrative to the point where we are driven to contemplate another 'formula':

> Natural variation acted upon by (, , , ? . . .) leads to natural novelty (e. g. new species).

The rhetorical force is irresistible and we find ourselves making Darwin's great conceptual step ourselves. The unknown and unobservable mechanism of speciation must be natural selection.

The reasoning is analogical, and, as the theory-family develops the limits of the analogy need to be examined through explicit statements of the positive and negative components in the analogy relation. Darwin systematically deletes some of the common implications of the term 'selection' from his scientific concept. His deletions include volition and any personifications of the natural forces involved.

The basic structure of the theory-family derives from the exigencies of explanation. In a great many cases the use of analytical analogues reveals patterns among phenomena, for whose explanation the community may be at a loss. The deficit is made good by imagining causal processes which could produce them. But, in the first instance, these processes will usually be unobservable, if real, in that people could not experience them in the same way as they experience the patterns the existence of such processes would explain. Reference to unobservable causal mechanisms and the beings upon whose existence they depend must involve the use of terms which denote beings which are, at that time, beyond experience. In short the community cannot tell what is producing the phenomena of interest by looking, touching, or listening. Just to guess is to leave open too wide a range of possibility. It is to remedy the lack of 'microscopical eyes (and ears)' that the *controlled* imagining of what those processes and beings might be begins. The role of source analogues is essential to this cognitive activity. It is from these that the community of scientists draws the images and the conceptual systems, with the help of which the cognitive work of pushing the imagination beyond experience is achieved in a disciplined way.

Looked at this way the methodology of theorizing can be described in four steps.

1 Methodological step: an analytical analogue is used to elicit a pattern or patterns from nature.
2 Theoretical principle: observed patterns are caused by unknown productive processes, and the clusters of properties that mark putative kinds are manifestations of unknown constitutions.
3 Theoretical principle: an analogue of the observed process can be thought (imagined, for instance) to be caused by some analogue of the real but unknown productive process.
4 Methodological step: the analogue of the real productive process or 'inner' constitution is conceived (imagined) in conformity to the source analogue.

I illustrate how this activity creates a semantically organized theory-family for the explanation of an observed process, a patterned sequence of event-types. Within this structure there are three analogy relations:

(i) An analytical analogy between the analytical analogue and the observed pattern.
(ii) A behavioural analogy between the behaviour of the analogue of the real productive process and the behaviour of the real productive process itself (which we already know, since it is revealed in the observed pattern).
(iii) A material analogy between the nature of the imagined productive process and the nature of the source analogue.

The behavioural and material analogies control the way the community conceives a hypothetical generative mechanism or process which would, were it to be real, produce the patterns revealed by the use of the analytical analogue. It is important to see that hypothetical generative processes so conceived are, strictly speaking, analogues of whatever the real productive processes might be. We know from experiment and observation, within the conceptual possibilities constrained by the analytical analogue, how the real productive

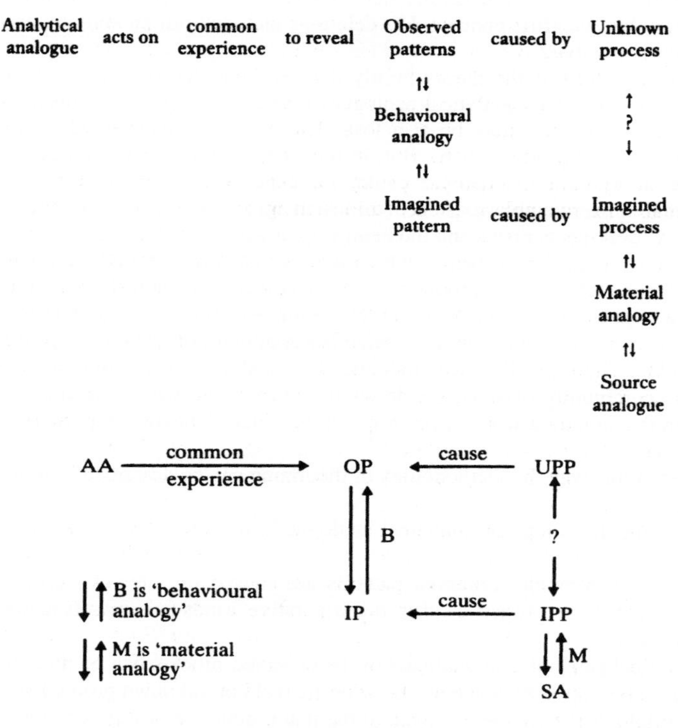

Figure 1
Schematic representation of a theory-family

mechanism behaves. We imagine, through the joint constraints of the behavioural and material analogies, what that mechanism or process might be like. These relationships can be summed up in the following schema, representing the structure of a theory-family. In the schema the double arrow represents a real-world relation; all the other relations are conceptual. In some treatments the conceptual entity I have called an imagined generative mechanism or process is called an 'explanatory model.' Schematically the structure of a theory-family can be laid out as in Figure 1.

The realist reading of this account of theorizing is created just by adding a fifth step to the methodology.

5 Epistemological claim: the hypothetical productive process or mechanism, conceived with the aid of the constraints embedded in the relevant theory-family, represents, to some degree, the nature of the real productive process or mechanism, when the theory-family is in a condition such that the theory which describes that moment in its evolution is plausible. The justification for picking out certain observable clusters of properties as something like the nominal essences of natural-kinds by reference to the relevant constitutive (and unobservable) micro- or macrostructures as real essences follows exactly the same pattern.

The next step in the analysis will be to give an account of plausibility and implausibility of theory-families, by reference to their momentary condition. I must emphasize that I am defending policy realism, not truth realism. It is no part of my account to suppose that the plausibility of a theory justifies the claim that the hypothetical productive mechanism or process it describes is just like the real one, or that implausibility would ground the dismissal of such a claim. Rather I argue that in the condition that a theory is plausible it represents a moment in the history of a theory-family, when the policy of undertaking a search through the appropriate referential realm for exemplars of the entities imagined (conceived) in the act of theorizing makes good sense.

The above schema is to be read in the iconic mode, that is it refers to patterns, properties, things, processes, and so on, real and imagined. A corresponding 'discourse' schema could be constructed for a science, in which each element in the above schema is replaced by a description. Such a discourse schema could be used to analyse scientific publications. When the explicit formal discourse of the scientific community is matched against this schema it becomes clear that only a very small part of it is reproduced in normal scientific writing. Usually only the observed patterns and the hypothetical generators of those patterns are described. The rest of the discourse is taken for granted, with some notable exceptions. When great scientific writers such as Darwin or Hales are writing up their work, much more of the implicit discourse of the scientific community comes to be laid out explicitly. I believe that for expository purposes it is better to describe the components of theory-families and their interrelations in the iconic mode, since the complexity of a discourse which did justice to the implicit analogies and their interrelations in a discursive mode would be formidable.

A theory-family develops in response to two kinds of external pressures. There is the need to accommodate new experimental results, which refine our knowledge of the manifest patterns of behaviour of the real causal mechanisms operative in some field of phenomena. These are accommodated by adjustments of the behavioural analogy which spark off adjustments of the material analogy. But there are also changes in the theoretical background to the theory-family which come about by further developments of the source analogue. These lead to adjustments in the conception of the hypothetical generative mechanisms and processes at the heart of the theory-family, through the material analogy. And in their turn they suggest new domains of research through the behavioural analogy which links their imagined behaviour to manifest experimental or observational patterns. I reserve the detailed exposition of examples of these processes of adjustment to the section on plausibility and implausibility.

Any theory of theories that is to merit attention must account for the meanings of theoretical terms in the scientific discourse, in particular the way in which theoretical terms have an excess of meaning over that which accrues to them simply from the empirical consequences that follow from their incorporation into a theory. I propose that the cognitive processes (mainly judgements of likeness and difference) which are involved in working with structures such as that sketched above determine the meanings of the lexical items that appear in the corresponding discourse. Ideally the etymology of theoretical terms should parallel the way theoreticians come to conceive of the hypothetical mechanisms, processes and constitutions which the terms are used to describe. Since the hypothetical generators of observed patterns are conceived by analogy with known generators of known patterns, the source analogue of the relevant theory-family, the terms descriptive of those hypothetical generators should be thought to acquire their meaning by parallel processes. The tropes of simile and metaphor would seem to be the obvious candidates. Both are linguistic devices which create new meaning from within the resources of a lexical system, and make no use of ostension to extralinguistic exemplars. In the case of simile, extralinguistic input is required for the literal meaning of the term to be used, but its use as a simile does not depend upon a point-by-point comparison between the first and second subject. Rather it creates that comparison. It invites the reader to look at the second subject in such a way as to emphasize certain aspects of it. Metaphor too extends the contexts of use of terms already having literal meaning, which may indeed have been based on extralinguistic exemplars, but it is used for just those occasions when we do not possess the linguistic resources to express what it is about the second subject that has struck us. It is not a comparison, but a catachresis.

So far I have left the source analogue unanalysed. The role of a source analogue in a theory-family is to provide and maintain a set of natural-kind rules within which hypothetical entities are to be designed. But for a realist construal of those entities in terms of a material practice of seeking concrete exemplary instances of such beings, a mode of reference must also be given. [For] instance 'pointing to a spatio-temporal location', 'testing a material substance for its ability to display a certain disposition in appropriate circumstances' and so on, are bound up with the implicit metaphysical component in the structure of the intensions of such concepts. Source analogues as the progenitors of natural-kind rules should also be structured so as to incorporate the necessary metaphysics to prescribe this or that determinate mode of reference. It turns out that this is indeed just how source analogues are structured. Domestic selection as the source analogue for natural selection constrains the metaphysics of speciation within a 'material process' categorical framework. Its one-time rival, as a source analogue, creationism, constrains the metaphysics of speciation within an 'act' categorical framework. Each source analogue further constrains the hypothetical entities involved in the theory of speciation with regard to the kind of process or kind of act hypothesized. The metaphysical component of the source analogue determines what kind of demonstration is required to establish a physical relation between the entity in question and an embodied human scientist, the necessary link for a proof of the existence of the beings proposed in the theory. The natural-kind rules determine, in a general way, the features to be looked for in deciding whether a putative specimen should be recognized as an exemplar of the kind of being in question. The metaphysical component then plays a central role in the setting up of the search procedures that are consequential on a policy-realist reading of a moment of equilibrium (plausibility) in the evolution of a theory-family.

Finally I need to show that the content structure of Realm 2 discourse, theorizing aimed at working out what must be the characteristics of beings which could, with technical advance, become objects of human experience, is just the structure that would make theories considered as descriptions of the state of a theory-family at some moment of equilibrium, mapping functions of the Aronson type. That is to say, theories become devices by which phenomena are mapped on to aspects of some common ontology. I owe a neat formulation of this point to Craig Dilworth. The two analogy relations, which are the main structuring relations of the theory-family, create the mappings. The behavioural analogy relates the hypothetical generative mechanism to the nominal subject of the theory, say 'overt bodily characteristic', which would be the Aronsonian 'phenomenon' to which genetics as a theory-family is directed. The material analogy relates the hypothetical mechanism of inheritance to the source analogue. This is the concrete form that the aspects of an Aronsonian common ontology for genetics would take. In this way there is created a mapping to aspects of the nature of genes, as complexes of chemically defined units, which are the real subject of the theory. Aronsonian mapping is the abstract or formal structure of theories considered as expressing or representing moments of temporary equilibrium in the unfolding of a theory-family. They play the role that deductive-nomological structures played in the logicist-empiricist view of theories.

2. PLAUSIBILITY AND IMPLAUSIBILITY

Judgements of the relative plausibility of theories are based, I believe, on a sense of the structure of the implicit content of the theory. This content is the current state of the cognitive object underlying theories of that kind, the theory-family. There seem to be five main aspects of a theory-family that influence asessments of plausibility and implausibility. I set them out as successive necessary conditions for making such a judgement.

1. The strongest condition necessary for a theory to be judged plausible is that it should represent a moment in the history of a theory-family at which there is a balance in the behavioural and material analogies. Precisely what is meant by a 'balance of analogies'? The idea is this:

(i) Behavioural analogy: the better the imagined behaviour of the hypothetical generative or productive mechanism simulates the behaviour of the unknown real mechanism which actually produces the observed patterns, the more plausible is the theory which represents that moment in the development of the theory-family. The worse the simulation, the more implausible the theory.

(ii) Material analogy: the more fully the imagined properties of the hypothetical generative or productive mechanism match the essential properties of the source model, that is those properties that define the natural-kinds it represents, the more plausible is the theory. This condition on the material analogy ensures that the reality-determining natural-kind rules which express the ontological commitments of this theory-family are conserved.

Balance has to do with the way the behavioural and material analogies are restored when a theory-family is disturbed either by new empirical discoveries or by theoretical innovations, or both. This can be illustrated with the later history of the gas laws.

When Amagat discovered systematic divergencies in the behaviour of gases at high pressures from those predicted on the basis of Boyle's law, the simple point molecule conception of the mechanism responsible for the behaviour described by the law had to be revised. Amagat realized that gas molecules would be more plausible existents if they were thought of as having volume as well as position and momentum, since they were modelled on Newtonian particles which are essentially extended. At high pressures the volume of the molecules themselves ('a') would reduce the effective space in which they could move to 'V − a'. By so modifying his conception of a gas molecule, adjusting the positive and neutral components of the material analogy, Amagat changed the theory-family in such a way that its new state could be represented in a theory which took the form of a revised gas law '$p(V - a) = k$'. This law was deductively consistent with his results. In short he restored the behavioural analogy between the way the real but unknown structure of the gas manifested itself in the new conditions he had created, and the imagined behaviour of the hypothetical nature of gas as expressed in the theory. A theoretical 'gas', imagined now to consist of spatially extended molecules, would behave more like real gases had been shown actually to behave in experiments. The analogies had been rebalanced, restoring plausibility to the modified theory as it represents a moment in the life of the theory-family.

The history of the discovery of the positron (described in detail in Hanson 1967) is a good example of the re-establishment of equilibrium after a theory-led disturbance in the relevant theory-family. Dirac's theoretical work led him to postulate the possibility of positively charged particles of the same order of magnitude as electrons. The actual story of the eventual matching of Anderson's experimental results with Dirac's theoretical concept is rather complex, but from the perspective of this analysis it represents the restoration of the balance of behavioural and material analogies in that theory-family. This example is borderline in that both electrons and positrons are perhaps best considered to be Realm 3 entities, not of the same metaphysical status as ordinary kinds. It is worth a reminder that what I have in mind in this discussion is not the demonstration of the existence of a hitherto unknown kind of being, but the restoration of the key analogies, so that the behaviour of a hypothetical generative mechanism, which theory describes, is analogous to the behaviour of the real corresponding mechanism of nature, which experiment has revealed.

Since the body of data upon which intuitive judgements of plausibility and implausibility are based includes both observed patterns of phenomena and the content of the source analogue which specifies the natural-kinds of the proposed hypothetical entities, and since plausibility is adjusted to the strength of the analogies based upon both the observed patterns and the source analogue, the assessment of plausibility and implausibility is fully determined by the data. In this scheme the difficulty which Quine called the underdetermination of theory by data does not occur. It is easy to see now why, if data are confined to experimental or observational results, there seems to be underdetermination.

The first component in the concept of 'plausibility of a theory' can be set out as follows:

A theory is plausible in so far as it represents a condition of a theory family in which the material and behavioural analogies are 'in balance'. A theory-family is in an imbalanced state, and the theory representing that moment is implausible, if either a behavioural advance has not been remedied by a change in material analogy (Amagat) or a change in material analogy has not been remedied by an advance in the behavioural analogy.

2. A theory-family in balance is the more plausible as its component analogies are strong. Various intuitive formulae can be developed for the representation of the comparative strength of analogies. The strength of an analogy depends on the relations between its positive component (likenesses between source and subject), its negative component (differences between source and subject) and its neutral component (those properties of the source whose likeness or difference from properties of the subject has yet to be explored). Generally an analogy is the stronger as the positive analogy outweighs the negative analogy, and both outweigh the neutral analogy. There are various matters of philosophical interest in defining 'measures' of weight and I postpone a detailed discussion of the assessment of the strength of analogies for a separate section. For the moment it will be enough to suggest that the more the comparison between source and subject, the two terms of the analogy, has been explored, and the more of the aspects studied that have

turned out to be likenesses, the stronger is the analogy.

Part of the difficulty that geologists found with the early versions of Wagener's hypothesis of continental drift and plate tectonics (Hallam 1973) seems to have been the fact that the analogy between continental masses and floating bodies had not been thoroughly explored. The neutral analogy was too large.

3. A theory-family which at a certain moment in its development is in balance, with strong analogies, is the more plausible in so far as the material analogy preserves the natural-kind rules for beings from whom the source analogue of the material analogy is derived. When Pasteur used analogical reasoning to develop his theory of disease he made an explicit comparison between suppuration and fermentation. Using the natural-kind conservation principle he inferred that if suppuration is like fermentation then there must be micro-organisms involved (bacteria) that are like yeasts. The causal agents of the infection of wounds are of the same natural-kind as the causal agents of the fermentation of liquors. And so the theory of disease as caused by micro-organisms is the more plausible. But when the attempt to balance the behavioural and material analogies calls for changes in the natural-kind rules characteristic of the source analogue in defining the entities of the theory the theory-family can slide into a state of implausibility. The putative reality of the imagined entities is undermined. It is this kind of difficulty that made it seem unreasonable to undertake a search for biological entelechies.

4. Since there are two independent sources of concepts for any theory-family, namely the analytical analogue (or model) and the source analogue (or model), the question of their interrelation can be raised. This provides us with a fourth feature that seems to go into judgements of relative plausibility, namely the degree of co-ordination between analytical analogue and source analogue. A theory-family in balance, with strong analogies, preserving natural-kind rules, is the more plausible in so far as its analytical and source analogues are co-ordinate, that is drawn from the same general conception of the empirical realm to which they apply.

For instance in ethogenic psychology the analytic scheme for picking out relevant social patterns is based on an analogy between social action and staged performances, the dramaturgical model. A co-ordinate source analogue would be 'person as actor following a script'. An ethogenic account of a social event is the more plausible in so far as the explanatory theory, developed for the problematic behaviour in question, uses concepts based on those of actor and script, such as 'knowledge of role'. A measure of implausibility would infect an explanation for which the material analogy for the generative process was a socio-biological source, with concepts like 'gene selection', while the analytical analogue continued to enforce an analysis of behaviour in terms of concepts appropriate to the description of a staged performance.

This component of (im)plausibility can be given a Kantian turn. It is equivalent to the principle that the schematisms by which experience is ordered should be in one-to-one correspondence with the categories in terms of which one's whole cognitive apparatus is organized. But in the building of a plausible scientific theory it is by no means sure that such a correspondence can be achieved. It is a desideratum which has to be actively pursued.

The four components of the concept of '(im)plausibilty' could be mapped on to a four-dimensional space somewhat as in Figure 2.

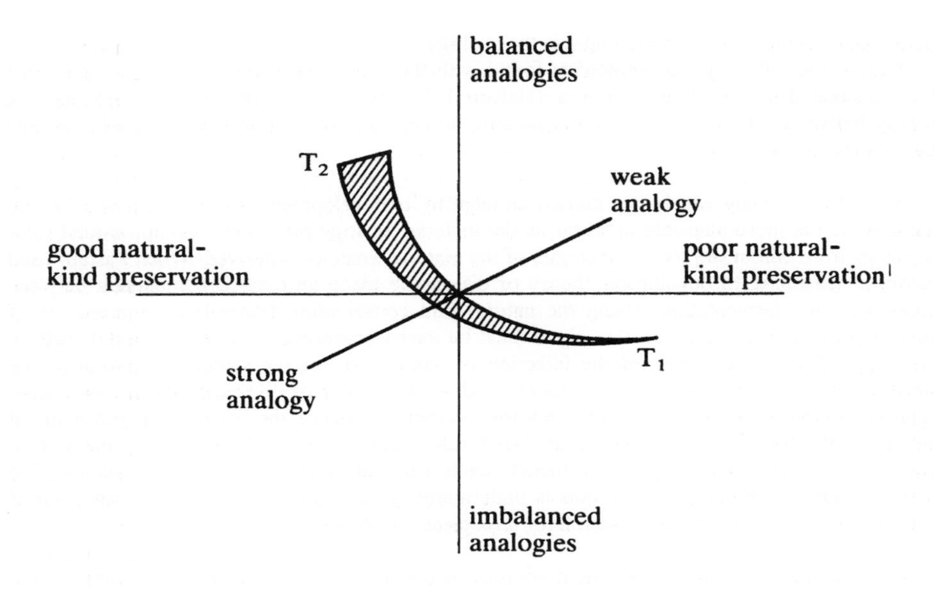

Figure 2
A graphical representation of the concept of plausibility

The breadth of the trace represents the degree of co-ordination between analytical and source analogues. But there is a further consideration.

5. Popper and others have pointed out that our confidence in a theory grows with its successful passage of tests of various kinds. This idea was originally proposed for a deductivist conception of theories and their testing, but the same principle can be applied to the more elaborate structure described here. Meeting the above four desiderata can be thought of as passing tests. If a theory-family has the resources to develop so that it still meets the four desiderata late in its history so much the more plausible must that late moment of equilibrium appear as a particular theory. The fact that the kinetic theory of gases has survived successive modifications, but all within the basic overall scheme defined by its analytical and source models, is, I believe, a ground for holding that the most recent version merits our assent as a plausible theory. I offer these five conditions for plausibility as contributions to spelling the useful but the rather vague terms, 'disadvantage' and 'advantage'. . . . But one must bear in mind that each successive 'theory' is a representation of a moment of equilibrium in a continuous history of adjustments within a theory-family. The Darwinian evolutionary theory-family and the theory-family developed around the concept of 'electron' can easily be shown to have evolved along the same lines as the kinetic theory-family.

If all five desiderata have been met, something more can be said about the nature of the

the concepts in the theory-family at this or that moment in its history, shows that a theory-family is a *structure*. The changing balance between the behavioural and material analogies preserves this structure against the disruptive effect of certain disturbances. There are internal mechanisms of adjustment which maintain the integrity of the whole. The existence of these mechanisms suggests that a theory-family is not only a structure but could be usefully thought of as a *system*. This suggests the possibility of further study of theory-families directed to a more detailed analysis of their system properties. It may also be the case that the social structure of scientific teams that work with theory-families mirrors the structure of those theory-families, and that the conversations within the team apropos their researches have system properties.

With the strict system of theory assessments displaced to the ethics of science, and the deductivist scheme of analysis reduced to a reflection of a mere heuristic and pragmatic aid to using parts of theory-families more effectively, plausibility, implausibility and complex structures of analogical relationships must fill the gap. But there are several ways in which (im)plausibility judgements differ from truth and falsity. For instance the grounds for plausibility judgements cannot be mapped on, to either of the grounds implicit in the traditional theories of truth. (Im)plausibility is not measured by degree of correspondence with the facts, nor is it an expression of the degree of coherence of the theory in question with other theories. Relative (im)plausibility assessments partake in some measure of each. There is a kind of correspondence in the use of observed patterns as a test for the behavioural analogy between the unknown real generators of those patterns and the hypothetical productive process imagined in accordance with the content of the theory-family. There is a kind of coherence in the requirement that our conceptions of the hypothetical mechanisms assumed in theories should be constrained by a material analogy to a source analogue which represents the kind of world we think we are exploring.

An important aspect of the concepts of truth and falsity is their use in defining the logical particles or truth-functions. If theory T_1 is true and theory T_2 is false then their logical conjunction T_1 *and* T_2 is false, and so on. If T_1 is true then not-T_1 is false. And there are many other familiar combinations. If 'true' and 'false' are transferred to the moral universe of the social order of scientists, what sort of meaning are we to give to the logical particles in theory-family discourse? I owe to Jonathan Bennett the idea of testing the way 'plausibility' and 'implausibility' work, and of more carefully establishing the meaning of 'and', 'or' and 'not' in so far as they are used to form complexes of theories.

Certain useful qualifications of the use of these concepts immediately become obvious, when Bennett's suggestion is pursued. Time must be taken into account. Two successive states of equilibrium of the same theory-family may be equally plausible, since their plausibility will be relative to the state of empirical knowledge and theoretical sophistication at any time. But to form a conjunction of the two theories, except in the purely historical sense that perhaps the scientific community once held the former and now holds the latter, makes no sense. Boyle's law and the Clausius model and Amagat's law and his model cannot be conjoined, since when the interface is 'zipped up' contradictions appear. A molecule cannot both have and lack volume. This points up the fact that, in this way of looking at science, and the same goes for Lakatos' view as well, successive theories are determinates under the same determinable. The fact that Boyle's and Amagat's laws cannot be conjoined is the same kind of fact that blue and red cannot be jointly predicated of the same thing at the same time.

However, determinates can always be disjoined, and indeed one way of interpreting the determinable under which they fall is as the disjunction of its determinates. A disjunction of plausible theories within the same theory-family is that theory-family. And if the disjunction contains at least one plausible theory and it is the latest in the development of that family, then the family is plausible, by condition 5 above. Again a temporal condition must be imposed on the 'molecular' structure.

If we try to form conjunctions or disjunctions of theories drawn from different theory-families, in any other sense than as mere catalogues of what the French Academy, say, believed in 1775, we run into further complexities. A conjunction of plausible theories could be assessed for epistemic value, and the infection of an implausible conjunct represented only if there were some melding of source analogues (in Aronsonian terms, some degree of mapping on to a common ontology). Consilience (non-contradictory conjunction of empirical hypotheses on a combined data base) would increase plausibility of the conjunction over any of its conjuncts taken separately only if ontological melding had occurred. There would be an increase in implausibility if there were an ontological clash. Further pursuit of these minutiae does not strike me as likely to prove fruitful, since the discussion so far has brought up two extra-logical matters which seem to exert a determining influence over how our intuitions should go. These are the temporal order of theories, as they reveal the way a theory-family has unfolded; a theory-family might unfold in various ways. The second matter is the degree to which a common ontology (common source analogue) can be created out of the theory-families relevant to each conjoined or disjoined theory. And this last seems to me to be a matter of content not form.

Finally, is the negation of a plausible theory implausible? It will depend a good deal on how much of the engendering theory-family is expressed in the theory, and thus to what parts of the theory-family structure negation is to be applied. If it is no more than that which appears in the printed scientific paper, not much can be said. The negation of a plausible theory is nothing. However, if the source model is contradicted (molecules are not point masses) then it might seem that something like the truth-functional relation between affirmed and negated theories does hold. At least the negation, in this last sense, of an implausible theory, may be plausible—*but only if the content is right.* It certainly now looks very unlikely that the meaning of the logical particles, as used in a metatheoretical discourse is particularly illuminated by the study of the relationships between plausibility and implausibility.

3. THE STRENGTH OF ANALOGIES: A METRICAL METAPHOR

Assessments of (im)plausibility, I have suggested, depend on a theoretician's intuition of the balance between the strength of the behavioural and material analogies of a theory-family. I offered a mere sketch of the idea of 'strength of an analogy' above. It can usefully be filled out a little. The strength of the behavioural analogy is greater if the positive analogy—that is, the similarities between the imagined behaviour of the hypothetical generative mechanism and the actual behaviour of the real generators of natural patterns—predominates over the negative analogy (the dissimilarities). But the more aspects of the possible behaviour of the hypothetical generative mechanisms which have been tested, whether they turn out to be similarities to or differences from the behaviour of the real generators, relative to all the ways these behaviours could be compared (the neutral analogy), the stronger the overall behavioural analogy. These relations can be expressed in an algebraic metaphor. Let pA, nA and tA 'measure' the positive, negative and neutral analogies.

Consider two behavioural analogies A_1 and A_2, say a gas behaves as a swarm of point particles behave, and a gas behaves as a random ensemble of 'plump' molecules behaves. We might compare them for strength by reference to the preponderance of likenesses and differences. There are more differences and fewer similarities in the former analogy than in the latter, so we believe. Thus

$$pA_1 - nA_1 < pA_2 - nA_2$$

expresses our intuition that the latter analogy is stronger than the former.

But we might also be impressed by the extent to which the neutral analogy of some model had been explored, compared with another. For instance some of the opposition to Wagener's hypothesis was the degree to which features of the 'floating continent' idea (A_2) were unexplored. Thus

$$pA_1 + nA_1 - tA_1 > pA_2 + nA_2 - tA_2$$

These formulae are not intended to be the basis of a serious calculus for 'strength of analogy' measures. For one thing they do not take into account the quality of the similarities, differences and undetermined properties of the relata. Similarities in behaviour may be few, but involving important properties, while dissimilarities may be many involving only unimportant properties. How is the relative importance of properties to be decided?

Any suggestion of a numerical measure of the relation between source analogue and subject presupposes criteria for individuating and identifying properties. For instance, it might be important for comparative purposes to treat degree of heat and hue (determinate colour) as distinct properties in assessing the balances of likenesses, and differences. But whether they should be treated as distinct will depend upon the level of analysis. In some contexts, such as metallurgy, hue is used as a measure of degree of heat. Relative to some underlying conception of molecular behaviour, hue and degree of heat might not be considered distinct. This is another sort of case illuminated by Aronson's concept of 'mapping on to a common ontology'.

But this suggests a way of ordering properties by relative importance. The greater the number of distinct superficial properties that can be grounded in a deeper property (mapped onto an item in a common ontology) the more important the deeper property. Thus, electron configuration of the subatomic structure of the atoms of an elementary substance is an important property, while atomic mass is of lesser importance. But one should not be tempted into trying to devise a quantitative index of importance on the basis of what seems to be a purely ordinal notion. And how do we pick out 'distinct superficial properties' from one another? How many of these are there in the colour spectrum for instance? I have touched on the problem of imposing a discrete parsing on a property continuum in discussing Mendel's researches. Again there is no general solution to the problem of how this can be done. In Mendel's case the basis was the atomistic concept of genetic factor, and this overrode the continuity characteristic of the varieties of any property found in nature.

Assessment of the strength of the material analogy is more complex. There are certain minimal conditions an entity must meet to be said to have a certain mode of being. Different ontologies will set different metaphysical limits to the possibility of existence. For instance a Cartesian materialism requires that a material being be spatially extended. Dynamical materialism, in the manner of Boscovich, required only that the centre of a field of force emanating from a material being be located at a point in space-time.

In assessing the strength of the material analogy one must take into account the requirement that the negative analogy between source analogue and the basic entities presumed in the physical groundings of a productive process does not transgress the minimal conditions for the existence of things of the natural-kinds of the source analogue. The fewer of the properties that define the minimal conditions for existence that appear among the similarities and the more that are in the neutral analogy and so undetermined, the more likely it will be that in further developments of the theory the metaphysical limits of the source analogue will be transgressed in the specification of the analogy in the associated hypothetical productive process. The worse the transgression, the weaker the material analogy and the less plausible the theory. For instance, the original kinetic theory of gases was weak, *vis-à-vis* the material analogy. Molecules, though analogues of the material particles that formed the source analogue, were imagined as mere points, and it was undetermined whether they should be thought of as having volume. The addition of volume to their properties in the later theories strengthened the material analogy relative to the Newtonian source analogue.

The relation between properties whose status has been determined and those so far undetermined has a similar effect on our intuitions concerning the strength of a material analogy as it does on intuitions about behavioural analogies. While much of the neutral analogy remains undetermined, it is full of potential surprises. The larger the neutral analogy relative to the positive and negative analogies, the weaker the analogy overall, and the less plausible the theory.

4. A PLACE FOR THE 'SEMANTIC' THEORY

The relationship I have called the 'behavioural analogy' can be represented as a similarity relation between the forms of laws. One law sentence describes the behaviour of the natural causal mechanism, the other that of the hypothetical or imaginary causal mechanism which is the heart of a Realm 2 theory. Thus in physics we often have cases of similarity between empirical laws like those describing the structure of the electromagnetic spectra of incandescent elements and laws describing a 'model universe', for instance those Bohr deduced from a description of his planetary model of atomic structure. There are very many such cases, for instance the phenomenological general gas law PV = RT and the formula derived within the Maxwell-Clausius kinetic theory $pv = 1/3nmc^2$. (Or Stefan's law and Planck's description of an ensemble of mechanical oscillators).

It is to this small corner of the sentential expression of the cognitive structure of Realm 2 theoretical science that the Sneed-Stegmüller account of the nature of theories in physics applies, so far as I can see. This becomes clear when we look a little more closely at the relationship between a pair of laws which represent a behavioural analogy. The similarity relationship between laws can be described in set-theoretical terms as a mapping from one set on to another. The first set is the ordered list of observational results representing the behaviour of the natural causal mechanism under study, relative to the imposition of a conceptual system on its field of behaviour powerful enough to engender data. The second set is the set of pairs, triples, etc., represented by a function which is the mathematical core of the hypothetical law of the imagined behaviour of the hypothetical generative mechanism which is the cognitive core of the theory. Any set of experimental results is, of course, a member of the 'Sneed' set of intended applications of the theory. So the 'semantic' account of theories is just another (and, I am inclined to say, more obscure) way of describing the situation that obtains when the behavioural analogy at a moment in the history of an evolving theory-family is good (cf. Sneed 1971, pp. 250, 252-3).

From my point of view theory-families are maintained by preservation of main features of the intensional structure of the source analogue. I have identified these main features as rules specifying the natural kinds of that universe. They are the specific form that the ontology of *that* theory-family takes. For Sneed (1971) and Stegmüller (1976) something roughly like a theory-family is created by structural morphisms (mappings) between specialized laws (the set of which is the totality of (possible) intended (and intendable) applications), say $F = m\omega^2r$, and the mathematical structures of the 'core', including in this case $F = ma$.

Both Sneed and Stegmüller (and I) deny that theory-families have a top-down structure—for instance axioms and boundary conditions leading deductively to specific law forms. We share the view that the structure is, so to speak, 'side-by-side'. But my account allows for a great many importantly different side-by-side relationships that could emerge as morphisms of mathematical structure. So far as I can see Sneed certainly, and Stegmüller probably, are committed to the view that their mappings are mediated by sets of numbers. By adopting the set-theoretical terminology Sneed and Stegmüller are unable to make the crucial distinction between extensional and intensional structural isomorphisms. Multiple analytical analogues coordinate with the source analogue, correspond to Sneed's 'I', the set of intended applications. But in my account these analytical analogues are not generally mediated by quantitative techniques, even in physics. When they are it is a very special case. Finally it is worth pointing out that the philosophically subordinate character of Sneed and Stegmüller's 'core desideratum' is acceptable only as a surrogate for a content rich-account of what is at the heart of a theory.

5. NETWORKS OR HIERARCHIES?

It should now be clear why the 'network' image as an expression of the way conceptual systems or scientific discourses are organized will not do. Conceptual systems as embodied in the cognitive objects that underlie particular moments of explicit theoretical discourse are indeed structures of interconnected parts, but their organization is hierarchical. In the Aronsonian formal version of this account of theories they are mappings from real beings to real beings. This mapping imposes a hierarchical order on the concepts that would be needed to express the mapping discursively. Two hierarchical relations emerged in my discussion of Aronson's formal account: part to whole, and cause to effect. There may well be others. These relations are asymmetrical and transitive. As such their presence within the cognitive object from which theories are drawn off ensures a hierarchical inner organization to the conceptual core of the theory-family.

Hesse's (1974) account of the network image (echoing Quine 1953, pp. 42-3) runs as follows (p. 28): 'any predicate may be more or less directly ascribed to the world in some circumstance or other, and that none is able to function in the language by means of such direct description alone.' This is precisely the consequence that would follow for the theory of scientific predicates from the dialectic account of the growth of theories (and kind concepts) brought about by the interplay of relatively practical and relatively theoretical contexts in governing the use of a term. But construing this in the *network* image suggests that there is no hierarchy in our beliefs. A much better image is that of a dialectical development, an image we owe to Whewell. The network picture tends to suggest that there is no difference in principle between the way a concept like 'Halley's comet' functions and the way a concept like 'intermediate vector boson' is used, while both differ from the way a concept like 'Africa' is used. With suitable equipment a human being could stand in the same physical relationship (or very nearly) to Halley's comet as they can to Africa. There could be no circumstances in which an embodied human being could become aware of an intermediate vector boson by landing on it, that is by establishing an unmediated physical relationship to it. Since the core of Halley's comet has turned out a roughly cylindrical lump, about 12 km. by 5 km., about the size of the island of Sark, making a landfall is only a technical problem, and will no doubt be achieved next time it comes round.

There is a second dimension of hierarchy. The concepts by which hypotheses about possible generative mechanisms are formulated are controlled by the natural-kind rules implicit in the source analogue of the theory-family. A discursive expression of that source analogue, say in an explicit statement of the metaphysical presuppositions of a certain theory-family, Boyle's *Origins of forms and qualities* (1666), for instance, must make use of concepts which subsume but are not subsumed by the concepts of any particular theory of this or that phenomenon. Boyle's generic mechanical concepts, 'bulk', 'figure', 'texture' and 'motion', take on specific characters in the theory of colour and others in the theory of chemical composition.

CHAPTER SEVEN

CREATIVITY IN SCIENCE

To create is to produce or generate what did not exist before, and most importantly, it is to produce not only an individual which did not exist before but one of a new and hitherto unknown kind. In science the most obvious product of creativity is a sort of discourse, the flow of theory. But theory is itself a secondary product, a description of potent things and products which produce the phenomena we experience. And yet, at least initially, the potent things and processes described in theory are not part of that experience. It is in our conceiving of ideas about them, by imagining possible potent things in which and among which causal activity occurs, that creativity is exercised.

But if theory is to provide understanding, it must be intelligible, and that intelligibility must derive ultimately from the intelligibility of the novel entities and forms conceived in the creative scientific imagination. So novelty must be tempered by connection with the known, or at least with that amongst the known which we take to be intelligible. What it is for something to be intelligible will emerge in the course of the discussion. But the only possible connection that would allow both intelligibility and novelty is that of analogy. New things and processes must be like known things and processes in some ways, but must be unlike them in others. The forms of unlikeness may be very various. Unlikeness may derive from the absence of some common property, as photons have no rest mass, or it may derive from a combination of a set of properties never found together in ordinary experience, as the spatio-temporal continuum must be both continuous and infinitely divisible. Sometimes, as in the latter case, the resolution of the consequent paradoxical intuitions is achieved only by proofs of consistency created in the formal domain of mathematics.

In this [chapter] I shall follow the creative scientific imagination in some of its acts and examine some of the constraints and disciplines which have developed to banish fantasy from the theorizing of scientists. These, we shall see, derive from stricter and stricter ideas of what is possible. And in doing this we shall be pursuing the philosophical issue of realism, since we shall soon confront the problem that the realm of the real seems to extend far beyond the realm of the experiencable.

1. FALSE THEORIES OF CREATIVITY

The importance of creativity will vary with different views of science. In a simple inductivist or positivist view of science, the passage from fact to theory is achieved by a purely formal addition of generality to observed fact, and logical axiomatization of the "laws" so derived. Scientific creativity is then, at bottom, no more than formal invention. To someone taking science seriously as it appears to be, namely an intellectual enterprise in which ever-new content is added to the revelations of experience, the creativity involved in the invention of new concepts can only be an illusion of the advancement of possible experience for an inductivist. On the positivist view this burgeoning of new concepts can have at best a literary purpose, providing a kind of attractive appearance to propositional structures and linking concepts that have only logical importance as bridges from one observation statement to another. Indeed the creative act is strictly dispensable, since all that is essential to science is contained in descriptive concepts and the logical and grammatical particles that are the connectives of their structure as a discourse, if we follow the inductivist-positivist view.

According to realist views of science, the concepts of theory do refer to possible real processes in real structures of things, and thus the introduction of a novel concept enlarges the possibility of experience and adds something to our conception of the world. But realist philosophers of science differ fundamentally on the issue of whether the appearance of a new theory or of its component concepts in the scientific community, and presumably once in the mind of an individual scientist, is a process which is susceptible to rational analysis, and then to reduction of the obedient following of certain canons. Thus some realist philosophers hold a purely psychological (and thus serendipitous) theory of scientific creation. According to Koestler in *The Sleepwalkers* (1963), scientific invention is some kind of accidental occurrence mediated by psychological processes, mysterious to the inventor himself, and amenable only to the kind of analysis he provides of Kepler's mind: part history, part biography, and part psychology. But his analysis of Kepler's intellectual life lays bare certain preparations, which anyone must undertake if he is to hope for a "discovery," in the sense of the invention of a new concept bringing order to the data. Kepler *already* had the ellipse, as a form, both geometrical and analytical, before he could creatively apply it to the problem of making sense of the orbit of Mars. There is then a two-stage process, the invention or acquisition of the required concept, and then the novel use of this concept bringing out something new in the data. In just this way in microsociology, once we have acquired Goffman's concept of the "with," the symbolically displayed pairing, etc., of people in public places, the world is suddenly full of "withs" (Goffman 1972). And what is more, not only must the specific form of the image or concept be present to the mind of the creator, but it must inhabit a mind prepared for that kind of form, as Kepler's was prepared for *an* orbit of harmonious proportions, by his adherence to a general metaphysics, or supra-theory, according to which all the processes of nature were compounded of processes involving forms having harmonious proportions. As I shall argue in detail later, Kepler's situation corresponds precisely with the scheme for rational analysis of scientific invention which merges from the theory I shall propose. Kepler had already studied a highly specific analogue of the orbit of Mars, namely the geometrical ellipse, and was thoroughly acquainted with its properties, and with the various consequences which followed from such properties as the ratio of semi-diameters. And, from his earliest studies, he had been convinced of the verisimilitude of a world picture of the universe, in terms of which the attempt to find a regular geometrical form to correspond to and represent the orbit of Mars made perfect sense.

In the same vein, the fallibilist theory of Popper rightly lays stress on the role of invention in science and the role of the imagination in that invention (Popper 1963). But the discipline under which that faculty works, making it the imagination of a scientist, rather than the fantasy of a crank, is left unanalyzed by Popper, and assigned to the "psychological." This unfortunate move is clearly connected with Popper's very narrow view of rationality, identified by him throughout his works with adherence to the canons of deductive formal logic. Of course the way invention occurs in science must be a topic for psychological study alone, and can conform to no schema, and have no canons of rationality, if rationality is confined to the principles of deductive logic. But one must take great care to distinguish the bogus claim to rationality of the inductivist, who purports to pass beyond experience in the dimension of generality, and the genuine claim of the realist, who sees the imagination of scientists generating conceptions of things, properties and processes that pass beyond any actual experience, not because they make some claim to universality, but rather provide an inkling of the way the world is here and now in those regions like the very small and the very distant, to which we have neither sensory nor instrumental access. The defender of the rationality of creativity seeks the canons of reasonableness in accordance with which such imaginative constructions of conceptions of the unknown can be rated proper or improper, plausible or implausible.

2. SCIENCE AS ICON OF NATURAL STRUCTURES AND POWERS

An adequate theory of scientific creativity can come only from a properly constituted view of what is being created. As in most issues in the philosophy of science, all will depend upon how far one regards the analysis of theories to be primarily a matter of laying bare their logical form. This article is written from very much the view that little of interest to the understanding of science and its modes of thought can be found by a search for logical form. Scientific thought cannot be understood in terms of the content-free principles of formal logic, be they deductive or inductive, if indeed there are any content-free principles of the latter sort. So if we regard theories, primarily, to be considered as formally ordered structures of propositions, we shall look only for the sources and means of creation of those structures. We have already seen how this leads to a stultifying psychologism.

In my view, a theory must be considered as it conforms to certain principles of content, and it must be analyzed for philosophical purposes, so as to bring out the various sorts of propositions it involves, classified by reference to the kind of thing they assert about the world. The logical form of such propositions and the logical structure of the discourse within which they appear is not, of course, irrelevant to our understanding of them, but, I hold, is far from exhaustively determining all that a philosopher might want to say about them (Harré 1972). Thus, for example, a causal proposition cannot be identified by its form alone, say that of a conditional, but is only truly causal if it explicitly or implicitly refers to an existing natural agent potent to bring about the causation when unconstrained and suitably activated. Thus, "ignition of petrol causes combustion" is intelligible as a causal proposition on two counts, neither of which can be dispensed with: it has the form "If *i* then *c,*" and it refers to "petrol" an inflammable liquid. Our understanding of the proposition as causal depends upon our understanding of "petrol" as "inflammable," that is, as something which naturally tends to bum in natural conditions when lighted.

Let us first ask in the most general way, but with more care than is usually applied to these matters: What is the content of a theory? We shall avoid, for the moment, the few very general, very atypical theories, that one finds in fundamental physics, be it classical or modern, and stay with the kind of theory that is typical of chemistry, or medicine, or physiology, or social psychology (of the reformed, ethogenic, sort). Commonly then, in addressing a theory, we confront a discourse which seems to be describing some arrangement of things with certain definite properties, the modulations and changes of which are responsible for the phenomena we are theorizing about. It might be that the distribution and form of animals and plants, both geographically and geologically, requires explanation. Darwin and Wallace produced theoretical discourses describing a process, which, repeated billions of times, was responsible for the phenomena, as they saw them, in the light of the theory.

Already much has emerged. Notice first how readily one slips into speaking as much of an explanation as a theory in this sort of example. I shall return to this point. Notice too that the way the observations of naturalists present themselves to the great biologists, the form of order they saw in them (such as spotting the gradation of the size and form of the beaks of Galapagos finches) was a product of holding the theory. This is not just a psychological observation, and we must pause to examine it further, for here a crucial moment of creativity occurs, one liable to be overlooked in our awe in the face of the invention of the grand design.

Recall, if you will, the discussion of the inductivist view of science and the narrow margin of creativity it allowed, doing science simply being the generalization of regularly recurring observational correlations. Mill, the hero of the inductivist view, describes this process as generalizing over similar cases. The important point to notice here is that he takes similarity for granted. The development, skeleton by skeleton, from eohippus to horse, is similar, for evolutionists, to the development, beak by beak, of one species of finch into another. But why should they be regarded as similar? And in what respects? Their similarity for evolutionists is problematic, and needs to be explained. Taking it for granted, Mill sees scientific method as imposing only a purely formal concept, generality, upon our scientific knowledge.

We must turn to the hero of neo-Kantian realism, Whewell, for a resolution of what is problematic in the inductivist view (Whewell 1967). For Whewell, scientific method is creative not only as to logical form, generality, but also as to content. A science, he holds, is produced by the existence of an "idea," an organizing conception, which is brought to the phenomena, and creates both the possibility of perceiving similarities, which bring the phenomena into the same class, and supplies the generalities, all at once. In controversy with Mill, he cited Kepler's discovery of the orbit of Mars, involving the ellipse, not as a generalization of the known positions of the planet in the star charts, but as a prior idea which provided organization to those positions (in some cases this may be sufficiently sensory to be called a gestalt) under which they became moments in the career of a planet along an elliptical orbit. So the gradation of the beaks of finches is both a product of, and evidence for, the idea of continuous evolution of species one from another.

But the Darwin-Wallace theory was not just the idea of gradual but accumulated change, the idea of organic evolution, which had been held in other forms prior to their development of it. Their theory purported to describe a process occurring over and over again in nature. But neither had observed this process; indeed, it has scarcely been observed today, outside the glass walls of drosophila boxes. (Ford's butterflies and industrial mechanism are recent exceptions.) How could Darwin and Wallace describe the process of evolution in such detail if they had never, and given the time span of the process never could, observe it? What were they describing? What stood between them and the almost untold series of minute changes in millions of species on millions of hectares of the earth's surface that they reduced to order? Clearly they shared an imaginative conception of the organic history of the earth and the natural forces and processes that shaped it. I am not interested, for the moment, in the question of from whence they derived that conception, but in the role of the conception itself.

It is this conception which stands between their limited experience of organic biology, and the utterly out-of-reach organic history of the earth. This again is not just a psychological observation about the personal thought forms of a couple of Victorian giants. By looking very carefully at the form such conceptions commonly take, we shall see how the necessary intermediary between ignorance and the unexperienceable is created, and so how theories can come to have content beyond the description of experienced phenomena. The conception which lies between ignorance and reality I shall call an *icon* of that reality. It is not a delineation since that reality is usually not known, though it may come to be known. In general, a theory describes an icon as a representative and surrogate for that reality. I choose the word "icon" in preference to the more commonly used "model," since the latter has long ceased to be univocal. I also want to draw attention to the frequent sensory or imaginative character of the bearers of conceptions of unexperienced reality, though as we shall see these develop away from the sensory into abstract forms. Darwin and Wallace "formed a picture" of the organic process, a picture which by their description of an icon of the organic process they convey to us. Of course, psychological idiosyncrasy is such that no psychological generalizations about how their conception of nature was present to their minds or ours, their readers, is intended. In particular, I do not intend to suggest that they or we must literally visualize the icon of reality at the heart of a theory. But by keeping a sensory connection, by the use of this word, I want to emphasize that nature must be conceived as a process in time, and a structure in space, of individual organisms, geographical and meteorological forms, ecological interrelation, and the like. Darwin and Wallace attempted to conceive of the reality and to convey their conception to us, though neither they nor we can experience that reality.

What then of the supporting evidence that these great men cite in their works? What relation does it bear to the conception of unexperienced processes responsible for all the phenomena of organic change? We know from the arguments of the Humean tradition that it can provide no logical support for the generalization of the theory or for the claim of the theory to universality within its domain. To understand the role of evidence in science we need to take a radically different view of it, more radical than Popper's fallibilism. I shall try to show by examples that a great scientist cites supporting evidence, not as premises or even as evidence in the legal sense, but as anecdotes, illustrative of the power of the theory to make certain widely-selected phenomena intelligible. Conceiving of the citation of evidence as anecdote brings the explanatory power of the theory to the fore, and raises the philosophical problem of what it is to make the phenomenon intelligible. So Darwin's account of the gradations of beak shape among finches from different islands in the Galapagos group appears not as a premise from which his theory might be inferred, together with all the other available evidence, but rather as an anecdote illustrative of the power of the icon to make the phenomena intelligible.

This can be shown in other cases in just as striking a fashion. In an influential work, *Asylums,* Erving Goffman sets out a theory of institutions, based upon the idea that an institution should be conceived both as a device for fulfilling its official functions, i.e., as a hospital is a place where the staff cure people, and as a setting for the staging of dramas of character, where personas are created and defended, so that a hospital is a place where people perfect dramatic performances as "surgeon," "nurse," and of course "patient," learning to conform to and excel in these dramaturgically conceived "roles" (Goffman 1961). In his discussion of those closed institutions he calls "asylums," Goffman cites instance after instance of people doing things that become intelligible only if conceived on his dramaturgical model, his icon of the nunnery or the barracks or the hospital as a theatre. Each citation is an anecdote, in which the power of the dramaturgical theory to make phenomena intelligible is illustrated. At the same time, it becomes plain that certain structures and textures of life in such places become visible, stand out from other phenomena, only if that life is examined by someone with that icon of the institution in mind.

I can sum up this theory of science in the phrase "icon and anecdote," bound into a single discourse by the explanatory power of the icon, its power to make our experience of the world intelligible, since above all our scientific icons are depictions of the productive processes which bring the patterns of phenomena into being. We must also keep in mind that our experience is only wholly what it is when we conceive of the world that way. Thus, understanding its genesis makes the ordering of experience in classificatory systems intelligible. In science, for every phenotype we find convenient to extract from experience, we must conceive a genotype, for every nominal essence we use in practice to select and identify things and samples of materials, we must conceive a real essence. I was pleased to read in Levi-Strauss the following elegant statement of the icon and anecdote theory.

> . . . Social science is no more founded on the basis of events than physics is founded on sense data: the object is to construct a model and study its property and its different reactions in laboratory conditions in order later to apply the observations to the interpretation of empirical happenings which may be far removed from what has been forecast (Lévi-Strauss 1973).

Creativity, then, must be at its most seminal in the origin of conceptions of the unexperienced, icons of the reality beyond but productive of our experience.

But why not just study the discourse, and never mind about the icon? First, it is *assumed* that a discourse can contain all that is in the object of the discourse, so that all relations relevant to the internal structure of the discourse are supposed to be somehow present within it. But there may be relations in the discourse, relations which we know and use, say between predicates, where two predicates permeate each other's sense, or relations of analogy between predicates deployed in an argument, which are dependent on prior icon relations, appearing in the icon as coexisting properties, or likeness and unlikeness between things. Thus, the copresence of a pair of properties in an icon becomes the source, in a diachronic process of meaning assimilation, of synchronic internal relations between two predicates, relations which may be quite crucial to understanding how the discourse is structured and how its development would be justified. Such relations can never be reached by the analysis of a discourse into its logical form, since a *fortiori* logical form extracts from all material relations between predicates, and yet it may be just those relations upon which the coherence of the discourse as a scientific theory rests.

3. TWO EPISTEMOLOGICAL REGIONS

"Beyond our experience," you say. "But surely traditional epistemology teaches us that we have no knowledge of what is beyond our experience." There are two epistemological barriers which scientists regularly overleap in their practice. The one is that erected by positivists between actual and possible experience, the other by the critical philosophy between all actual and possible experience and the realm beyond all possible experience. I shall try to show that our powers and techniques of creativity are such that we can, at first tentatively and cautiously, and finally boldly, trust ourselves to pass beyond them, and dwell in thought, in worlds accessible only to our creative imagination.

The realm of actual experience is limited in two ways. We are confined to the senses we actually possess, and invariants we can express between them, so hand, eye, and ear conspire to provide our experience of a bell. But without conception, perception is blind. And what we can actually identify within and across our sensory fields depends upon the sort of concepts with which we are mentally prepared for the world. As Kant had it and contemporary psychology confirms, our experience is a product of schematic ordering and supplementing our sensations, the schemata being of conceptual and perhaps even of a linguistic origin (Gibson 1968). All this is commonplace. But it is not difficult to imagine things, and structures and processes and properties, that are too far, too small, too fast, or too slow, or even too big to be experienced by us as we are presently constituted, though the Gibsonian invariants in the objecthood of these entities are just the same invariants as in the objects of ordinary experience. Locke talks of what we might see with microscopic eyes, or hear with a more acute sense of hearing, and Geach has speculated on the colors we would see were our eyes sensitive to a broader spectrum of electromagnetic radiation. In this way we can conceive of a realm of possible experiences, and populate it in our imagination with objects undergoing processes which we do not, but might, experience.

But theories describe icons of structures and processes which would be experienced within that very realm. Kepler conceived of a fine structure for the snowflake which would be a natural form of packing for minute ice particles (water molecules, if neat little spheres) and which would, if much repeated, yield the universal hexagon. Kekulé conceived of a structure in space in which the carbon atoms of benzene would form a stable ring, and Harvey completed his sanguinary plumbing with minute imagined vessels, closing the hydraulic circuit, while Van

Helmont conceived of disease as the invasion of the body by an army of invisible minute organisms. These men populated a world of possible experience with fabulous creatures of their imagination just as Mohammedans filled the air with djinni, Descartes ignited a furnace in the heart, and Velikovsky supposed a radically different history for the earth. So entering a realm of possible experience in the imagination is fraught with hazard, for the imagination is as capable of fantasy as of sober speculation. I shall return to examine the discipline to which it is subjected in science and from which we shall extract the fragments of a creative criterion. It is because of the existence and acknowledgment, and ultimate justifiability, of this discipline that we can override the extremes of positivism which would have us conceive of science as no more than the "mnemonic reproduction of facts in thought." Let us call the activity of thought in populating the realm of possible experience the work of the reproductive imagination.

But the scientific imagination does not confine itself to the same realm of creation as is continuous with the realm of perception. It attempts to conceive of the structure of the world beyond all possible experience. Scientific thinkers are driven to attempt this ultimate barrier to knowledge by two factors. The first is simple. Our ordinary experience is full of instances of phenomena whose effects are inexplicable by any work of the reproductive imagination. Electric and magnetic phenomena are the most striking examples. We are simply not equipped with sense organs sensitive to the magnetic influence.

First attempts to solve the problem of the mechanism of the magnetic influence involved the work of the imagination at its reproductive stage, leading to a proliferation of magnetic fluids, particles and the like, clearly denizens of the realm of possible experience. Norman and Gilbert went further. Gilbert's imagination leapt the barrier of all possible experience, leading him to postulate the *orbis virtutis,* a shaped, structured field of potentials or directive powers. A structured field of *powers* was something, though spatially extended and temporally enduring, that was clearly not, as such, an object of possible experience, given our lack of magnetic sensibility. Eventually Faraday offered to the visual sense a picturable icon in the lines of force technique for representing the field.

Not only is the perception of the magnetic influence beyond the range of any of our sensory fields, but magnetism, like electricity and gravity, forces the creative imagination to the transcendental stage, since there is a conceptual, not just a contingent, difficulty about the perception of fields. A field is a distribution of potentials, and though we speak of the energy of the field at a point, that energy is not manifested in any kind of action. Thus a field is a *fortiori* imperceptible, its existence known only from its manifestations, and from the presence of field generators, like conductors carrying a current, iron atoms, lumps of matter, and the like, the laws of structure of whose produced fields we have discovered by examining the manifestations of the field on other similar occasions. The icons which represent fields, like rotating tubes of moving, elastic fluids, are not representative of objects or processes in the realm of possible experience, and to suppose them to be so would be a major epistemological error.

But the matter is more complicated. The transcendental imagination is required to generate not only distributed potentials, but to conceive an intermediary between potential and action, not as far beyond possible experience as the potential, but in its realm. At some definite point in the room there is a gravitational potential. But at this moment no test body is at that point. Suppose we now bring a test body to that point, say an alabaster egg, and support it there on a platform. There is still no action, but since action is now possible if the platform is removed, we are obliged to postulate another entity—the tendency to fall acquired by the alabaster egg from the gravitational field. We know that the egg acquires the tendency from the field, since on the moon, for example, we have good reason to think that the egg's tendency to fall will be much diminished, so the tendency is not an intrinsic property of the egg. And we can easily check whether the egg has indeed acquired the tendency by removing the platform and seeing if, indeed, the egg does fall. If we replace the platform by a hand under the egg we might claim to be experiencing the tendency of the egg to fall, but we could hardly claim that it is an experience of the potential of the field, since that potential will be there when the egg is removed, and then we feel no tendency for downward fall. The tendency, then, is not so far from the edge of the boundary of all possible experience as is the potential, and may indeed be held to within the realm of possible experience if the experience of weight and pressure to which we apply our powers of resistance, frustrating action, is accepted as the experience of a tendency, and I cannot think of any argument that would oblige us to deny this.

Contemplating the egg has led us to a complex icon of the whole situation, and in supposing it to be a delineation of the real, the subsequent fall of the egg becomes intelligible, as does the dent it makes in the velvet cushion upon which it usually rests. There are two material things, the earth and the egg, and two immaterial things, the gravitational field, our conception of which is certainly a product of the imagination in its transcendental phase, and the tendency caused in the egg by the power of the field, a thing arguably on the border of the realm of possible experience.

4. TWO ACTS OF CREATIVE IMAGINATION

I shall try to show that though the creative imagination of scientists is, in a certain sense, free, and indeed we shall come to see in exactly what sense, nevertheless, analytical schemata can be constructed to represent the dynamics of concept-construction, in an idealized form, and from which canons, exemplified in rules, can be abstracted. The essential point to be grasped, in considering the acts of the imagination in its reproductive phase, is that in producing an icon of a possible reality, the imagination is not modelling something known, but something which is in its inner nature unknown. We know how the mysterious structures and things behave, since they have produced the patterns of phenomena we wish to make intelligible, so at least our icon must depict a possible being which behaves analogously to the unknown real being.

But simply to conceive possible realities in terms of their behavior is no advance on positivism, since we already know the behavior patterns of things and express these in the laws of nature. The task of the reproductive imagination is deeper, for it must enable us to generate a conception of the nature of the objects which behave exactly like, or in various degrees analogously to, the real things actually in the world. The forms of thought involved in this act of the reproductive imagination can be idealized and schematized by distinguishing between the subject of the conception, what the icon is a model *of,* and the source of the conception, what the icon is modelled *on.* Since the nature of the subject is unknown in the sense of beyond all actual experience, the relation between subject and icon can be mediated only by likeness of behavior: a swarm of molecules behaves like a real gas behaves, a person in a shop behaves like an actor playing "person in a shop," the evolutionary process produces results like a plant breeder produces. But the creative act of the reproductive imagination is to produce an icon of the unknown nature of the real world, and this icon must be at least capable of being recognized and understood as a plausible depiction of a possible generating mechanism for the patterns of behavior whose explanation is our problem. So if we imagine an evolutionary process as consisting of minute variations in form and function generation by generation, and certain of these variations leading to greater reproductive rates in their possessors, we have conceived a mechanism (or at least the rough outlines of a mechanism) which would generate the pattern we call "the evolution of species," which is itself a product of an act of structuring upon the individual bits of knowledge about form and structures of individual organic specimens.

But the natures of things imagined and the generative mechanisms they severally constitute must be plausible, both as generators of the observed patterns and as possible real existents.

Thus, not any sources will do. Sources of models, as icons of reality, must conform to two criteria:

1. They must be the kinds of things, structures, processes, and properties the current world picture regards as admissible existents—gases rather than imponderable fluids (1800), material atoms rather than atmospheres of heat (1850), electric charges rather than solid atoms (1920), neural networks rather than mental substance (1960), and so on. The metaphysics of science consists in the discussion of the coherence and plausibility of the world pictures, literally conceptions of structures, that occupy space and endure for a time, and any possible systems they may form, which serve as general sources for conceptions of possible realities, though, in one of the examples I have cited, that of elementary electric charges, we are on the borders of possible experience.

2. Having thus conceived proper kinds of things, we need to be able to imagine plausible laws for their behavior. And we find such laws, or closely analogous laws, in existing science. Electrons obey, *for us,* Coulomb's law, the law obeyed by the charges on suspended pith balls. Gas molecules obey Newton's laws of motion, the blood in Harvey's imagined tubules obeys the ordinary laws of circulating fluids. It is thus that the disciplined imagination works in reproducing a version of experienced reality in the realm of possible experience. This kind of representation may, in certain cases, turn out to be not just a representation to fill a gap but a depiction of reality itself.

But what about the human imagination in its transcendental employment, its transcendental phase so to speak? The first point to notice is that the behavioral constraint is the same in both phases. The world as conceived beyond all possible experience must behave just as the real world behaves, or in a very similar manner. So the dispositions we assign to reality in our imagination must be closely analogous to the dispositions we find the real world to have. But if these dispositions are to be grounded, that is, to be powers and liabilities, dispositions grounded in the nature of things, must we not try to conceive of natures of things the details of which *must* lie beyond the boundary of all possible experience, and if that nature is beyond all possible experience, how are we to conceive it? Yet it is my contention that physicists, cosmologists, and psychologists both can and do achieve creative acts of the imagination in the transcendental phase of activity, and that we can follow them. How is this possible?

The first point to notice is that the world beyond all possible experience can share one kind of attribute with the objects of the world of possible experience, namely structure. It is this attribute that makes for the possibility of intelligibility of conceptions of that world, and of course for its mathematical description. Objects in the experiencable world must have both sensory qualities, or considered in themselves, at least the power to manifest themselves sensorily, and they must have structure (structure which need not be, though it often is, spatial), while processes in that world, to rise above the dead level of imperceptible uniformity, must have both powers of sensory manifestation within thresholds that allow us to say they have structure in time. (Compare the difficulty that was once experienced in knowing whether the Australian aborigines had melody, when the multi-toned structure of their tunes is within a semi-tone, our normal unit of melodic differentiation.) In the world beyond all possible experience, there are, a *fortiori,* no powers of sensory manifestation, though two connections must remain with the world of experience. In the one there must be an actual causal connection, in that the objects in that world must have powers to affect objects in such a way as to change or stimulate or release their powers to manifest themselves or changes in themselves to us. And secondly, more germane to the issue of conceivability, both synchronic and diachronic structure may be attributable. For example, in Medieval chemistry, the four principles which had the power to manifest themselves in warmth, coldness, wetness and dryness, though not perceptible themselves as such, nevertheless were imagined to be present in things and materials in definite proportions (non-spatial synchronic structure) which might be changed, and so change the nature of the substance, a process, if intelligible, having diachronic structure.

The role of icons in conceiving qualityless structure thus becomes both clear and shows itself to be equivocal: for icons, if based upon the reproductive power of the imagination, but passing into the transcendental phase, will constrain conceived structure to the structures of possible experience. The imagination, in its transcendental phase, must proceed to acts of abstraction and generalization to pass beyond this constraint, and of course the acts of the imagination in this phase are identical with abstract mathematical creation. In the very last analysis icons of the world beyond all possible experience *may* be required to have the character of abstract mathematical structures. But of what?

We have already noticed how, in science, dispositions are grounded in hypotheses about the natures of the individuals which manifest these dispositions. If the individual is a source of action such as a material thing as the ultimate source of the gravitational field, or an acid, relative to chemical analysis at the molecular level, then the grounded disposition is a power. But if, as we move into the transcendental phase, we have left behind all properties other than structural with which we might ground dispositions, then those structures in which all secondary dispositions are grounded must be of primary or pure dispositions, that is, of ultimate powers, that is, in the elements of the most fundamental conceivable structures, powers, and dispositions must coincide. In natural science, fields are the best example of structures of pure powers. Icons, such as Maxwell's tubes of fluid, may be required for conceiving structures of potentials, but here at least they are but dispensable models of our abstract conceptions of reality. They are, at best, a system of metaphors for holding onto the sense of the abstract objects, and it is to this role that another aspect of the sense of the notion "icon" is directed. An icon as a religious painting is not just a picture of some worthy person, but is a bearer of meaning, generally abstract with respect to that which it depicts. In their transcendental employment the models generated by this phase of the imagination are truly icons.

5. DISCIPLINING THE FANTASY

Common sense would have it, no doubt, that the test of the imagination in conceiving objects in the realm of possible experience is whether, when our senses are extended by the development of some device, such as the stethoscope or the microscope, the hypothetical object or process appears, that is, it is a matter of whether and to what extent reality, when it is revealed in experience, matches the icon. But common sense needs defense. There is a tradition in philosophy of casting doubt upon the authenticity of what is perceived, by insisting that only the existence of and properties of the immediate elements of various sensory fields involved are known for certain. Happily, as far as stethoscopes, probes, microscopes, telescopes, and slow motion film are concerned, one can establish a gradual transition from the objects and processes of unaided perception, to the sounds, shapes, colors, motions, and so on, brought into our experience with the help of instruments. One can hear the same sound with or without the stethoscope, but one can also hear clearly sounds heard only faintly or not at all without it. Thus, we establish a continuity of the existence of percepts. By this achievement, the world of possible experience penetrated by instrumental aids is made one with the world of actual experience, so the extended world is no more nor less dubious or inauthentic than the world of unaided perception, the ordinary world. And this is all we need for the control of creativity at the reproductive phase of the work of the imagination. Philosophers may continue to argue about the epistemological and metaphysical status of material objects, but their disputes and distinctions cannot detach bacteria from bodies, nor galaxies from ganglia.

But discipline in the world imagined to lie beyond all possible experience cannot be based wholly upon instruments. However, there is a kind of penumbral region, wherein structure is simply spatial, where structures whose elements are beyond all possible experience may nevertheless be displayed. I have in mind the photographs of molecular structure obtained by field ion microscopes, or the tracks of "particles" observed in cloud chambers. While the phenomenal properties of the structure are linked to its elements only by long and sometimes ill-understood causal chains, the structure so projected is at worst isomorphic with the structure of the thing or process being examined, at best that very structure itself. However, if we consider cases deeper into the inexperiencable, only reason can come to the aid of the creative imagination, and that only a *posteriori.* At the deepest level, the best that we can do is show by argument that the structure of elementary powers we have imagined as the ultimate structure of the world fulfills certain necessary conditions for the possibility of our having the kind of experience we do have, and there may be a still more general form of argument which would link certain structures (and certain powers) to the possibility of any experience at all. In fulfilling these conditions, the world as we experience it is made intelligible.

A process or structured object becomes intelligible if the following conditions are met:

1. From the imagined fundamental world structure the form of the process or object can be deduced, i.e., from the tetrahedral distribution of the valencies of the carbon atoms the observed form of the diamond can be deduced with the help of certain ancillary hypotheses, that is, the structure of the valencies provides a reason, *relative to accepted physics,* why diamond has the form it has. Form, as we may say, is inherited from form. The intelligibility of the form of diamond comes not just from the fact that the proposition expressing this stands in a certain logical relation with some other propositions, but that among those other propositions are some descriptives of some underlying and fundamental form, that is a structure of units or elements that are, for that case, not further decomposable. Thus, to cite structure is to make intelligible, and by linking structure via the deductive link, which has the effect of preserving content, that intelligibility which derives naturally from citation of structure alone, is transferred to the form of the diamond. But sensory qualities, like color or timbre, cannot be so made intelligible, only their associated forms, wavelength, or harmonic structure, can be referred to more fundamental forms and so acquire intelligibility. It is a prime rule of science that qualitative difference should be explained in terms of structural difference, and in so doing, the only final form that an explanation could take is achieved. In short, only form or structure is intelligible in itself. Philosophical argument for this proposition could do no more than take the form of the analysis of all satisfactory explanatory forms and the exhibition that their satisfaction derives from that feature.

2. In the nodes of the imagined world structure, there are agents, that is the structure is a structure of sources of activity, as for example a complex formed from repressed traces of disagreeable experiences, a concept which represents a state of the world beyond all possible experience, is a structure of agents, its elements having power to effect changes in behavior, so that the complex itself becomes a structured agent, the source of the pattern of neurotic or compulsive behavior, making it intelligible by showing that the form of the behavior is a direct transformation and manifestation of the form of the complex.

6. SOCIETY AS CREATED ICON

But when we create icons in the pursuit of the social sciences, we cannot take it for granted that there is a real structure, some independent world, of which that icon, however imperfectly, is a representation. Indeed, both the existential status of society and the significance of societal concepts is highly problematic and cannot be taken for granted. We speak of the nation, the army, the middle class, as we were speaking of the island, the Thames valley, and so on. Our power to create societal concepts is, as we shall see, a creativity of another kind.

I shall approach this difficult problem through two examples, illustrating different facets of the role of societal concepts, and their associated icons in our lives. Both examples will illustrate how we are unprepared to live in an unintelligible environment, that is, an environment which does not either exhibit structure or clearly manifest an underlying structure. Imagine a large complex of buildings unified by a boundary wall, and a common calligraphy in the labels displayed at various entrances (that a hole in a wall is an "entrance" is also a social, not a physical, fact, so in this analysis an underlying and unexamined ethnomethodology is being taken for granted, but at least by me, knowingly). People move in and out, some prone on stretchers, others arrive with every mark of respect in Rolls Royces. Inside, uniforms are much in evidence. Many people are in bed, and even some of those who are walking around are wearing pajamas and dressing gowns. What on earth is this strange place and what is going on? The innumerable momentary interactions and sayings of individuals, or rather some part of them, are made immediately intelligible by the hypothesis that this is a *hospital,* that is, the whole begins to exhibit structure. The introduction of this concept is strictly comparable to, though much more complex than, the introduction of the concept "galaxy," which made the appearance of the night sky intelligible by referring its observed form to an underlying and aesthetically pleasing structure, the spiral form of the stars in the galactic plane. As ethnomethodologists have insisted, there is *always* an everyday problem of intelligibility, which it behooves sociologists to contemplate, since it is nearly always solved by those involved. Sometimes the continuous everyday solution needs supplementation by a stroke of scientific genius, as when Goffman made us realize that many flurries of activity, unintelligible within the official theory which glosses "hospital" as "cure-house," become intelligible within a single supplementary theory in which the institution is glossed as a setting for dramas of character. Each theory generates a rhetoric, a unified theory of explanatory concepts, with an associated grammar (in which rhetoric do we put the socio-grammatical rule that the superintendent can refer to the hospital as "my hospital," and to whom?). Rhetorics are drawn on in accounting sessions, in which in the course of talk the momentarily mysterious is made intelligible by allowing itself to be so described as to find a place in a structure, this time of meaningful activities within semantic fields recognized as legitimate in the rhetoric. Finally, one should notice that the official theory may find expression in what is literally an icon or icons, diagrams on suitable walls, in which structure, as officially conceived, is laid out. Nowhere on such charts appears such power and influence structures as that of the janitor in a school, where the boiler-room society over which he provides is the apex of the counter-hierarchy, and in which such officially-defined figures as the headmaster carry very little weight.

The position implicit in the example above, which it should be clear is not at all the same as the old theory of methodological individualism, can claim at least one systematic exposition in the past, namely, that of Tolstoy in the sociological chapters in *War and Peace.* As a mark of my admiration for his formulation of the theory, I have called it the "Borodino Theory," since he broaches it explicitly in his analysis of that battle and in his recurring theme of the contrast between the manner of generalship of Napoleon and of Katusov. As Tolstoy sees it, the Battle of Borodino is a middle-scale social event within an inexplicable, very diffuse, and very-large scale human movement, the periodic movement of very large numbers of people from West to East and East to West. This migratory oscillation has no name, and is not mentioned in historians' accounts of the affair. They are concerned with nations, armies, generals, governments, and the like. And their role, according to Tolstoy, is to impose order and intelligibility upon meaningless eddies in the groundswell, such eddies as the Battle of Borodino. The battle becomes, in *War and*

Peace, both an instance of Tolstoy's theory, an anecdote showing its power to make phenomena intelligible, and a kind of model for the analysis for all middle-scale human events.

The battle is joined by accident in conditions which prevent either commander getting a clear view of the battlefield and its changing dispositions. Both commanders are surrounded by eager staff officers and constantly receive messages from the officers on the field. But by the time the message, usually in garbled form, has reached the commander, the situation it described has changed. Napoleon nevertheless issues detailed orders throughout the day based on the "information" he receives — but these orders rarely reach their destination, and even more rarely are intelligible to the commander they are intended for, and even more rarely still enjoin courses of action still possible by the time they arrive. But on the French side, a great flurry of command goes on. Katusov, on the other hand, does nothing, he believes no messages, he issues no orders. He waits for the issue to be decided. But as a final irony, no issue is resolved, at least in the final dispositions of that battle, though as Tolstoy points out the French losses that day turned out to be fatally weakening to their army.

But, asks Tolstoy, what do historians make of that battle? They *impose* order upon it. They represent the haphazard movements on the battlefield, enjoined by the exigencies and impulses of the moment as splendid tactical moves, flowing from the genius of the commanders, and brought about by their orders. "The Battle" as a structured, ordered, hierarchical social entity, is a product of retrospective commentary; in the technical language of the new sociology, it is an account. A series of flurries, intelligible as the mutual actions of individual men at the microlevel, become elements in a larger structure, by an act of the creative imagination. In terms of this, commentaries upon and explanations of actions are contrived for happenings which are no longer conceived as closed entities, but as elements in a larger structure having relations with other elements of that structure, for example, with the thoughts and orders of the commanders. But that larger structure has its being only in the imagination of those who share the theory, a theory, of course, which any member can hardly fail to share. Only one who follows at least one step on the phenomenological path, one who, like the ethnomethodologists, wishes to subject the natural attitude and its products, "battles," to scrutiny, can come to this. Ironically, it is in the social sciences that the positivist theory of theoretical concepts has its only plausible application, since in the social sciences the "Borodino Theory" would counsel us to treat societal concepts as serving only the interests of an imposed intelligibility, and not being referential terms pointing outside the theory to real existents.

The creative imagination of the social scientist is the most potent of all, for he can create an icon whose close simulacrum of a real world is so potent that people will live their lives within its framework, hardly ever suspecting that the framework is no more than a theory for making the messy, unordered flurry of day-to-day life intelligible, and so meaningful and bearable.

7. EVOLUTIONARY EPISTEMOLOGY

But sciences and societies have a history. And the question as to why a particular form *appears,* makes itself visible in various manifestations, at a particular time and in particular circumstances, must be tackled. To get clear on the basis for a diachronic analysis, one must distinguish the productive process of a "next stage" from the sequence of those stages. Only by clearly separating them can the problem of their several intelligibilities be solved. In general, I would wish to claim that just as in the sequential stages of plant and animal life there is no pattern from which a law of those stages can be inferred, that is, they have no intelligibility as a progression, so there is no pattern in the sequence of stages of sciences or societies. Patterns *are* discerned and described, but I would wish to argue that these reflect current ethnographies and current obsessions—God's will working itself out in history, economic determinism, and the like—the projection of which on the sequence of stages is the source of historicism. But that does not mean to say that the process of historical change cannot be understood, and that it cannot be made intelligible. I follow Toulmin in his claim that the *general form of all historical explanations* was invented by Darwin and Wallace, a form which allows for the intelligibility of a historical sequence without falling prey to historicism.

The form of our understanding of the diachrony of social and scientific creativity will be evolutionary in the natural selection mode. Thus the origin of new forms, be it animal, vegetable, or structures in thought, will be taken to be (relatively) random, with respect to the environment in which those forms will be tested. Thus, in the moment of inception, all novel forms will have the character of random mutations, and thought forms, fantasies, will be taken to be innumerable. We shall return in a moment to the important issue of how far the inception of thought forms is disconnected from their environment, and we shall find that it is not quite so clearly free as organic mutation.

But by what sort of environment are they selected? We must acknowledge the complexity of that environment. New ideas are contemplated, deliberated upon by people, and in the course of these deliberations are accepted or rejected, or sometimes merely forgotten, or abandoned because of the appearance of a novelty more in fashion. Sometimes they are tested, as to what further intelligibility they lend to what we think goes on, and sometimes even as to what they lead us to think there is. Sometimes ideas are rejected out of hand, as silly, threatening, "unintelligible," obscene, and so on. How is some order to be brought to this multiplicity?

The credit for the introduction of the basic idea of evolutionary epistemology must go to Popper, and hindsight, I feel sure, will regard this as his great contribution to philosophy. Effectively, Popper proposed to bring order to the selecting environment by the use of a principle of formal logic (Popper 1972). The instrument of natural selection upon ideas whose appearance is serendipitous with respect to that environment, and of only psychological interest, is the principle of *modus tollens,* that a proposition which has false consequences is false. By itself, "falsification" is just a logical principle, but in Popper's works it is uncritically coupled with an epistemological principle — "rejection" — that is, whatever is falsified must be rejected as knowledge. Popper's particular version of the evolutionary natural selection theory comes to grief on that coupling, since it cannot be taken for granted and it turns out to rest on two levels of theory, one metaphysical and the other scientific.

In order to pass from falsification to rejection, one must suppose that the falsified principle, hypothesis, or theory is not worthy to be accepted as knowledge. This requires recourse to an assumption about the stability of the universe, a metaphysical assumption that the universe will not so change in the future as to behave in such a way that the falsified principle is then true. But this is a principle to which Popper cannot have recourse, since it is a form of the general inductive principle of the uniformity of nature, the negation of which leads to an evolutionary epistemology for science in the first place.

But even if the passage from falsification to rejection be granted, say, as a principle, itself a mutation surviving in a hostile environment by virtue of its power to make scientific method intelligible, the application of the principle depends upon assuming some embracing scientific theory as true. In general, falsification is itself an interpretation of a yet more fundamental relation, namely contradiction. The product of a "testing" is a contradiction. "All A is B" is in contradiction with "This A is not B." To assign "false" to the principle, to make this *its* test, requires that we assign "true" to "This A is not B," that is, the assignment of a truth-value to the general proposition depends upon a prior assignment of a truth-value to the particular. That is what makes this a *test* of the general proposition. But it is notorious that the result of any experiment is very far from being a "brute fact," and we might as easily have assigned the truth-value the other way. In practice, "This A is not B" gets "true" or "false" in priority because it is embedded in a more embracing or otherwise more attractive theory than "All A is B," which is, for the purpose of the test, detached or isolated in some way. Thus the passage from contradiction to falsification is not unequivocal. (Popper did, of course, attempt a "basic statements" theory to anchor truth somewhere, but has been forced to relativize it, which is to give it up.) It seems that Popper's own attempt to give body to the general theory of which he was the originator is too much dependent on logicist assumptions as to what is rational, and on the uncritical acceptance of the transitions contradiction to falsification, and falsification to rejection.

Toulmin, by contrast, is prepared to include a much wider range of items in the selection mechanism, and furthermore, makes an important and interesting concession to a mild teleology, unthinkable in the Popperian theory (Toulmin 1972). The appearance of a mutant idea, for Toulmin, is not wholly detached from what has gone before in the realm of ideas, not unconnected with the tests and trials to come. As we get the idea of the kind of tests an innovation is to face, we censor what one might call "first fantasy," so that only plausible ideas are offered for the community to test. And synchronic creativity, as I have described its structure in Part I, involves a disciplinary feature in the very process of creation itself, which "close-couples" the new creation to the old. Mutation occurs, then, within a very narrow range.

What of the selection mechanism itself? Institutional and social factors become prominent once we move away from a simple logicism. Clearly, an idea will have a better chance of discussion and consideration if it is proposed by someone in a certain place in the institution, be it the society of scientists or the board of a company or the general meeting of a commune. And greater effort will be made to make the world conform to the idea before it is rejected. This latter feature is very prominent in the natural history of political ideas, where the "world" in which the idea will run its course is a human construction which can, within certain limits, still unknown, be reconstructed so as to preserve the idea unrefuted.

Toulmin's particularization of the theory is not without difficulties as well. He chooses to discuss the problem, not in the propositional ontology of Popper, but with concepts as his individuals. However, this leads to considerable difficulties and unclarities in his statement of the theory, occasioned by the problem of individuation of concepts. The problem is central, since he treats "population of concepts" as strictly analogous to "population of organisms." But what is the "individual" concept? Is it the concept of an individual person, so that my concept of the atom is a different concept from yours, even though our concepts may match throughout their semantic fields? This would seem to be the natural interpretation, so that a concept would reproduce itself by being more and more replicated in the minds of others. But he also speaks sometimes as if it was the concept we shared, which is the individual which is naturally selected, and its progeny, not as its replicas in other people's minds, but the further logical and conceptual descendents that it spawns. It seems clear that this second interpretation must surely be a mistake, and that one must stick to the individuation of a concept as my concept, and apply the species notion to link, under the same phenotype, my concept and yours when they are, as concepts go, alike.

I hope I have said enough to indicate that both the diachrony of theories and societies can be understood in a general way by the strict application of the evolutionary analogy and the idea of selection, but that the final account of the balance between rational, societal, and other factors in the selective environment has yet to be struck.

CHAPTER EIGHT

METAPHORS AS THE EXPRESSION OF MODELS

1. THE INTERACTIVE CONCEPTION OF METAPHOR

For those concerned with the use of metaphor in scientific theory two questions immediately present themselves: How do metaphors work? and Why are metaphors necessary? The first of these is one that may never be fully answered, not at least without a theory of meaning and a theory of mind at present far beyond us. The second may be answered simply—We need to use metaphor to say what we mean—since in the course both of literary composition and scientific theorising we can conceive more than we can currently say. Discomfort at having in the first case only a partial answer, and in the second a vague one is relieved by seeing the ways in which these two questions may illuminate one another.

Despite the difficulty of giving a fully adequate theory of the way in which metaphors work (and who, in any case, would be qualified to give it - philosophers, linguists, neurophysiologists?) we can note strengths and weaknesses in the various theories generally presented, and this itself points in the direction of a more adequate account. Transfer is implied, perhaps unfortunately, by the etymology of the term metaphor, and it is as a study of transfer or comparison that traditional studies of metaphor have developed. What are often referred to as the 'theories of metaphor' are for the most part theories as to the nature of the transfer or the comparison which a metaphor effects. They are theories of the way in which metaphor gives us, in Dr. Johnson's words 'two ideas for one' (cited by I. A. Richards 1965, p. 11).

These theories comprise two main groups: substitution and Gestalt theories. Basic to the former is the idea that metaphor is another way of saying what could be said literally, 'a sort of happy extra trick with words, . . . grace or ornament, or *added* power of language, not its constitutive form' (Richards 1965, p. 90).

The shortcomings of the substitution theory are legion. Taken at face value it reduces metaphor to the status of a riddle or word game, and the appreciation of metaphor to the unravelling of that riddle. To assume the ready availability or even the necessary existence of a literal substitute renders metaphor, on this interpretation and especially for the purposes of philosophical or scientific reasoning of any sort, almost useless.

A slightly more plausible variant of the substitution theory is the idea that metaphor is a kind of comparison, a condensed simile. Metaphor is treated as the merely ornamental comparison of similars. The comparison theory, though implying a more active mode of cognition than the simple substitution theory, fails to identify the most interesting sort of metaphors. These involve a use of terms, not merely to compare two antecedently similar entities of whose attributes the author must already be appraised, but enable one to see similarities in what have previously been regarded as dissimilars.

Furthermore, as we shall show, if metaphor is just comparison, then the content of scientific assertions involving metaphor will be confined to material concerning the realms of actual and possible experience, since comparisons are essentially rooted in experience. But most sciences are, for reasons we shall develop, inclined to include assertions about those features of the world that are beyond all possible experience.

All versions of the substitution theory share the conviction that metaphor is a way of saying what could he said literally. It is with this that Gestalt theorists disagree. It is basic to their position that what is expressed by metaphor can be expressed in no other way. The combination effected by the metaphor results in a new and unique agent of meaning. A majority of modern commentators would accept some such position, derived ultimately from I. A. Richards' analysis, in which the modifying term ('vehicle') gives to the primary subject ('tenor') an extended meaning (Richards 1965, p. 90).

Max Black has been the main philosophical exponent of Richards' ideas, proposing in an early article, 'Metaphor', what he has called the 'interactive' view of metaphor (Black 1962). Black's contention is that each metaphor has two distinct subjects, principle and subsidiary, and that the principle subject acquires new meaning through its involvement with the subsidiary one. The subsidiary subject 'organises' one's thought about the principle subject in a new way and this operation makes metaphor irreducible to any one literal formulation. Metaphor is neither a kind of substitution nor a function of simple comparison, which is the province, Black suggests, of simile.

In Black's original interactive theory both primary and secondary subjects bring with them their own 'systems of associated commonplaces', though he offers no account of those systems, in particular of their structure. So the metaphor 'man is a wolf' depends upon certain shared knowledge and assumption about the nature of man and of wolves — wild, ruthless, relentless, and so on. In the metaphor, the *two* systems of implication interact (hence 'interactive metaphor'), and produce a new, informative and irreplaceable vehicle of meaning. Both primary and secondary subjects are illuminated by this interaction. It is the cognitively irreplaceable status of such metaphors which Black wishes to stress, and which makes them radically different from mere comparisons.

Black uses various metaphors to develop his theory of metaphor, one of which is 'filtering'. In organising one's view of man, the wolf metaphor 'filters' one's understanding, suppressing some details and emphasising others. Black compares this with looking at the night sky 'through a piece of heavily smoked glass on which certain lines have been left clear'. However, the implications of filtering are in fact inconsistent with Black's suggestion as to the twofold character of interaction. It is difficult to see how the smoked glass is affected by interaction with the night sky. The notion of filtering has been criticised in other ways: How does this filtering take place? and what controls are exercised on it? Why are some commonplaces accepted, and not others? Indeed, while the early interactive view has met with general acceptance and has come to be regarded in English language philosophy as a basic text, Black's terminology has met with criticism, In particular, 'interaction', 'filtering' and 'screening' are objected to as being neither fully explanatory nor fully explained.

By way of reply Black has written 'More about Metaphor' (Black 1977). The major change is his recognition of a closer bond than he has previously allowed between models and metaphor, and between metaphor and analogy, and this constitutes his reply to the difficulties of filtering. He says that between primary and secondary subjects, or more precisely between their two implication complexes, there exists an isomorphism of structure. 'Hence every metaphor may be said to mediate an analogy or structural correspondence.' This provides an answer to the question of how filtering may be controlled, but sounds surprisingly close to the comparison theory which he earlier rejected. Although Black insists that his position is not to be confused with those in which metaphor is identical to comparison, his talk of isomorphism, analogy and structural correspondence belies his claim.

Black's interaction view is further undercut by his new contention that it is only the secondary subject (for example, the wolf) which is to be regarded as bringing with it an implicative complex. Interaction is now a less appropriate term for what is going on, and Black's efforts to meet criticism have resulted in retraction of most of what made the original interactive theory interesting.

Contradictions are inevitable given Black's continued insistence that the first principle of his interactive view should be that each metaphor has two distinct subjects. This, and his continued reliance on examples such as 'man is a wolf' and 'Nixon is an image surrounded by a vacuum', make his theory applicable primarily to those metaphors which involve two nouns, and inevitably suggests comparison. Furthermore, the whole of Black's interaction theory rests upon the idea that it is the two *subjects* of a metaphor, or their systems of implicature, which interact, a notion which, if discredited, discredits his theory, despite the evident correctness of his idea of the role of the 'system of associated commonplaces'. We shall show, in our own theory, how, by generalising some ideas of Saussure, we can retain this idea without any commitment to a comparison of two subjects.

It has been mentioned that a great deal of the modern discussion of metaphor is indebted to the study of metaphor found in one small book, I. A. Richards' *The Philosophy of Rhetoric,* and it is Richards who may be credited not only with originating the interaction view but with putting it in its most consistent and illuminating form. Richards says: When we use a metaphor we have two thoughts of different things active together and supported by a *single word or phrase,* whose meaning is the resultant of their interaction (our italics). Max Black dismissed this notion of 'two thoughts working together' as psychological language and as an 'inconvenient fiction', perhaps because Richards' phrase suggests the ideational theory of meaning which he put forward with C. K. Ogden *in The Meaning of Meaning* (1923/1972), but if one allows, by a principle of historical charity, that the thoughts that one is dealing with here will be primarily couched in language there is no reason why Richards' suggestion is any less convenient than Black's 'two present subjects'.

Richards wants to emphasise that metaphor is an intercourse of thoughts, as opposed to a mere shifting of words or crude substitution, as suggested by the ornamentalist view of traditional rhetoric. It is not substitution of terms, but, in his phrase, the 'interanimation of words'. The notion that two thoughts interact is at the root of his distinction between 'tenor' and 'vehicle'; the tenor is the underlying subject of the metaphor and the vehicle is the terms which present it, so in this quotation used by Richards:

> A stubborn and unconquerable flame
> Creeps in his veins and drinks the streams of life.

The tenor is the fever from which the man is suffering, and the vehicle is the flame which drinks his life. Here it is important to note that what Black would call the primary subject, 'fever', is not explicitly mentioned in the passage. We talk about fever by using the word 'flame', and its associations determine what we mean by so doing. Since 'flame' occupies the centre of a different semantic field from that occupied by 'fever', the use of the term 'flame' enables us to say things about fever different from those we could say by using the word 'fever'. This supports Richards' contention that it is thoughts (associated commonplaces) and not words which interact. In section 2 we shall show how this can be worked out in more detail by the use of Saussure's conception of *valeur.*

Part of Max Black's dissatisfaction with 'ideas interacting' arises because he has not grasped Richards' distinction between tenor and vehicle. When Black gives account of these terms he uses examples such as 'man is a wolf', in which 'man' would be tenor and 'wolf' vehicle. Not only does this miss Richards' more subtle insight that tenor and vehicle may be co-present in one word or phrase ('That wolf is here again') but it prompts Black to state that in the interaction metaphor 'two distinct subjects are present'. This contention is perhaps related to a residuum of extensional or referential theories of meaning. In the above example, 'flame' is being used to refer to fever, not to flame. So it cannot be the referential meaning of 'flame' that is in point. But it is mistaken, both if one means by present 'there in the utterance', as can be seen from the above description of the fever, and mistaken if one means by 'present', 'there in the mind'. While the primary subject 'fever' may be absent from the text but present in the mind in a metaphor such as the one cited above, it is by no means true that two distinct subjects are present, even to the mind, in a metaphor such as 'the giddy brink' or 'eddying time'. Black's account cannot, in fact, deal with such metaphors. Richards is not so limited for he can say, for example, that the tenor is the 'brink' and the vehicle 'giddy', and the metaphor works through the associations one has with giddiness.

This misunderstanding of Richards' distinction between tenor and vehicle is responsible for many of the inconsistencies of Black's interactive theory. The stipulation that each metaphor have two distinct subjects was responsible for the confusing notion that in the metaphor both subjects were illuminated, a notion which had to be abandoned in the later article. Stripped of this form of interaction, yet hampered still with two subjects, Black's analysis in 'More about Metaphor' drifts inevitably towards the comparison views which he had been at pain to denounce.

It is only by recognising that the tenor and the vehicle may be co-present in the one word or phrase, since tenor corresponds to reference and vehicle to 'sense', and that a metaphor has one true subject and a vehicle which is used to illuminate it, that a full interaction theory is possible. The insight of an interaction is not that two subjects and their commonplaces interact but that, in Richards' words,

> The vehicle is not normally a mere embellishment of a tenor which is otherwise unchanged by it but that vehicle and that tenor in co-operation give a meaning of more varied powers than can he ascribed to either.

This is 'interanimation of words' and not comparison of two subjects, because the referent of the old word 'fever' has been enriched by being described with a word having the sense of 'flame'. Hence we have a richer conception of fever and soon, perhaps, a richer semantic field for 'fever'.

The illumination of one subject through the interaction of tenor and vehicle can be seen in this metaphorical description taken from Virginia Woolf's, *To the Lighthouse:*

> Never did anybody look so sad. Bitter and black, halt-way down, in the darkness, in the shaft which ran from the sunlight to the depths, perhaps a tear formed; a tear fell; the waters swayed this way and that, received it, and were at rest. Never did anybody look so sad (Woolf 1977, p. 31).

What is being spoken of here is not both a grief and a shaft of some kind, but simply some private, sickening grief which is uniquely illuminated by being spoken of in terms appropriate to a shaft. The excellence of the metaphor is not that this is a new description of a previously describable human condition, but that this subject, this particular mental state, and these particular connotations are revealed as such only through this metaphor. This 'interanimation' of terms has uniquely identified the state so that the metaphor is not an adornment to what one already knows, but a vehicle for a new insight made available by this interaction of terms, leading to an increment to the psychological description. And here, through discussing how metaphor works, we arrive at an answer as to why metaphor is needed—we need metaphor because in some cases it is the only way to say what we mean since the existing semantic fields of the current terminology referentially related to the subject *in* question are inadequate to our own thought.

2. SAUSSURE'S *VALEUR* THEORY

The theories we have discussed in section 1 are all adequate to some instances of metaphor (there are, for example, sonic metaphors which do function as little more than ornamental substitutes for what could otherwise be literally stated): however, all involve a premature fusion of what metaphor is with how it is that some metaphors work. The most satisfactory theory, we have suggested, is an interactive one developed along the lines suggested by Richards, and for a working definition of metaphor it is best to choose a formulation which retains the insights and possibilities of Richards' 'interaction of thoughts', yet avoids suggestion of an ideational theory of meaning, or indeed suggestion of any mechanism or process at work in metaphor. The following is suggested: metaphor is a figure of speech in which one entity or state of affairs is spoken of in terms which are seen as being appropriate to another.

The motivation for wishing a well-designed theory of metaphor to accompany realist philosophies of science should now be more clear. The theoretical sciences experience crises of vocabulary. If the progress of science is seen to require the attempt to describe real things, some of which are beyond all possible observation, then one must concede the need to give an account of the terms which are used to describe these beings, their properties and relations not available to experience. Boyd has made the important point that these terms are essentially incomplete and improveable a *posteriori*[1]. One must ask under what conditions such terms can be introduced into a language *so that they may be intelligible.* We have some notion of the point of reference of our terms but since these entities are beyond experience the experiential terminology is semantically unsuitable, taken in literal application, that is, relative to its original source of meaning.

Clearly, a term that can be used in accordance with these demands must be:

(i) Meaningful to a user of the language *without recourse to further experience.*
(ii) And yet, somehow imbued with novel meaning.

---

[1] From Boyd, R. 1979, pp. 371-372:

These programmatic features of theory-constitutive metaphors—the tact that they introduce the terminology for future theory construction, refer to as yet only partially understood natural phenomena, and are capable of further refinement and disambiguation as a consequence of new discoveries—explain the fact that repeated employment and articulation of these metaphors may result in an increase in their cognitive utility rather than in a decline to the level of cliché,

What is significant is that these programmatic features of theory-constitutive metaphorical expressions are, in which, typical of theoretical terms in science . . . normally, we introduce terminology to refer to presumed kinds of natural phenomena long before our study of them has progressed to the point where we can specify for them the sort of defining conditions that the positivist's account of language would require. the introduction of theoretical terms does require, however, some tentative or preliminary indication of the properties of the presumed kinds in question.

[Metaphors] provide an especially apt illustration of ubiquitous but important features of scientific language generally ... there exist theory-constitutive metaphors in abundance, and. . .a non-definitional account of reference of the sort advanced by Kripke and Putnam can be employed to defend the view that the metaphorical terms occurring in theory-constitutive metaphors actually refer to natural kinds of properties, magnitudes and so on . . . which constitute the non-literal scientific subject matter of such metaphors . . The use of theory-constitutive metaphors represents a nondefinitional reference-fixing strategy especially apt for avoiding certain sorts of ambiguity.

The former condition implies that somehow the new term is drawn from the common stock, and that whatever is novel about it is created by a process internal to the language. The latter condition requires that the term be controlled by novel rules of use, and involve intentional content which is distinct from any other term, or that term in any other usage.

These conditions can be met by only one kind of linguistic phenomenon. While neologicism could meet condition (ii), they could not, in general, meet condition (i). But metaphor (and perhaps some other figures of speech) could meet both conditions. A neologism fails since mere location of a lexical sign in a new network of laws of nature, for example, implying a unique set of rules of use, is inadequate to fix new meaning, since such a condition *can* guarantee no new intentional content—there could be tenor (that is, a novel referent), but no vehicle (that is, no sense).

It seems clear that any account of meaning which correlates meaning with sensory experience in acts of observation by ostensive reference is bound to be incapable of explaining the meaning-creating power of a process internal to a language, such as metaphor. Worse, taken seriously, it would lead to the reduction of all sense-extending figures of speech such as metaphor to comparison. In a comparison, the term through which the similitude is drawn is unaffected in meaning by standing in that relation, since it is precisely its literal meaning which is partitioned into likenesses and differences. Indeed, comparisons could work only if the predicated term remains unaffected in meaning. And these meanings are, in general, given in terms of existing dimensions of experience of the actual world.

Thus, meaning theories which are essentially ostensive *in* character are excluded from a role in giving an account of that kind of metaphor in which new meaning is created. At most they could explain how old meaning is reshuflled. Since ostension theories entail that there are no real metaphors, only comparisons, the demonstration of the existence of irreducible metaphor would be a *reductio ad absurdum* of such theories. In order to understand metaphor we shall have to turn to a different way of conceiving meaning from 'whatever' a term refers to. The most powerful theory that provides a thoroughgoing alternative is that of Saussure. With our development of Saussure's *valeur* theory we shall show how Richards' theory of 'interaction of thoughts' can be given a quite concrete interpretation.

If metaphorical description is a process internal to language, then we ought to find an account of it in terms of Saussure's theory of *valeur,* that aspect of meaning by virtue of relations that a term bears to all other terms of a language, that is, the internal relations of the language as a system. I remind the reader, *valeur* can be represented graphically as follows: a horizontal axis represents all the structured forms into which a term may enter. It could be a list of the well-formed sentences of a language in which a term 't' appears. This is the syntagmatic axis, and it could be thought of as generated by a set of rules, the grammatical rules of the language. At every point at which the term 't' occurs one could imagine a set of vertical axes, each representing a category of possible meaning-preserving substituents for the term 't' at that point. Distance from the horizontal axis would represent the likelihood of the substitution.

The permissible set of alternatives at any occurrence could be thought of as generated by sub-categorical rules, representing the metaphysical and even the empirical status of the term 't' relative to the other terms in the sentence. For instance, the sentence 'My cat likes to lap cream' contains an instance of the term 'cream'. Paradigmatic axes at that occurrence might include close to the horizontal axis such terms as 'milk', 'water', 'mouse' and 'blood', but not such terms as 'the circuit at Brands Hatch'. In short, 'lap', in that sense, specifies categories and classes of terms which are admissible substitutes at that occurrence of 'cream'. Let us call the rules that specify the *set* of possible objects of the verb 'to lap' *in that sense* the sub-categorical rules, relative to the sub-categorical rules for 'cream', that is, that it is a liquid, comestible dairy product.

We define a metaphorical use of a term as a use which violates the sub-categorical rules of the lexical items in a sentence. To insert 'the circuit at Brands Hatch' in the sentence above would be a violation of the sub-categorical rules associated with 'cat' and 'lap'. Contemplation of 'lap' would relate our sense of what 'lap' means in this context to the sub-categorical rules governing the set of verbs of which 'cat' could be the subject, in that sense of 'cat'. Graphically these rules could be represented by the sets of items on paradigmatic axes erected at the point in the syntagmatic axis where the term 'lap' occurs.

But the term which is used metaphorically, that is, functioning as the vehicle of the metaphor, has with it its own Saussurean grid, with its own syntagmatic and paradigmatic axes. If the metaphorical use is accepted as intelligible, it must involve a reshuffling of the items on the paradigmatic axis of the term so used, relative to its *valeur*. In short, a metaphorical use of a term involves an interaction between the set of Saussurean grids representing the *valeur* of all the terms involved in the sentence in which the metaphorical use occurs. The mathematical representation of such an interaction would require matrix algebra. Fortunately, since the axes of the Saussurean grids can be generated by the repeated application of sets of rules, grammatical for the syntagmatic axis amid categorical amid sub-categorical for the paradigmatic axes, we can express the matter succinctly as the principle that whenever a genuine metaphorical use has occurred there has been a violation of a sub-categorical rule of the set of rules for substitution in the relevant grammatical categories relative to the other terms. Metaphor is *interactive* because in *valeur* terms, the effect of a metaphorical use will be to alter the order of the set of items that lie on the paradigmatic axes at that point. But to reshuffle those items will necessarily be to alter the sub-categorical rules associated with them, in their appearances in syntagmata. To construe a usage as metaphorical is then for a native speaker to employ tentatively modified sub—categorical rules, so that the usage is not rejected as a category mistake. In so doing, the native speaker allows himself, as it were, a richer intension than the referent of the term in the metaphorical employment currently sustains.

Thus if the term 'wave' is used metaphorically for the causes of luminiferous phenomena, in its new use it must he associated with some differences in the items on the paradigmatic axes representing the literal use. So 'creamy white horses' is not a proper meaning-preserving substituent for 'waves' in the sentence, 'Light is propagated by transverse waves'. But if 'waves' is being used in its metaphorical sense, there must be **a** reshuffling of the items paradigmatic to 'light' so that, for example, 'flame' is no longer an admissible substiuent for it.

Why do we call the use of 'wave' metaphorical and not the use of 'light'? To speak of 'light' in this sense must also involve changes in the previous sets of sub-categorical rules governing this term, though these reorderings may not always engender metaphor. Sometimes the effect correlative to a metaphorical predication is an alteration in the extensional scope of the subject term, that is, the domain of its application becomes wider. Compare 'I see the cat' with 'I see what you mean'. Neither use of 'see' could properly be called metaphorical, nor do these uses depend upon a covert model.

3. THE DEMAND FOR METAPHOR IN THE SCIENCES

For the purposes of discussing metaphor in scientific theory we have noticed that it is necessary to distinguish metaphors from models. A metaphor, we have said, is a figure of speech; a model is a non-linguistic analogue. An object or state of affairs is said to be a model when it is viewed in terms of its relationship to some other object or state of affairs. The relationship of model and metaphor is this: if we use the image of a fluid to explicate the supposed action of the electrical energy, we say that the fluid is functioning as a model for our conception of the nature of electricity. If, however, we then go on to speak of the 'rate of flow' of an 'electrical current', we are using metaphorical language based on the fluid model. Strictly speaking 'rate of flow' has not the same sense when used in the context of electrical phenomena as it does when speaking of liquids, say. The model 'spins off', as it were, a number of metaphorical terms (flow, quantity of electricity, condenser, resistance, and so on) which we apply in formulating electrical theory, but clearly without the intention of a point-by-point comparison between liquids and their behaviour, and electrical energy. So models and metaphors may be closely linked; we can have the latter when we speak on the basis of the former.

Simile, on the other hand, resembles metaphor in being a figure of speech. It is customary to identify simile syntactically by the presence of 'like' or 'as'. The analysis of simile identified in this way has led a number of theorists, including Black, to reject the idea that metaphor is simply simile without the 'like'. Objection to this equation is usually made on two grounds: first, that simile lacks the rhetorical impact of metaphor; and secondly, and more importantly, that simile as simple 'same-saying' cannot rival the richer and more complex interactive meaning of metaphor. Both these arguments stand only if one takes as examples uninspiring similes such as 'these biscuits are like cement', or 'the sun is like a golden ball' where the range for comparison is narrow. But if one takes a striking simile, such as this one from Flaubert, 'Human language is like a cracked kettledrum on which we beat out tunes for bears to dance to, when all the time we are longing to move the stars to pity' (Flaubert, *Madame Bovary*, cited in Platts 1979) the comparison is by no means obvious or flat, nor would it be improved by deleting the 'like' to make it a metaphor. Not all simile is same-saying of a trivial sort. Simile, like metaphor, may be the *modus vivendi*, or comparisons of two kinds, the comparison of seeming dissimilars. For this reason we are justified in saying that metaphor and simile are overlapping categories, but differ in grammatical form. This does not denigrate the novelty of metaphor, but recognises the full capacities of simile, nor does it deny the stigma that some similes are mere comparisons.

There is, however, one role which metaphor performs and which simile, precisely by virtue of its grammatical form, cannot. This is to supply a term where one is lacking in our vocabulary, the process of catachresis. Catachresis, so defined, took place when the lower 'slopes of the mountain were called its foot, or when the support of a wine glass was called its stem, because no satisfactory straightforward term was available *in* the lexicon for this purpose. In the language of the linguists, catachresis is the activity of filling lexical gaps. Simile cannot, for reasons of syntactic form, be easily used in catachresis. One may say of the voluble guide 'He's just like Cicero used to be', but the catachretical form will be 'He's a cicerone'.

It is the role of catachresis which is, in an indirect way, the reason why metaphor is so very useful in scientific theory-making, for, as suggested earlier, it is not the model in itself as heuristic device that makes models indispensable in creative theory-making, but the fact that the model gives rise to, 'spins off' a matrix of terminology which can then be used by the theorist as a probative tool. Speaking metaphorically on the basis of a model, a scientist is enabled not only to posit but to refer to theoretical entities by the use of terms which transcend experience in that their semantic context is not fully determined *a priori* by the empirical conditions for their application. Meaning is not exhausted by the conditions of assertability.

The demand for a defence of metaphor in science is not unconnected with the view one might have as to the representational quality of scientific theories, as to whether they could be taken to be possible descriptions of the states and processes in the real world, the world that exists independently of men. The strong realist position, transcendental realism, requires that there be referential terms and descriptive predicates that refer to and serve to describe states of the world which could be forever beyond the bounds of possible experience. Furthermore, defenders of that position would also claim that it would be quite naïve to expect the denizens of such realms to be like those with which common experience makes its acquainted, even as these are themselves much influenced in their manner of manifestation by the conceptual system with which we perceive and understand them.

The argument for transcendental realism runs as follows:

I. The stratification condition: the natural sciences explain regularities of behaviour and coexistence of properties at any one level of the natural order by reference to beings, their properties and relations at some other level of order. Levels can be defined epistemically by reference to our ways of knowing them, and correspondingly, metaphysically, through the use of various hierarchy creating concepts such as whole-part and collective-individual. People are the individuals in social collectives, and electrons the parts of atoms. The simplest cases are where explanations of regularities manifested in the behaviour of materials in common experience are achieved by reference to microstructures, as in chemistry. Sometimes explanation is achieved by reference to the macrostructures in which the entities in whose properties we are interested are embedded. The regularity of day and night, the seasons and so on, as observed on the earth, are related to structural and dynamic properties of the system of planets within which the earth is a component.

2. There is no *a priori* reason why the entities, properties, processes and relations of levels of the natural order remote from that revealed, however concept relatively, in experience, should be capable of description in the same vocabulary, having the same sense as that in use for common experience.

Taken together, these two conditions imply that a naturalistic exegesis of the import of scientific practice requires a theory of meaning which would allow the creation of new meaning independent of actual experience. Or so it would seem in order to argue that point successfully, it will be necessary to deal with another way of describing the unobservable.

3. A third feature of the natural and socio-psychological sciences alike is routinely to substitute dispositional attributions for occurrent. Thus the predicate 'hard' is treated not as the term for an occurrent property, 'hardness', but is to he read dispositionally as meaning 'capable of resisting penetration, etc.'. So the phenomenological attribute 'hardness' is replaced by the dispositional attribute 'capable of resistance', with the important consequence that in common with any dispositional attribute it may be said to be a property of a material being when it is not actually being displayed.

Combining this feature with the stratification aspect characteristic of theorising in natural and social science, we find that occurrent properties at one level are replaced by dispositional properties, at that level, which are themselves replaced by structured complexes of more fundamental dispositions at the level immediately above or below the level of actual human experience. Thus for example, actual lay-offs, bankruptcies and unemployment are explained by reference to the dispositions of an ordered and structured social system of firms and other institutions. Or the actual combinations of substances in chemical interaction are explained by the dispositions of the component atoms, clusters of which are molecules, to attract or repel each other, and so on.

Properties can be attributed to the unobservable on this model. Thus we may wish to attribute a disposition to some unobservable entity which is such that its antecedent and consequent are both descriptions of observable states of the world, but the disposition is ascribed to an unobservable. So one may wish to ascribe the curvature of a flight path observed in a trail of condensation produced from an observable hot filament and an observable cloud chamber by reference to a disposition, or structured field of dispositions, of an unobservable electric field, and the dispositional properties of an unobservable charged particle.

It is to meeting the objections to this treatment of scientific discourse that the theory of metaphor can contribute. Consider a repetition of the above explanatory move. The dispositions of substances are explained in terms of the dispositions of constituent atoms and the structures they form. Atomic dispositions are explained in terms of dispositions of their constituent protons, neutrons and electrons and the structures they take tip. The dispositions of sub-atomic particles are to be explained in terms of the dispositions of their constituent quarks, and the structure of their interaction and so on. Either this regress is infinite or it terminates. There is no reason to suppose that the universe is infinitely complex. So the former alternative is radically indeterminate. To focus any argument we must turn to the second alternative. Consider the possibility of the termination of a regress such as that above. It must terminate either in a simple disposition or in a structure of simple dispositions. But the principle governing the use of dispositional terms was that every disposition must be grounded in a deeper level. It was that principle that led us to formulate a theoretical explanation for the observable dispositions in the first place, and so to initiate the regress. Hence the regress cannot terminate without abandoning the very principle of its construction.

The one possible solution, if one does not wish to contemplate an infinitely complex world, and does wish to treat science realistically, is to introduce some other form of predicate. But it must be of such a kind that it is intelligible, and its intensional content is richer than any predicate whose content is exhausted experientially. The argument of the preceding sections suggests that of all catachretical possibilities, metaphor meets these requirements particularly well. The metaphorical employment of a term brings about a reordering of its semantic field, as well as those of the term with which it is used, so generating new intensional contents, most of which are yet to be explored. So, returning to the main line of the argument, it seems that a realist construal of science requires predicates, of which those created by metaphorical usages of existing empirical predicates are the very exemplar.

CHAPTER NINE

MODELS AS RHETORICAL DEVICES

Among the most potent rhetorical resources of contemporary discourse are the terminology and even the results and theories of the natural sciences. Of course, there are many ways of looking on the use of scientific terminology and theory other than its use as a persuasive rhetoric. But, I believe, it is as much to their role as rhetorics that we owe the use of fragments of the standard vocabularies and theories of science within Greenspeak documents and speeches as it is to their use in reporting matters of fact in professional journals. I shall refer to the use of a scientific vocabulary outside its usual area of application as 'scientism'. The very work 'science' itself can be used scientistically as in 'library science', 'Christian science' and so on.'[1] . . . . [The] identification of the deployment of a fragment of a scientific theory, of a measure of an atmospheric constituent, of the effects of certain industrial processes, as persuasive rhetoric does not imply that the science so deployed is false or suspect. There have been occasions when the anxiety to prove a point has overcome the natural caution of some group using the techniques of science. But, by and large, the uses I will discuss are *bona fide* science. There is one striking way in which rhetorical conclusions go beyond scientific premises, and that is in the expansion and compression of timescales. I shall have much more to say about this temporal feature of science as Greenspeak in this chapter.

In antiquity, rhetoric was studied as part of the training for legal and political debate. Teachers of rhetoric were well aware of certain techniques in the use of language that would help to persuade an audience to favor one account or view of some matter—say, the character of an accused—over another. We can understand in what way a rhetorical device works only if we look at it in light not only of the beliefs that have been successfully promoted but also those it has made unattractive (Billig 1987). The traditional home of rhetoric is in the dialectic of debate in adversarial contexts. The use of science as a rhetorical device presupposes an implicit contrast with the irrationality of other ways of looking at the world. There is a rhetoric of science as well as a rhetorical use of science by others. I shall not be concerned with how scientists persuade each other of the belief-worthiness of their findings (Latour and Woolgar 1979).

We can illustrate the persuasive or rhetorical use of scientific terminology with a news item from the July 1, 1995 issue of the London *Times.* The headline says that the temperature at Wimbledon reached 110^0 F. The usual way of presenting air temperature in the United Kingdom, for example used invariably in the London *Times,* is in degrees Celsius, not Fahrenheit—in this case, 38° Celsius. Both '110° F' and '38° C' refer to the same degree of agitation among the Wimbledon molecules. Why report the temperature in degrees Fahrenheit in a headline? It seems to us that the larger numerical expressions of the same physical property, 'heat', is a good deal more dramatic. This suggestion is supported by the fact that when reporting very cold weather the Celsius scale is always used, having the rhetorical advantage. One never finds 16° F used instead of -10° C! . . Both temperature readings are accredited as science, but they differ in their powers of expression of subjective experience.

---

[1] In the 17th century, the language of Euclidean geometry became the popular rhetoric for the physical sciences. Gilbert's *De Magnete* of 1600 uses no Euclidean terminology, whereas Newton's *Optics* of 1726 is set out in terms of theorems, corollaries, axioms and soon.

1. SCIENCE IN GREENSPEAK

I shall use excerpts from a number of documents to demonstrate the way in which the voice of science is employed as the voice of authority in Greenspeak, [to] show how the use of certain rhetorical characteristics of scientific discourse in general reappear as familiar devices for canvassing rhetorical support in environmentalist claims and debates. Of course, I do not wish to deny the importance of environmental science in diagnosing problems and suggesting solutions. It is by just such research that it was found that algae develop more rapidly in iron-rich water and so fix more carbon dioxide. Perhaps increasing the iron content of the earth's water would balance some of the emissions of 'greenhouse gases'. In their campaign against the oceanic disposal of the Brent Spar platform, Greenpeace claimed to have a *scientific* case for their campaign against the dumping of the rig. This case was later shown to be flawed. The viability of the scientific case for and against some practical program is less important than the rhetorical power of a discourse shaped by the discursive conventions of the natural sciences. I am concerned not with the *scientific* question of whether Greenpeace were right or wrong in their factual claim but, rather, with their use of scientific discursive conventions in presenting it.

'Science' is both a source of knowledge and a resource in the shaping of public opinion. It is to its use in the latter task that our analysis is directed. One must also bear in mind that papers published in mainstream scientific journals also display this duality, since their authors not only wish to report their findings but to secure the belief of their readers in those findings. Again, in displaying the use of concepts, theories and measures from the natural sciences in Greenspeak, whether by conservatives or reformers, as persuasive, I do not mean that Greenspeakers who, for example, recalibrate temporal parameters are guilty of fraud or dishonesty. Language is a kit of tools used for purposes. We can only assess the use of any tool by examining its relation to the task which it is used to perform. When scientists are reporting how they believe the world is, we understand their use of the dialects of science in one way; when they [or someone else] are using the language of science in support for a program of political action, we must understand their use of those discursive devices in other terms—namely, as to how far they have the power to persuade. In this study, our question is: Why does science have this power?

2. SCIENTIFIC RHETORIC AND POLITICAL WORK

Let us begin with an example of Greenspeak in which the rhetorical use of 'science' is quite plain to see, since in some respects it has lost its moorings in science proper. From the writings of Teddy Goldsmith (1992). I shall demonstrate the way in which the voice of scientific authority is employed to close the gap between the scientific evidence that is drawn on by the author and the political response he wishes to encourage in his readers. I shall contrast the political judgments he expresses—that the commitment of some government to sound environmental policies is 'superficial and half hearted'—with the 'six scientific points' presented by this author as, so to say, 'swamping' the way in which the politicians express themselves. From the May 30, 1992 issue of the London *Times* I take an example of disaster rhetoric in the following quotation:

> Global warming, ozone depletion, desertification, large scale pollution and species loss were all threatening to combine with runaway poverty and hunger in the Third World in one crisis which could destroy 'the security, well-being and very survival of the planet'. It was the most frightening analysis possible yet it was not dismissed as exaggeration. (Goldsmith 1992, p. 17)

Why did it have this privileged status? Well, because Goldsmith, author of the article quoted, claims a unanimity and seniority for the authorities he cites, *as speaking with* the voice of science. This unanimity and seniority was of course a spin-off from the presentation of the report in a scientific mode. In this paragraph, a conditional conclusion is offered on the basis of what amounts to a utilitarian argument. If the practices of the industrial nations continue unabated, there is going to be a massive crisis in which the Third World will be impoverished *and the planet* will be destroyed. Strong words indeed.

Interestingly, the argument is set out in terms of the situation for human beings, roughly in terms of what would be conducive to the greatest good of the greatest number of people. There is a more general utilitarian argument associated with the "deep ecology" movement (Devall & Sessions 1987): The morally privileged position of human beings, assumed in much Greenspeak literature, is brought into question. This is a point to which I shall return from time to time: the unargued assumption that the health of Earth in some sense is to be identified with the continual presence on it of human beings and the quality of life defined exclusively in terms of human well-being. We must bear in mind that in the use of science it continually interacts with contestable assumptions about the moral place and role of human beings in the biosphere.

3. THE STRUCTURE OF SCIENTIFIC DISCOURSE

To fully understand the force of science as rhetoric, we must look closely at the originating context, the discourse of ordinary science. Consider the way in which the content of a scientific discourse is organized. Generally speaking, . . . to make an investigation possible at all, there must be an abstraction or idealization of the phenomena of interest. This procedure is controlled by our adherence to a general assumption about the kinds of things, substances and processes that there are in the world. For example, 17th-century physicists, such as Robert Boyle, as a way of clearly describing the behavior of gases classified them as 'elastic stuffs'. Thinking of gases as elastic stuffs suggests a way of studying their properties. We study elasticity of springs by comparing their deformations under different test weights. Boyle decided to follow up this idea by constructing and experimenting with a gas 'spring'. This was the famous Boyle apparatus, with which he and his assistant Hooke discovered the first general law of gas behavior: that the volume of an enclosed sample of a gas—in this case, air—was inversely proportion to the pressure exerted on it. By experimenting in many other ways with different versions of this gas spring, physicists were able to arrive eventually at the most comprehensive of all gas laws:

$$PV = RT,$$

where P is the pressure exerted on the gas, V is the volume occupied, T is the gas temperature and R is a constant.

But *why* is it that a confined sample of gas behaves as if it were a gas spring? Why are gases best thought of for scientific purposes as elastic stuffs?

To build an explanation, a later generation of scientists tried to imagine what the real nature of gases might be. They proposed a mechanism that would simulate the experimentally discovered behavior of gas springs. We can call such an imagined mechanism an explanatory model. An explanatory model is not a free invention. It is constructed by reference to some general assumption about the natural kinds that make up the world, even those aspects of it that we are unable directly to observe. In the case of gas theory, the favored natural kind was 'material particle in motion'. So a gas was imagined to be a swarm of Newtonian particles, moving about in a confined space, colliding with one another and with the walls of the confining vessel. These imagined particles were called 'molecules'. A mathematical study of the way these imaginary particles should behave yielded the formula

$$pv = l/3nmc^2.$$

The two formulae derived, the one from the use of an explanatory model and the one from experiments with an apparatus that represents a 'stripped down' version of the world, are formally similar to one another. The law of the behavior of gas springs is $PV = RT$. The law of the behavior of an enclosed swarm of gas molecules is $pv = l/3nmc^2$. It is relatively easy to establish rules for interpreting the relevance of each in terms of the other. P (pressure) is p (momentum change as particles bounce off the walls of the vessel), V and v are both expressions for volume, and T (temperature) is interpreted as energy (a function of nmc^2). It is the similarity of the laws that allowed Clausius, Maxwell and Boltzmann to offer the behavior of molecules as a tentative explanation of the behavior of gases.

Looking at any discourse that proclaims itself to be scientific, we must be alert to identify the natural kinds that are controlling the construction or conception of the models involved in the experimental and theoretical research programs. Only when we have identified these correctly do we have a clear understanding of the content of the discourse and its standing as science. Part of the persuasive power of science comes from the plausibility of the assumptions that lie behind seemingly objective descriptions and explanations of the phenomena of interest.

4. THE IMPERIALISM OF NATURAL KINDS

In every scientific discourse, assumptions of natural kinds as the sources of explanatory models and controlling the idealization of phenomena that make experiments possible are the ultimate sources of intelligibility and of the meanings and structural characteristics of the discourse itself. But it should be pointed out that sources compete for hegemony in any particular branch of science. For example the concept of 'elastic stuffs', which plays such an important role in the science of gases, has two possible competing forms. In the way Stephen Hales used this model it was sufficient simply to identify elasticity as the key property of gases, whereas in the hands of Lavoisier it was the molecular features that were salient. Compare the way these two great scientists explained the famous bell jar experiment. In this experiment, a burning candle floating on a cork is enclosed in a bell jar with a finite amount of air. The candle eventually goes out after the volume of air in the bell jar has decreased by one fifth. According to Hales, the bell jar experiment shows that combustion causes air to lose a proportion of its elasticity. According to Lavoisier, combustion removes some of the molecules, indeed one fifth of them, from the mixture of active and passive components in the air. It removes what I would now call the oxygen molecules from the original nitrogen/oxygen mixture. If the molecular account wins out historically over the generalized elasticity account, we are strongly inclined to treat the phenomena that we used the concept elasticity to explain as *actually* molecular phenomena. This is a point of the very greatest importance as we shall see when we come to analyze Greenspeak discourses from the point of view of the explanatory models they invoke. Some explanatory models, however, are simply aids to thought and are not taken seriously as depictions of a reality independent of human beings. Deciding between a realist and a heuristic interpretation of explanatory models is neither easy nor secure. However, if a model is to serve as the basis of a program of action it is obviously of great importance to know whether it is being used as an aid to thought or whether it is an adequate representation of how the world really is.

However, our beliefs about the nature of the world generally give priority to some basic natural kind on the basis of which the classification of everything else is set up. So in biology we have 'molecules', 'cells', 'organs', 'organisms', 'ecologies', 'biosphere'. Each natural science has its 'ontological hierarchy of kinds'. Such hierarchies explain the choice of explanatory models in each discipline (Aronson, Harré, and Way 1994). Changes in such hierarchies inevitably are linked to changes in explanatory models. Historically, the influence may go in either direction. Again, it should be obvious that practical programs ride on the back of assumptions about the nature of the world.

[Consider by illustration] the application of this way of analyzing the content of scientific discourses by a brief examination of the way in which Darwin constructed models in his famous analysis of the natural origins of the diversity of plants and animals we know today and that is revealed in the fossil record. We could see this as the building up, step by step, of a pattern of interlocking analogies under a basic choice of type on which all his pictures of evolution were to be devised. Here is an example of 'metaphors doing a persuasive job'.

First of all is the abstraction or idealization of nature in which Darwin pays great attention to genealogies or lines of descent, the 'bloodstock' conception with which all the farmers and horse breeders of his time were imbued. This device leads to an analytical model, 'nature as like a great farm'. The very same generic type of phenomenon also can be thought of as the controlling source of Darwin's explanatory model, the idea of natural selection. On the farm, stock breeders make a systematic selection of breeding animals. Transferred to nature as the source of an explanatory model this forms the basis of Darwin's theory of the origin of species by natural selection.

Darwin's presentation of his theory can be summarized in accordance with the model structure outlined [above]. The source of his controlling model is the behavior of farmers, pigeon fanciers and other stock breeders and the way in which varieties are developed in domesticity. There is domestic variation, generation by generation, of animals and plants in farm and garden. Then there is domestic selection of breeding stock. Darwin points out that this process yields a vast variety of novel forms of plants and animals, breeds as domestic varieties. This entire structure is then transferred to the natural world, to the wild, to nature as self-originating and self-managing.

According to Don Schön's (1963) treatment of scientific metaphors, the explanatory power of a science grows when concepts are displaced from one context to another. They take some of their original meanings with them, but their meanings are modified and transformed in the course of insertion into the new context. Darwin describes natural variation in the wild, such as the diversity of the forms of the beaks of the finches of the various Galapagos Islands. His problem is to account for natural novelty. If the finches had a common ancestor, what process produces the *new* shapes of their beak? His explanatory method is the same as that used in the domestic context: Namely, he employs the concept of selection of a breeding stock. In transferring this concept from the domestic context to the context of nature, of the wild, the concept is subtly transformed.

The term 'selection' in the phrase 'natural selection' is to be understood in a somewhat different way from the same word as it is used in the domestic context. Farmers and plant breeders are overtly active of selection. There is no intentional breeding in nature. 'Selection of a breeding stock' then does not describe a process common to farm and forest but serves to bridge the gap between intentional and nonintentional processes of selection by pointing to a functional similarity. For the rest of the book, Darwin develops and differentiates the concept of natural selection from that of domestic by the deletion of various unwelcome features that are part of the concept of 'selection' as it is used in its original domestic context. He needed to find a pattern of nonagentive, nonteleological, nonhuman causality that would perform a similar selective *function* as that performed by the human agent as selective breeder. In this way we can understand how the content of the theory of natural selection is created and though the use of the 'breeding' model is attractively and persuasively 'packaged'.

5. SOME SCIENTIFIC MODELS IN ENVIRONMENALIST DEBATES

5.1. Thermodynamic Models I: Cycles and Balances

I turn now to an analysis of the models that appear in the first chapter of a well-known collection of environmentalist essays (Southwood 1992). In this chapter we are presented with a general analysis of environmental issues within the framework of a conventional scientific discourse. To understand the force of what is being advocated, we must extract the models being deployed and try to identify their sources. Analysis reveals two striking discursive devices. There are thermodynamic models galore, and these are interwoven with subtle recalibrations of time. Together they lock into place a powerful system of metaphors with apocalyptic implications.

What is the generic source of models of basic ecobiophysical processes that is at work in Southwood's chapter? Well, patently it is thermodynamics. It is used for analyzing, idealizing and extracting conceptually manageable patterns from what we know of human life and its impact on the environment. The leitmotif of many of the models found in Southwood (1992, p. 6) is the transformation of timescales through the shrinking and expanding of which the thermodynamics of a generalized biology are recruited to the rhetoric of Greenspeak. The basic thermodynamic model that serves to simplify and schematize the relations between people and their environment, subtle and complex as they are, appears in the guise of an idealized and abstracting formula:

$$I = (P \times E) + (P \times E \times N),$$

where I is the human impact on the environment, P is the number of people, E is the energy used per capita, and N represents the nonrenewable energy use. This looks for all the world like a law of nature, say,

$$PV = RT.$$

Where does the abstract formula of the thermodynamic model fit onto the world as we know it? Darwin's abstract model of natural selection appears concretely in the patterns of distribution of the beaks of finches. The law of molecules is matched to the law of gas springs by various identity relations. In Greenspeak, the identity is forged between an aspect of the thermodynamic model and something apparently nonthermodynamic, namely food, that comes to the fore as the argument develops (Southwood 1992, p. 12). It is in that moment that the model and its subject are tied together. We are invited to consider the impact of human life on the environment in general thermodynamic terms, but terms that have been temporally adjusted and rendered human.

Southwood begins his account with a striking model of the early state of affairs on planet

Earth. This model takes the form of an imagined scenario of the early history of the planet. The central concept of the model world that Southwood describes is the photobiont, the first "photosynthesing microbe that produced oxygen" (1992, p. 6). Of course, this is a construction, an invention, a picture of the way in which the origins of the organic world as we know it might be understood. But from the point of view of rhetoric, it is also a picture of the way an organism can transform the entire global atmosphere.

Interestingly, the way in which Southwood employs his model involves the deletion of features of the biology of those photobionts we can study in contemporary environments and the substitution of other features that fit his model organisms for their role in his virtual model Earth. The Southwood model works in the following way:

> In due course these changes [to the environment] were to drive the anaerobic organisms that originally populated the earth to take refuge in unusual environments such as sulphur streams or the guts of other animals (p. 6).

These events occurred over eons of time by our standards. Our knowledge of the biology of photobionts is drawn from the observed behavior of such organisms in sulphur streams and other unusual environments over timespans of at most a week or two. In the essay we are analyzing, the biological phenomena of the model world is based on the anaerobic biology of the sulphur streams projected over eons of time. In the model, the concept of photobiont and its effect on its environment have been subtly transformed. Its temporal characteristics have been adjusted to fit its role in the virtual world of the imagined Earth history.

But the analogy upon which the use of photobiont biology in contemporary Greenspeak is set up in another way: between the effect of the activities of 'homo sapiens' on the global environment and that of primitive 'photobionts'. This requires a second recalibration of the temporal concepts of photobiology. In the model world there is imagined to be a very slow rate of change, compared with the later history of bio-evolution. The human time scale is ultrashort compared with that of the photobiontic model. *Homo sapiens* threatens to transform the atmosphere in decades in a way comparable to the effect of photobionts in eons. This suggestion depends on a rhetorical use of a double temporal recalibration of processes described in the terms of the natural sciences.

In highlighting the *rhetorical* force of the compression and decompression of timescales I am not impugning the scientific validity of inductive reasoning from limited domains of evidence. Similar recalibrations can be seen in discussions of population growth in relation to the exhaustion or overexploitation of resources. Again, our focus on the rhetorical force of such recalibrations should not be interpreted as if it were an attack on demography. In many instances, the argument or analysis displays the human population growing exponentially by decades, with corresponding atmospheric changes mapped onto a similar time scale, but by inferences from data from the ice ages that cycled over hundreds of thousands of years. By inserting human activities into the 'evaluation' we have explicit recalibration of the temporal parameters of atmospheric change. In effect we end up with a discourse in which tens of years of human history and tens of thousands of years of geological history are subtly mapped on to a common calibration. There is a perfectly respectable pattern of inductive inferences lying behind the apocalyptic conclusion of such reasoning.

The point, again, is not to impugn the scientific respectability of a distinguished scientist's analyses but to highlight their role in a discourse presented as a contribution to environmentalism. As such, it does not just report but must aim to persuade. It may well be that human beings can do in decades what photobionts achieved in eons. The persuasive power of the parallel, I argue, comes in large part from the rhetorical recalibration of time. Southwood does not take the time to explain the parallel. Indeed, he never states it explicitly. The recalibration of time, as a rhetorical trope, will play an increasing role in our analysis as our studies of Greenspeak develop.

5.2. Thermodynamic Models II: The Greenhouse Story

The most important thermodynamic model, from the point of view of the public perception of environmentalism as science, is based on the analogy of Earth to a greenhouse. This image functions as an abstract and idealizing model in a great many environmental discourses. The rhetorical use of the model is familiar to everyone who reads a daily paper or listens to the radio or watches TV. To gauge the rhetorical role of this model with its apocalyptic concept[2] of 'global warming' I shall analyze a measured presentation of the physics of the atmosphere by Mason (1992, p. 87).

The opening paragraphs of that paper are couched in the familiar mix of science and prophecy. Thus we have the claim that "the concentration [of CO_2] is now 27% higher than that which prevailed during the industrial revolution. . ." coupled to the apocalyptic prophecy that "higher temperatures will be accompanied by . . . a rise in sea level" (p. 60). Mason's model of Earth's physical situation shows the quantity of incoming radiation balanced exactly by the quantity of outgoing radiation. The question for the atmospheric physicist is how *exactly* this balance will be perturbed by the effect of 'greenhouse gases' in the atmosphere.

The pivotal point at which prophecy and science meet occurs in the following passage:

> It is virtually certain that the troposphere is warming very slowly in response to the continually increasing concentrations of CO_2 and the other 'greenhouse' gases but the signal is yet too small to detect above the large natural climate variations, partly because it is being delayed by the thermal inertia of the oceans (Mason 1992, p. 90).

How can a signal that is 'too small to detect' establish that something is 'virtually certain'? The confidence in the claim must derive not from observations but from the model within which the discourse is framed. In the context created by the model, this speculation is endowed with the authority of its discursive environment. This is not to deny, of course, that it might turn out to be supported by finer-grained measurements. Interestingly, a new general hypothesis to explain global warming has recently been proposed. It turns on the effect of bursts of cosmic rays from distant novas on the solar wind. To carry public conviction, this thesis would need to find a model as 'user friendly' as the humble greenhouse.

[Again] rhetorical devices play two different persuasive roles in lay and scientific discourse. In some cases, the rhetoric persuades one of a conclusion for which, in a more generous exposition, a rational argument could be provided. In other cases, and Mason's (1992) presentation seemed to be one of them, model-based rhetoric is used to close a gap in the discourse, for which at the time no bridge course be established. Embedded in a discursive environment that is marked by all the devices of the presentation of scientific reports, the distinction is easily overlooked, perhaps even by the authors of environmental position papers that draw heavily on the results of scientific work.

Our comparison between the opening and closing paragraphs of Mason's argument shows just how the figures, graphs and equations in which the argument is presented conceal a

[2] The widespread coupling of 'global warming' and 'rise of the sea level' in disaster stories, such as the scenarios in which densely populated low-lying areas are flooded, justifies, we think, describing the discourse that makes use of these expressions as 'apocalytic'.

somewhat speculative discussion. But what is in still a rather inadequate working model of the Earth's atmosphere is a powerful image, potent as a rhetorical device. Again the technique of closing gaps in this way is not peculiar to Greenspeak. It can be found in the most hard-core physics and chemistry.

However, it is important to notice that these are intended as realist rather than merely heuristic models; that is, what they picture are systems and processes that could exist. They are virtual worlds, one or more of which might resemble our real world in relevant respects quite closely. A closed car on a hot day will tell one all one needs to know to *appreciate* the metaphor of the greenhouse and its use as an explanatory model. What we lack, as lay folk, is an adequate basis of comparison between that car and Earth, a comparison which would be mediated by such pictures as that conjured up by the use of these models. The 'greenhouse' metaphor makes the *picture* intelligible, but does the picture make the state of Earth intelligible? We shall have to wait on atmospheric science to tell us. To add a touch of irony to the greenhouse story, there is a newly marketed gadget that uses solar energy to cool the interiors of the very same cars that sunlight has warmed up.

5.3. Gaia: The Organismic Model of the Earth and Its Cosmic Environment

A third virtual world offered as a scientific model of Earth is presented in Lovelock's famous 'Gaia' concept (Lovelock 1987). Lovelock offers the Gaia hypothesis in contrast to two other model worlds that stand in polar opposition to one another. In one, life and the physical environment are coupled but loosely, the sort of virtual world we have come across in discussions of the greenhouse story. The other model presents a virtual world in which the role of life processes in regulating the state of the planet is minimal or nonexistent. According to Lovelock, neither alternative is adequate. The physical and the biological systems are one system. The planet Earth, its physical structure, its biosphere and indeed the Sun as well constitute a whole system. This system is self-regulating in such a way that the conditions for life are maintained *at various different equibilibria.*

The important thing from the point of view of this study is not to attempt to adjudicate between these models as science but to look at them in their rhetorical role, their use in argument. The contrast between the Gaia model and the greenhouse model as persuasive images does not lie in a catalogue of discoveries, such as a possible increase in the proportions of $C0_2$, methane and so on in the atmosphere in relation to processes in the biosphere. It is in the root idea on which the Gaia model is based, the idea that the whole system is self-regulating. All self-regulating systems have boundary conditions that, if exceeded, will force the system to maintain a new equilibrium, and finally, if displaced far enough, to 'crash'. If we are living in something like this virtual world, we are very far from an environmental crash that would eliminate human life altogether from the planet. But we may be in the vicinity of a shift to a new equilibrium.

The significance of choice of model is underlined by the alternative role of the Brazilian rain forest in the Gaia model from its role in the 'standard' picture. In the Gaia model, the discovery that there is no net production of oxygen in equatorial forests is irrelevant to its plausibility. But there is a very definite answer to the question of why we should preserve them rather than replacing them with soy or something else of immediate utility. In the Gaia model, the forests do play a fundamental role to be understood by reference to the generic thermodynamic model that lies behind all three alternative pictures. The forests are important for their cooling effect. They are an enormous air conditioner, which reduces the tendency of the temperature of the biosphere to increase cumulatively.

The relation between the virtual world presented in the Gaia model and the moral duty of human beings is quite other than that portrayed in the apocalyptic visions of the near future conjured up in the simple thermodynamically inspired virtual worlds. Short-term human interventions are called for if the real world is best modeled by the 'greenhouse' picture. But according to the Gaia picture, the efforts of human beings are puny. We must withdraw from the hubristic post of stewards of Earth's estate to the more modest position of mere planetary doctors, and 'barefoot' at that, helping the patient's own immune system to resist infection. . . .

6. THE SCIENTISTIC USE OF A SCIENTIFIC VOCABULARY

The discourses of science are not only a source for the analytical and explanatory models used by Greenspeakers but also offer specialized vocabularies and other forms of symbolic presentation such as graphs and diagrams. In this chapter, I look only at the use of terms for numerical measures that carry rhetorical force. I have already introduced the distinction between scientific and scientistic discourse. In the former, the terminology of, say, thermodynamics has a proper place, and its use for persuasive purposes is based on well-grounded scientific research. In the latter, the prestige of the terminology is used without such grounding. We have already encountered a rhetorical use of recalibrations of temporal parameters in our exploration of thermodynamic models. Recall the simple case of temperature recalibration with which the London Times journalist brought home to readers just how hot it had been at Wimbledon. Instead of 39°C the temperature was reported as 100°F. Numbers attached to measures are the bearers of the rhetorical force. This feature of the use of scientific terminology in reaching doomsday conclusions is evident in the example taken from the article by Goldsmith (1992) from which I have already quoted. It is richly adorned with the characteristic rhetorical devices of scientistic discourse, that is, terminology not well grounded in respectable research.

There are some chemical formulae in Goldsmith's text, but above all there are statements of measures, some expressed in percentages and others in degrees:

> Even if emissions stop today, ozone loss would be of the order of 20-30% by the year 2000. A 1% loss is estimated to increase ultraviolet radiation by 2% and the incidence of skin cancer by 5% (p. 17).

As lay readers we are unable to interpret this prediction since we are not told how many Americans have developed skin cancer over the past two decades. At the heart of the argument is a statistical inference from unknown and perhaps unknowable premises presented in formal numerical terms.

Temporal recalibration appears in the following:

> Undoubtedly the most serious environmental problem is global warming. There has already been an 0.5 to 0.7°C increase in global mean temperatures since the start of the industrial age, approximately 1750, and it is predicted by different international and national agencies that emissions of greenhouse gases will lead to a 1.5 to 4° increase over the coming decades (p. 17).

We see here a characteristic example of shifting time bases. It occurs in the juxtaposition of an 0.5° difference in 200 years to a 1.5° difference in 10 years. Nowhere is the logical relation from one number to the other presented. The short way with temporality of this order is continued in the rest of the paragraph:

> The last Ice Age was triggered by a mere 1° drop in temperature. . . . Tens, if not hundreds of millions of refugees from the areas that are no longer habitable will swarm into those that still are (p. 17).

We are being asked to extrapolate now on an even grander scale but in the opposite direction. The temperature increase relative to Ice Ages has to be understood on a timescale of tens of thousands of years. The temperature increase that has, with a charitable reading, actually been detected is 0.2°, the predicted temperature with the collapsing of time into a decade is 1.5°, which is of course greater than that needed to trigger an Ice Age. But a decade is not 20,000 years. It is obvious when one looks closely at this paragraph just how the shifting from one time base to another functions rhetorically when the flagging of the shift is omitted. Once again we encounter the recalibration of time measures that we noticed in the previous section. To demand that *we* do something makes sense only in a time span measured in decades. It took Nature tens of thousands of years to bring about the Ice Ages. Economic conditions and political possibilities constrain any practical program to a fairly short-term implementation. In addition to this is the 'cultural attention span' problem. There is a strong 'fashion' element in attention to green issues that should be a matter for a sociological study of green movements at different historical moments.

Again, the use of scientific terminology in this article is rhetorical. It is not well grounded in the rational foundations of the temporality of geology and climatology. A scientistic description, that is, a description drawing on a scientific vocabulary but not well grounded technically, is offered as part of the rhetorical package. The most striking way in which Goldsmith's article displays its rhetorical character is shown by the fact that when the scientific evidence, such as it is, does not fit the rhetorical needs of his article, he attacks science:

> It is argued in particular by George Bush [senior] and the oil industry that there is no scientific evidence that global warming is occurring, but the concept of scientific evidence when applied to complex biological, social or ecological issues is largely meaningless (p. 17).

This quotation follows directly after the paragraph in which the scientistic claim that a 0.2° increase in mean temperature of Earth since the advent of the industrial age can be extrapolated to a 1.5° - 4° increase over a decade or two against the backdrop of the 20,000 or 30,000 years required for the coming and going of an ice age. There is a startling juxtaposition in the same paragraph of the two main attitudes to science one finds in the spectrum of speakers of diverse dialects of Greenspeak: deference and rejection.

The way in which the time dimensions of processes are adjusted for rhetorical purposes in some instances of Greenspeak reminds one very much of a similar rhetorical device employed in AIDS rhetoric. There have been many different predictions of the rate at which the 'unstoppable spread' of the lethal virus into the heterosexual population will occur. We have had predictions that range from a relatively mild epidemic to the claims of chat show hosts and hostesses that the dead in the United States alone would number some 20 million by the year 2000. Of course, not one of these predictions has come true. The heterosexual epidemic, predicted for the West, has simply not eventuated, and yet the rhetoric of unstoppable, incurable viruses has persisted in the apocalyptic presentation of the likelihood of the spread of the disease. Even the reports of AIDS in Africa are contentious. Our point is not about epidemiology but about the recalibration of time for rhetorical effect.

If Goldsmith's article had been presented as a scientific paper, of course it could not have been taken seriously. However, we must remember, and this example illustrates the point with great clarity, that the use of a scientific vocabulary as a scientistic rhetoric, the numbers, the degrees Celsius, the percentages, the chemical formulae and so on, do not necessarily indicate that the writing is a part of a scientific discourse. [Additionally,] persuasive rhetoric tends to draw from a locally prestigious source, and Greenspeak is no exception. This is how persuasion is done.

7. GRAMMATICAL STYLE IN 'SCIENCE' WRITING

The myth of 'objectivity' is immanent even in the preferred grammar with which scientists and those who imitate them write up their researches. The active and personal engagement of the researcher is written out of the story by the convention that the passive voice should be preferred: "Two drops of saline solution were added . . ." is preferred to "I added two drops of saline solution." It is as if Nature Herself brings forth the truth uncontaminated by the person of the scientist:

> For in the passive construction the actor has disappeared—the doer has disconnected—replaced by the deed itself, sterile and isolated, and apparently accomplished without human input (Kahn 1992, p. 152).

While the inorganic molecules of the virtual world conceived by chemists stand in no moral relations to their manipulators, the same is not true of the real world of biology. Of the living reality of the creatures with which biologists deal we need no virtual simulation. They are there for all to see. Again to quote Kahn (1992),

> It is indeed a passive, soulless [sic] voice which science presents in its literature on animal research, perfectly reflective of a mode of thinking that proceeds from outside the moral realm of active responsibility (p. 153).

One of the reasons why the natural sciences serve as a powerful source of rhetorical devices is that they incorporate within their rhetorics the idea of impersonal authority. In the dialects of Greenspeak highlighted in this section, Greenspeak itself appears as a dialect of the language of Natural Science. This appearance is rarely deceptive. Nothing in our analysis entails that Greenspeakers *qua* scientists never or only rarely produce or faithfully report genuine scientific findings, some sound, some not so sound, just like all other scientists. Sometimes, their zeal for their cause overwhelms their discretion. The scientific claims of Greenpeace concerning the environmental consequences of dumping the Brent Spar rig in the Atlantic seem to have been unsound. The vice of excessive zeal is not unique to Greenspeakers. The claims for 'cold fusion', made in all sincerity, have turned out to be ill-founded and the claimants to have been carried away by an understandable enthusiasm. The point of recruiting the language, structure and grammatical style of scientific writing in many of the dialects of Greenspeak is that not only is it one amongst contemporary discourses that tend to persuade, which is the focus of this discussion, but that much public policy turns on choice amongst competing models of the biosphere. Is it Gaia or is it a greenhouse? Is it a greenhouse or a mite bobbing on the cosmic wind? Which of these pictures one chooses to live by may make a huge difference to what one eventually dies by!

8. SUMMARY

At a first reading it would seem that Greenspeak is a scientific discourse endowed with the authority of the voice of the natural sciences. Natural science is a rich reservoir of terminology and models. In making use of physical science models, Greenspeak is not only a beneficiary of a borrowed scientistic rhetoric but is also a dialect of natural science. Environmental studies are an important branch of biology in relation to geophysics. We can see rhetorical uses of a borrowed terminology in the Aristotelian tradition of rhetoric conceived as general 'art of persuasion'. But throughout there is another use that coincides with the commonsense meaning of 'rhetorical' which is juxtaposed to 'substantive' or 'rational'. The rhetorical use of natural science and the use of the devices of science for persuasive purposes are not always driving in the same direction. If one were inclined to offer environmentalists advice it would be to be very careful indeed in drawing on natural science, whether for its bona fide results or for its prestigious discursive style. The natural sciences have derived their authority from the rigor with which hypotheses and models are tested experimentally. The upshot of this has been the almost continuous revision of what is accepted as well-established fact and plausible theory. It does not do the cause of environmental reform much good to be caught out in exaggerated claims and apocalyptic scenarios that do not eventuate.

SECTION FOUR

A METAPHYSICS FOR PHYSICS

SECTION FOUR

A METAPHYSICS FOR PHYSICS

INTRODUCTION

The discussion to follow is an exercise in analytical philosophy of science. My aim is to examine some scientific practices to reveal their conceptual structure. Such an enterprise partakes both of the history and socio-psychology of scientific work undertaken by a community, and of the traditional interest of logicians in extracting ideal patterns of reasoning. The enterprise is both empirical (What do high energy physicists do?) and normative (Is this pattern of reasoning such that it should serve as a model for other scientific communities pursuing their special interests?).

In this [chapter], I want to place the emphasis on the shift in the Philosophy of Science away from form, at least logical form, to content. Any naturalistic account of Theories and Theory Construction seems to necessitate our understanding of the content and its internal structures. Logical structures are to be found, but they are fragments within a larger whole, the overall organization of which could not be formulated using the apparatus of a traditional formal logic alone. Whether more sophisticated Logics can be developed is an open question. For instance, Aronson (1991) and Way (1991) have a new theory of verisimilitude which introduces a more sophisticated formal understanding of the content of theory-families. For the moment I want to draw a contrast between the naive view of theories, which advocates of the deductive nomological approach, such as Hempel and Popper have deployed, based on a rather simplistic formal logic, and the structured content that has been revealed in naturalistic analyses of theorizing by, e.g., Lakatos (1970). I have called the content-rich structures 'theory-families' (Harré 1986).

We should think of the structure of a theory-family in terms of the dynamic organization of content. To understand the dynamics of the process of development of such a family, say the family of virtual particle field theories, we need to look at the way theorists draw content from the source analogue, the 'common ontology' in Aronson's terminology, and to see how the development of a theory-family is controlled.

Understanding the nature of theory-families and their evolving structure involves the identification of a source model or source analogue, from which the theoretical concepts of the theory proper acquire their hypothetical empirical content and which control its development. In analysing a theory-family and its successive stages, we want to know what the hypothetical empirical but actual theoretical content of concepts such as 'molecule', in theories of gases, or 'social rule' in the theory of social behaviour might be at each stage of development. How do those concepts arise out of source models or analogues? By abstraction respectively from the Newtonian particle concept, and the concept of explicit, written rule. We can study in particular cases, how such concepts, which have been transformed during their displacement from source context to the explanatory context of the theory-family are to be analysed. The structure of the content of a theory-family is a crucial factor in understanding what is or is not an explanation. That content must include both the source analogue and the conceptually dependent explanatory concepts which are used to describe hypothetical generative processes which could have, were they real, produced the patterns of phenomena routinely observed. So we must give some sort of account of the source model or analogue, where it comes from, and what determines the selection of such a very important feature of a theory family in any of the sciences. Recently

Aronson has shown how the relations between the content of source models and theoretical concepts can be formalized through type hierarchies of natural kinds (Aronson 1991).

Aronson uses the idea of common ontology, in much the same way as I have used that of a source model. The construction of a whole host of theories in classical physics depended upon taking the Newtonian particle as instantiating the common material ontology, which was realized in all sorts of concepts in classical physics such as Avogadro's number, root mean square velocity and so on thus serving as the supertype at the apex of the natural kind hierarchies of classical physics. This aspect of the source model is absolutely crucial to understanding the organization and origin of the content of theories in any kind of scientific enterprise. To understand a branch of science properly one must identify the common ontology, and then one will see why the content of the theory is constructed the way it is.

The obvious question then is: why should we choose one common ontology rather than another? For positivists or Duhemian conventionalists, the answer to that question would be pragmatic and in some very deep sense, arbitrary. One can choose any common ontology one likes, provided it satisfies certain pragmatic considerations such as economy of thought. Focusing on these criteria presupposes that one is not interested in the question of the reality of the hypothetical entities, concepts for which were created by extracting content from the source model by displacement of concepts. So positivists and conventionalists are unlikely to find questions concerning the origin of source models particularly interesting. In fact someone like Hempel, given his treatment of models as merely heuristic, would regard it as irrelevant. At best a content engendering model is of psychological interest because it provides the scientists who use it with a way of thinking concretely about the abstract calculus which is, according to that point of view, the core of a theory. But for realists of whatever school the questions that I want to raise, 'From whence come the hard core concepts?' (Lakatos 1970) and 'what are the source models?' should be of central interest. One's attitude to the hypothetical entities one might take seriously as possible referents of the terms of a theory, concepts for which are derived in this fashion, is going to be influenced by what one thinks about the status of source models: are they a common ontology or a pragmatic convenience?

One view about this matter of choosing a common ontology has been that the question is to be treated as a grander form of a hypothetico-deductive procedure. We, the scientists of a community working in some field, think up a metaphysical scheme, an ontology. We try it out. The form of this 'trying out' would be more elaborate than a simple deductive structure, but overall it would look very much like a hypothetico-deductive test. If the schema begins to show us empirically interesting consequences, then by going step by step back through the structure of our 'trying out' of consequences, we would feel some confidence in the utility of the common ontology. I think that is part of the way that people do tend to think of metaphysical foundations, even if they are realists.

I want to use some examples from high energy physics which I think are suggestive of a rather different approach (Brown and Harré 1990), something other than the 'hypothetico-deductive' aspect of the business of choosing those concepts that are going to determine the content of a theory-family.

Quantum field theory is very complex, but for the purposes of this analysis it is possible to simplify it a great deal. The question to ask is, why is it that a certain way of thinking about the process of interaction between particles has become the heart of quantum field theory? Why, when we think of two electrons interacting does a certain way of interpreting the process, expressed in a certain *picture,* introduced by Feynman for thinking about an interaction, the picture of an exchange of particles, seem so appropriate? At the first glance a Feynman diagram seems to be a purely iconic device to help us think about the complex mathematical structure of the amplitude terms which would describe the interaction. One can identify, pick out, a certain algebraic structure—ig/q^2, which reminds one of a photon. We tend then to read the mathematics in the light of the icon, as if there were an imaginary entity, a virtual photon.

There are all kinds of ways, including non-pictorial ways, of reading the mathematics. Why should it be that quantum field theory, and the whole thrust of high energy physics, has gone in this direction? Why should we choose as our common ontology, the idea of a particle or particulate entity? Why should this be the common ontology, the source of our concepts, for understanding this and other more complex interactions?

I propose the following thesis: choice of common ontology is influenced by the techniques available to the experimentalists. Theoreticians look over the shoulder of experimental physicists to see what the latter can do. Experiments are so set up as to give a particular kind of result. Essentially, the result, whether it is achieved by computer simulation or photography or whatever, is some kind of representation of *tracks.* Apparatus is so set up to produce just this kind of result. It is no surprise then, that if that is what experiments yield, and physicists may have to work pretty hard using computer presentations to get the results to come out as tracks, they favour the sort of metaphysics the concept of 'track' suggests. It is, of course, a particulate metaphysics. If experimenters try to construct track-making equipment, then conceptually the community will tend to be on the side of the particulate interpretation of the mathematical structures developed by the theoreticians. Expressing this in terms of the idea of a theory-family we can say that it will include a source model which is highly particulate. When one examines the mathematics and sees that theoreticians pick out one particular bit of the algebra, out of the complex array of amplitude terms, and identify it with a photon, part of the explanation is that the whole of the experimental programme is set up to identify tracks.

This leads to a general point about the setting up of ontologies in the physical and in the social sciences. They are not arbitrary, nor are they assessed hypothetico-deductively. By looking at the style of the empirical work characteristic of the special science in question, we get some sort of idea as to what it is that is driving the conceptual scheme, not just from the inside, deriving from beliefs about the universe, but from the outside as well. This was implicit in the structure of Darwin's great work *The Origin of the Species.* He sets up a common ontology in terms of the idea of selection, first as an agricultural technique, and then moves the concept into the broader context of life in general. But, for Darwin, farming was not only a model for nature. It was a part of nature.

Darwin looked at the natural world as a seething mess of variations in organic forms. It is not that he started with the idea of domestic selection as an analogy, and then arrives at the idea of variation. The two concepts are parts of the same conceptual structure. He is interested in selection of variations, so it strikes him that one particular ontology is to be chosen rather than another. Stock breeding is a kind of experimental technique. In the case of high energy physics, the experimental techniques are all focused on tracks and so we get a particulate conceptual scheme, as a common ontology serving as a source model.

When Feynman drew those diagrams he insisted that people should not take them seriously as pictures of reality. For him they were just helpful devices for thinking about the energy exchanges in the interaction. . . . The last thing Feynman wanted anybody to do was to take the diagrams as expressing an ontology. But that seems to be what has happened. Once a common ontology is in place, something quite powerful and elaborate can come out of it, a pattern of reasoning that is much more complex than the simple cases I discussed in my analysis of the scientific role of iconic models (Harré 1988).

I shall call this pattern of reasoning the 'reversing analogy'. Even people who are not professionally acquainted with physics know that the photon concept is applicable not just in electromagnetic theory but also in the general account of the transmission of light. We can distinguish between free photons, those invoked in the current ontology for light, and virtual photons, the entities which are talked about in the above theory. The concept of 'photon' as the intermediate vector particle, which we are invoking in the theory, is the means by which we pick out one particular kind of term from the amplitude expansion. The intermediate vector particle [IVP] in our icon, was originally simply a device. We know how to write a description of a free photon. We have this concept in place. Then we analogise to pick out a mathematical entity. The analogy goes from the free particle to the virtual. We can talk about these as virtual photons which are not identical with free photons. And so we come to understand them. The upshot is that free photons and virtual photons are species of one genus.

Next consider the case of a 'weak' interaction, electron/neutrino, for example. Again theoreticians present a mathematical analysis, which shrewdly follows the pattern of that successfully deployed for the electromagnetic interaction. We have a diagram, a bit of mathematics, and an interpretation. Using a similar pattern of reasoning, we come up with the idea of another kind of intermediate vector particle, the *W*-particle. At this point it is only the name of a 'virtual entity' whose existence so far is entirely notional. It is just a feature of the drawing. The drawing is a representation of the mathematics. Again, we pick out the three algebraic fragments by extending the photon analogy. The weak interaction is to be analysed analogously to the way the community successfully analysed electromagnetic interaction. Now what? We have a way of reading this bit of mathematics.

But we have also arrived at the concept of a species of a new genus of IVP. *W*-particles within the same Aronsonian supertype—'photonic entity'. Should we not be able to create a situation in which we can observe or at least have some evidence for the existence of the free species? We conclude that we ought to be able to make an independent identification of *W*- and *Z*-particles, just as we can make an independent identification of photons. So we have the experimental programmes to seek the *W*- and the *Z*-particles.

The reverse analogy goes as follows. First the photon concept is used to interpret the Feynman diagram. Then the overall linkage is created by supposing that the weak interaction is analogous to the electromagnetic interaction. It is to be used in an analogous way for interpreting the mathematics. But there is a further step: in the original analogy mode, the free photon was the source concept from which the concept of the IVP was created by displacement. But in a new, analogous case, the intermediate vector particle, the virtual *W*, serves as the source model for the 'free' *W*.

This is a very complex piece of reasoning, but each stage is entirely intelligible as an analogical step within a type hierarchy.

The point that I want to bring out is that it is the sort of experiments that can be done that is driving this whole complex apparatus of reasoning. If the community favours experiments which yield tracks, one works extremely hard to make the results of theorizing come out as beings which generate tracks. There are other things to measure as well, but, I would argue, it is the track concept above all that is driving the whole analogical apparatus.

But there is a further question. What are we going to say about the world on the basis of the kind of experimental/theoretical complex that I have just described? There is a very delicate balanced system of analogies and it has produced some striking experimental results. But the tracks are not the particles. So what sort of concepts are we going to need to be able to say anything sensible about the world on the basis of these amazing discoveries?

To answer this question, another style of concept must be set up. I will try to show that it is just what we need for dealing with the kind of situation in which the experimental apparatus plays a very large role in determining how a theory is to be interpreted, i.e. in determining the content of the theory. The new concept, in this context, is that of an 'affordance'. It comes originally from the psychology of James J. Gibson (1979). But it is modified in making its way into the philosophy of physics. It is one of a family of concepts of which the concepts of 'disposition' and 'power' are typical members. It is characteristic of these concepts that we ascribe them to substances, individual or mass, as properties, but we ascribe them not only when they are actually manifested, but also when they are latent. We ascribe them independently of their manifestation.

Why do we have to talk of manifestation? Because these concepts all involve conditions. Their general form is this: if certain conditions are satisfied, and the substance in questions *has* such and such a disposition, power or tendency, then some effect, specified in the disposition will occur. We ground these dispositions, that is justify their ascription when merely latent, in the nature of the relevant substance.

There is some constitutive property which, if a material substance has that property and certain conditions obtain, the substance will manifest or cause to be manifest some observable phenomenon.

The use of such concepts is very widespread in the physical sciences. In classical physics we think that the apparatus we are employing for investigating a certain class of phenomena, is 'transparent', i.e. can be eliminated from the ascription of the property we are interested in. Temperature is measured with a thermometer. We can think of the temperature of a substance as a disposition of that substance to have a certain effect on the thermometer, and on lots of other things. The thermometer is a neutral vehicle for the manifestation of the disposition. We can simply eliminate it from the story and talk about the calorific value of the substance, measured quite independently of the existence of the thermometer. The thermometer is only an intermediary in the process of discovering how hot the material is. If we think of heat as a disposition, then it is a disposition which is independent of the apparatus employed to measure *it.*

In classical physics, generally speaking, all the dispositional properties are of that sort. In this they are not affordances. The concept of an 'affordance' is needed when we realize that the effect mentioned in the conditional clause of the ascription is manifested to human beings only in the apparatus that they have themselves constructed. It is as if there could be no temperature without thermometers. The conditions for affordances include a human construction. Only when that human construction is in place is the phenomenon manifested, because the phenomenon in question exists only as a state of that very human artefact. This concept is double-sided. Of course an affordance is a property of something in the world, but it is only manifested by the world as an aspect of a human artefact. Bhaskar (1978) contrasts the dispositions and powers that we ascribe to nature on the basis of the interaction between an apparatus of human construction and the natural world, and those dispositions and powers which are being realized, in all kinds of phenomena when the apparatus is not present. According to the argument of this [chapter], pure tendencies in Bhaskarian closed systems should be treated as affordances of the world/apparatus dyad, taken as a whole.

An affordance is a concept that incorporates a human practice. So if we say that soap affords cleanliness, we are indeed including the human practice of washing by rubbing saponified fat onto ourselves and dissolving surface contaminants, by virtue of the chemical properties of the material. The whole of that complex of concepts is accommodated in the meaning of the affordance.

There are two sides, so to say, to an affordance concept in physics. It is ascribable to the world as a material property but it involves essentially the idea of some human artefact or practice. Contrast simple 'power' concepts. Gravitational force is a simple disposition. Should this study be shaken by an earthquake, everything will fall to the ground, independent of it being a human construction. An apple will fall from a tree whether there are budding Newtons there or not. There are two kinds of dispositional concepts to be recognized. Those like 'gravity', and those like certain concepts in high energy physics, which must be treated as affordances. The latter refer to effects which are produced only in the circumstances of the human artefact. It is only the apparatus in conjunction with the world that affords these phenomena in which we are interested, in high energy physics. *W*-particles are affordances, that is, to say, they exist only as products of the world together with the *W*-particle-affording apparatus. We have become familiar with the notion that facts are conceptual constructions, 'laden with theory'. We must also acknowledge that in many interesting cases they are also 'laden' with the human artefacts in which a natural power or tendency is realized.

The apparatus is more deeply involved in the analysis of scientific activity than in the classicist logicist treatments, where it vanishes in favour of neutral 'data'. Ascribing affordances to nature is actually to ascribe them to a logically indissoluble combination of apparatus and world. A *W*-'particle' and (I would generalize this to include all the entities evoked in the discourse of high energy physics, even electrons) are affordances. Some of these affordances will be ascribable in cases in which no human construction is involved. Those are cases in which there are naturally occurring entities that are like human apparatus. For instance the Aurora Borealis is a discharge phenomenon afforded by the electron-wind, the earth's magnetic field and attenuated gases of the upper atmosphere that constitutes a naturally occurring 'set up' of which a Rayleigh discharge tube is another variant.

This analysis enables us to talk about the problems that are raised by the inclusion of a human component in the concepts of contemporary physics, in a way that avoids those queer ideas about consciousness which some philosophers and physicists have fallen into. The human involvement in quantum mechanics is not via consciousness, but via apparatus. It is the apparatus that is the human element. But I will not undertake a detailed treatment of this issue here.

Let me just take this argument one step further, by introducing the philosophy of physics of Niels Bohr (Honner 1987). Bohr calls the material outcome of an experiment a 'phenomenon'. At first glance, the concept of a 'phenomenon' does not seem to carry the sense of dispositionality/conditionality that affordance does, though it does involve the human artefact.

Nevertheless, if one follows Bohr's work carefully, it seems that the notion of phenomenon really is treated as a conditional notion very like the notion of affordance, if not identical with it. Bohr thinks that the quantum mechanical phenomena arise in art indissoluble blend of apparatus and world, and that one cannot draw a line and say that this is where apparatus ends and the world begins. It is the whole set up which produces these phenomena, and of which phenomena are properties. This allows Bohr to introduce one of his leading principles, the only one I shall discuss in this paper, namely his Correspondence Principle. With what concepts are we human beings to describe phenomena? As affordances they are reactions of humanly-built equipment which human beings must be capable of noticing. The Correspondence Principle amounts to the fact that all observational concepts must be classical, simply because they are the concepts appropriate to describe the states of any apparatus which can be simply observed by people.

This fits very well with the affordance notion, because the Correspondence Principle requires that the concepts which are to be employed in describing the apparatus must be affordances. They have to be concepts which include the reactions of humanly-constructed apparatus as the effects that are referred to in the phenomenal description of what happened. The condition clause in an unpacked affordance concept must be a constraint imposed upon a humanly-built apparatus. The effects have to be observable by human beings. We, as physicists, are ascribing all this to the 'nature of things'. But the subject of ascription is a totality, no element of which can be dropped out without destroying the conditions for the common meaning of the concepts employed. According to Bohr, these reactions are phenomena or according to me, affordances.

What are we going to say about the experimental results which they have produced at Cern? Clearly they are Bohrian phenomena. The Correspondence Principle must apply to them, that is, the tracks produced by how-so-ever sophisticated technology must be clear enough for people to see and to measure. They must be described in terms of spatiotemporal concepts with which we are familiar in describing any daily occurrence. We can say, on the basis of those descriptions, that nature plus the apparatus at Cern, affords us *W*-particles. It is quite another matter to say that nature minus the apparatus at Cern would afford us *W*-particles. I do not think we have any ground whatever for saying that. I hope it is clear by now that this refusal does not entail the claim that *W*-particles are artefacts—they could not be more real as affordances *of the world.* They are what the world affords when constrained in just this way.

The idea of the common structure of theory-families can be applied in two epistemologically quite distinct situations. One kind of application is to the case where the hypothetical entities, the concepts for which we have derived from the source model or analogue are either observable or observable in principle. I owe the following example to J. Aronson.

In Binghamton, New York there is a radio programme in which there is an expert motor mechanic, who can be consulted by telephone. One's car is misbehaving and one reports this in detail to the mechanic. The mechanic has a general theory of cars, and, using that theory, he proposes various existential hypotheses about routinely non-observed parts of the mechanism. 'Try cleaning the plugs', so one cleans the plugs but it still won't start. Gradually he works through the *ontological* possibilities. Finally one verifies some such hypothesis, by disclosing a faulty part by dismantling the machinery. It turns out *also* to be empirically adequate because the car now starts after the part is repaired. There is empirical verification of the existence of a hypothetical entity the concept of which was formed by this special kind of conceptual process: exploring the ontological possibilities. The theory of bacteria was crudely formulated in the seventeenth century, that bacterial infection causes disease. We had to wait till the nineteenth century for such technological advances as would enable us to observe these entities.

In these applications we have observables-in-principle, and so there is natural kind conservation, that is, the kinds of beings which would be observed by microscopes have to be the same sorts of beings that are observed by the naked eye. In microbiology we are still observing by seeing or hearing or touching so there is preservation of a generic natural kind, 'organism', in the analogical construction of the concept of 'bacterium'.

The metaphysical scheme that is appropriate to both levels of this ontology is 'substance' and 'attribute'. Medical research into AIDS, for instance, depends on the belief that there is a virus involved. Researchers must find an individual substance at the right place and time and see what attributes it has. There is not just a common ontology, but a common logical grammar for both discourses as well. However, in this [chapter], we have been looking at a second application of the analogical concept generating structure, where we are conceiving descriptions of hypothetical entities like intermediate vector particles. They occur in the kind of context where, under no circumstances that I can envisage, would anyone identify these entities, in themselves, as kinds of objects, if they are objects at all. To suppose they must be is a mistake in metaphysics since the entity track as I have argued, is experiment-driven, through affordances. There is no natural kind of conservation: We do not have to suppose that quarks are of the same generic natural kind as cannon balls, though medical research, at the other level, requires both bacteria *and viruses* to be organisms. The way of escaping the problems which would be raised by trying to apply the substance-attribute scheme to the 'entities' of high energy physics is to look for a different metaphysical scheme altogether, namely, the idea of located affordances instead of trying to say that there are electrons streaming from the sun to the earth, being drawn round the earth's magnetic field and so forth, causing the Aurora Borealis. We should say that there is a state of the universe, in the vicinity of the earth, such that, if we approach it with a certain kind of apparatus, it will afford electrons. The one thing that the beings of the world as it can be observed-in-principle have in common with the world of the unobservable-in-principle, is, I shall suppose, only spatio-temporal location. We seem to need the idea of the world which affords sub-atomic particles as the same spatio-temporal world as the world of macroscopic objects, when we talk about located affordances. The simplest concrete form of this ontology is the ordinary, unsophisticated field, as field-potentials are overtly affordances. We do talk about the location of the various potential levels, in the spatial field, changing in a temporal manifold: The joint spatio-temporal manifold can be used fully to locate the affordances. It would be a great mistake to ask what is the stuff of which the magnetic field is made and what are its properties. The idea of 'field energy' must surely be just a conceptual model. It is just such a mistake to take the *W*-particle as a kind of *thing,* an entity, and ask what are its properties. *W*-particles are afforded by certain kinds of apparatus in interaction with the world, by virtue of some conceptually quite indeterminate state of the physical universe.

If we think this through, in the metaphysics of high energy physics of Bohr's approach to physics, we can see that there is a very sharp break with the traditional metaphysical scheme, which worked very well in bacteriology. If we try to use it for theorizing in high energy physics, in developing the quantum theory, it gets us into all kinds of trouble. I believe that a very attractive resolution of problems like the Einstein-Podolsky-Rosen 'paradox' can be found if we shift away from an ontology involving things moving about and being spotted by detectors. Rather one should think of the whole apparatus as affording certain kinds of results. Then we are no longer faced with the need to admit the idea of bizarre, causal interactions which that sort of experiment seems to require.

Let me just very briefly summarize the analysis. When the content of theories becomes the central focus of philosophical interest, the sources of that content come into focus. One might suppose the content of theories comes from source models by displacement of concepts. But if one analyses some real cases, the role of the empirical techniques also seems to be important. Empirical techniques cannot be detached from the conceptual apparatus theoreticians have created. In classical physics it looks at first sight as if we can do so. But in high energy physics, where I want to follow Niels Bohr, we are not looking at the world through transparent apparatus, but at what is afforded by the apparatus-and-world as an indissoluble unit. In describing the results we are not ascribing properties to substances no matter how it may appear. We are locating affordances. We cannot fasten those affordances on to an apparatus-free universe in which there would be just pure goings-on. About these we would know absolutely nothing.

CHAPTER TEN

MODELING IN QUANTUM FIELD THEORY

1. THE ORIGIN OF THE CONCEPT OF 'INTERMEDIATE VECTOR PARTICLE'

The philosophical interest of quantum field theory, with respect to the interests of referential realism, centres around the interpretation of its major generic concept, the intermediate vector particle [IVP]. Where did the concept come from? The creation of quantum field theory can be thought of as the implementation of the following steps: Classical field theory assigns a number to each space-time location. This number represents the field strength at the point. Quantum field theory assigns a quantum mechanical operator to each space-time location. Thus, the theory engenders a probability distribution of possible states of affairs at each space-time point, that is possible values of the observables corresponding to the quantum mechanical operators. Interactions, say between two electrons, are described by a suitable wave equation. The exact form of the equation is determined by the requirement that it be locally gauge invariant. Amplitudes can be expanded into sets of terms, representing possible states and their probabilities. Each such term can be parsed so as to expose a particle style expression, analogous to the standard expression for the particle aspect of some already well established quantum mechanical entity, in particular the photon.

This act of parsing is nicely illustrated in Aitchison and Hey (1982, p. 29). They consider the lowest order amplitude term for electron-muon scattering. The term is parsed so as to reveal three components, those representing the electron and the muon, and a third 'mystery' component. It takes the form $-ig^{uv}/q^2$. How do we know what to *call* this term? Aitchison and Hey build its meaning by reference to photons.[1] Very significantly, it is called the 'photon propagator'. In the Feynman diagram representing the lowest order transition amplitude it is this term which is visualized as the 'intermediate vector particle', the exchanged photon. Taken literally the term represents not a photon, but another being of the photon *genus*. It is not massless, since the 4-momentum, q, is such that $q^2 \neq 0$. Each such interpretation is represented in a Feynman diagram. Each diagram represents a possible state of the system expressed in particle terms. Does a Feynman diagram have any ontological significance?

Feynman himself warns us against reading his diagrams realistically. Yet when one looks at the way certain experimental research programmes have been developed the corpuscularian model does seem to be functioning ontologically. The corpuscularian reading of quantum field theory provides a genus of beings, of which the 'virtual' particle is a species. The model quantum field theory uses for the intermediate vector particle of the electromagnetic interaction is the ordinary photon, thus ensuring the IVP is of the same genus as the photon, at least in

[1] The problem for a philosopher is at bottom how to understand the motivation for a certain kind of talk: photonic talk. The conditions for the mathematical forms of description typical of the physical sciences do not always fully determine how every aspect of a mathematical formalism is to be taken. A parallel example to the problem of understanding how the photon propagator comes to take that title, with all it implies for the origins of photonic talk in QFT, is Maxwell's 'displacement current' (cf. Whittaker, 1951). There was a feature of his mathematical description which did not have a direct interpretation within the framework of the existing conceptual structure, tied in as that was to the phenomena of electromagnetism. But by applying the ether *models* the displacement current interpretation falls out. Something similar is going on in QFT. I have called this 'parsing the amplitude'. I shall try to show that the real force of.Feynman diagrams lies in their being iconic devices for that act of parsing.

concept. We now have a situation in which there is a virtual and real species under the same genus. Extending this model of quantum field theory reasoning for weak and strong interactions would suggest that for every species of IVP there might (should?) be a real particle of the same genus. This line of argument seems to me to lie behind the design of experimental research programmes directed to 'looking for' such beings as the *W* and *Z* particles. The corpuscularian reading of the diagram enables the formulation of a project: the 'search for the "particle".' The search for the 'particle' consists in trying to find a *track* or *tracks* in a suitable recording medium. In this way the whole of quantum field theory might at first sight appear to lie within the framework of policy realism. The corpuscularian reading of a Feynman diagram does not, by itself, license any existence claims, but it has turned out to provide a reasonable guide for the policy of looking for a kind of particle of the genus adumbrated in the corpuscularian parsing of the amplitudes, and visually presented in the Feynman diagram.

But one can go a step further. It seems that the distinctive properties of the two species, IVP and free particle, under the common genus, can be explained by reference to their distinctive contexts. Perhaps, after all, it is not so absurd to think of there being a real exchange, and to think of a new status of being, and of 'virtual' as a new ontological mode. In the case of the electromagnetic interaction the concept of the IVP was modelled on that of the known free particle. In the case of the weak and strong interactions the concept of the free particle was modelled on that of the IVP. The next step must be look closely at what it is that counts, in experimental physics, as finding the free particle.

Finding it is *always* in a display of effects, not of intrinsic properties. The tie-up between tracks as displays and a particle reading of a Feynman diagram, however tricky the latter may be, is mediated by an important technical consideration. The functions describing quantum field theory interactions are superpositions not mixtures. This means that each possibility can occur, 'one at a time', rather than as in a mixture as simultaneous contributions to a combination of the total possibilities. Without this feature it would make no sense at all to take each term of the amplitude expansion, give it a particle interpretation and go on to an ontological (mis)reading of the diagram representing that possibility. The slide in the metaphysics of quantum field theory towards corpuscularianism also entails a complementary change in the implicit treatment of the probabilistic 'wave' aspect of any quantum mechanical treatment. Within the general framework of quantum mechanics, and quantum field theory is after all just second order quantum mechanics, a wave interpretation of the terms of an expansion of an amplitude is quite legitimate. But the exigencies of the experimental programme and the kinds of material practices that are possible within it have so highlighted the particle aspect of matter that, so far as I can see, the wave aspect has effectively been transmuted into behavioural propensities of particles, represented as the probabilities that this or that level of complexity of virtual particle exchange will occur. It is an implicit corpuscularian metaphysics encouraged by the existence of experimental techniques for recording *tracks* that seems to push towards a particulate ontology.

This is how *Science News* (January 18, 1986) reports the 'fifth force' suggestion:

> Fischbach and his coworkers relate this suggested force to a quality of matter called hypercharge or baryon number. The baryon number is related to the number of neutrons and protons, and therefore to the chemical composition of a material — thus explaining the differences in force for different materials. The researchers propose a formal similarity between this hypercharge force and electromagnetism. Just as electromagnetic forces are carried from object to object by intermediary particles called photons, so this hypercharge force would be carried by 'hyperphotons'. A number of experiments could test for the existence of the hypercharge force, including a direct search for the hyperphotons themselves. (D. E. Thomsen)

Of course this is scientific journalism, not a quote from a textbook. A textbook quote would be expressed in algebra. Hidden within that algebra would be the move by which amplitude terms are parsed in such a way that simulacra of photon forms appear, and from then on function within the framework of a corpuscularian metaphysics, particularly when the move from genus to species is made. It is that move that lies behind the very possibility of a research programme. When the wave aspect of matter does happen to be emphasized apropos an empirical domain, and hence relative to some possible experimental programme it is experimentally interpreted in terms of ensembles. This, idea is already built in to the Feynman images, since each member of the ensemble is represented by a separate diagram, each with its own repertoire of IVPs. So looking for the W particle, the details of which I shall attend to more closely below, depends on taking a member, usually that of lowest order, of the total set of Feynman diagrams, as the marker of the genus to which the other species, the free particle, belongs.

I am not suggesting anything as absurd as that the masters of quantum field theory are so naïve as to be simple-minded corpuscularians. Rather that the totality of cognitive and material practices of which quantum field theory is a part, involves a submerged preference for corpuscularian thought, which is evident both in the terminology and in the experimental programmes. Since there could scarcely be any other way of doing experiments in high energy physics except by recording tracks in various media, the exigencies of experimentation exert a gentle pressure towards that preference. In this I am making a similar point to that made by A. I. Miller to which I have already referred. Backstage in relativity is a notion of visualisability. Neither Miller nor I are making psychological observations. It is rather that within the conceptual structures operative in high-energy physics and particularly in quantum field theory, corpuscularian concepts exert a certain hegemony over the complementary picture.

. . . I would like to complete this brief sketch with a remark about the role affordances might play in explicating what could be meant by a claim to the reality of a virtual particle. Free particles are products, I shall argue, of the effect of apparatus of a certain design on the basic stuff. We are not entitled to say that the stuff consists of particles, only that it can afford particles when shaped up by our equipment. To say that the virtual species of a genus of particle whose free species has been found, that is has engendered tracks or the equivalent, is real is to say something like this: were we able to probe the region in which, say, an electron interacts with another, it would afford particulate phenomena in a piece of equipment, if we could have it, of the same kind as that which forces the basic stuff to display an affordance of that genus in free particle experiments. This is a counterfactual statement. It would be justified by reference to the free particle case. In that case equipment of the right sort has forced the basic stuff to appear as the free species of the genus under discussion.

2. CORPUSCULARIAN METAPHYSICS AND THE STRUCTURE OF QUANTUM FIELD THEORY

The use of covariance to select putative laws, that characterized the research programme of relativity, is matched in quantum field theory by the criterial use of local gauge invariance under well-chosen symmetry groups. This constraint effectively selects those laws which retain the same form no matter how the scales are set for assigning values to their constituent variables. The interactions between particles in the subatomic realm, including decay processes, appear experientially, that is are manifested in Realm 1, as permanent records of ionization tracks in various sensitive media, such as photographic emulsions. Sometimes the absence of a track is a significant event. The analysis of these events and the attempt to explain them within the framework of quantum field theory is the business of high-energy physics. This proves to be a second context in which certain features of the mathematical description of the episodes in question turn out to be ontologically significant. The recent rapid development of quantum field theory provides another useful instance of Realm 3 methodology. There is a constraint on the mathematical form of the expression of laws of nature, that they be gauge invariant and renormalizable. The coupling of this constraint with a strict set of conservation principles (to be called 'book-keeping rules') and the further constraint of the corpuscularian metaphysics creates the physical significance of the theory. Again we find, given empirical adequacy two necessary conditions for a realist reading emerging. Together they are not sufficient. But the constraining of the plurality of theories still further, by the metaphysics of corpuscularianism (which permits the application of the other two principles), does not enhance its putative realist reading though it makes a sufficient condition for selecting theories. The categorical framework of 'corpuscularianism' is somewhat weaker than the substantialist metaphysics that grounds relativity theory. The relationship between gauge invariance and renormalizability of the mathematical forms of the putative laws of nature applicable to this realm will need to be spelled out. I begin with a summary of the salient features of quantum field theory, as it strikes a philosopher.

2.1. The meaning of the gauge invariance constraint

In *Symmetries and Reflections* Wigner (1970) places great weight on the physical irrelevance of the absolute value of potential energy. Thus he introduces the idea of a global symmetry, general gauge invariance. Laws which involve differences in or differentials of potential energy are unchanged in form under the transformation

$$V(x') = V(x) + C$$

where V(x) is the potential energy term, and *C* is a constant, representing a change in bench-mark, common to all observers. But could a local change in bench-mark be tolerated, one which would represent a different change in measuring conventions for each observer? Suppose *C* were to be a function of position C(x). The observer O_i would recalibrate his or her measures of potential energy by adding $C(x_i)$, but the observer O_j, would recalibrate his or her measures by $C(x_j)$. Could there be laws which were invariant in form under the transformation

$$V(x') = V(x) + C(x)?$$

It seems that certain laws can be reformulated to be locally gauge-invariant, including Lagrangeans and wave equations.

For example in the wave equations of quantum mechanics the absolute value of the phase of the wave front is not relevant to the form taken by these equations as laws of nature. It is only phase difference that counts in quantum mechanics. This is a global phase invariance. But the Schrödinger equation can be modified so that it is locally gauge-invariant. A change of phase representation by each observer can be accommodated by making the phase a function of x and t. The overall form of the Schrödinger equation is preserved if, when we change ψ to $e^{ie\psi(x,t)} \hbar\psi$, we change *A* to $A + \nabla x(x, t)$. One could say that the requirement of local phase invariance has provided a reason for introducing the vector potential.

Quantum field theory is that branch of high-energy physics that is concerned with the creation of hypotheses about the exchange processes that stand for forces in the subatomic realm. As I understand it the current research programme in this sector of scientific research is to make the Lagrangeans describing the electromagnetic, the weak and the strong interactions, locally gauge-invariant. This has been achieved for quantum electrodynamics by adding the photon as a 'gauge particle' to compensate for local changes in the electric field. Though it would not be historically accurate to say that there was a coherent programme of extending this treatment analogously to all interactions, this is roughly what seems to have happened. The theory of the weak interaction was achieved by introducing the W^+, the W^- and the Z_0 particles to 'carry' the interaction (the conditions for this step will be tackled below), and to be the gauge particles which make the Lagrangean description of the weak interaction locally gauge-invariant, under some well-chosen symmetry group. The further generalization of the programme to the strong interaction depended on a number of further considerations, which amount to the need for the theory to entail the existence of very weak 'forces' at short distances. I understand that only non-abelian gauge theories behave like this. It will emerge as the argument unfolds that this empirical-cum-mathematical feature chimes in well with the metaphysical constraints that lead to the gauge theory programme in general.

2. 2. The basic physics of interactions

The essential step is the replacement of any explanatory apparatus that refers to forces acting between corpuscles with a scheme in which all interactional phenomena are mediated by the exchange of particles. Collisions between particles and decays of particles, as well as the forces that bind together the components of the atomic nuclei, are comprehended under the same scheme. This class of phenomena are represented in Feynman diagrams as in Figure 1. Such a diagram represents the simplest mode of exchange commensurate with the demands of the conservation laws relevant in each case. The contributions of more complex modes of exchange to the interaction are dealt with by a technical device, 'renormalization.' I shall discuss the relation between renormalization and gauge invariance below. The expressions which refer to intermediate vector particles are theoretical terms, whose place in theory is justified by their role in maintaining the 'book-keeping' conservation principles germane to the kind of interaction being described.

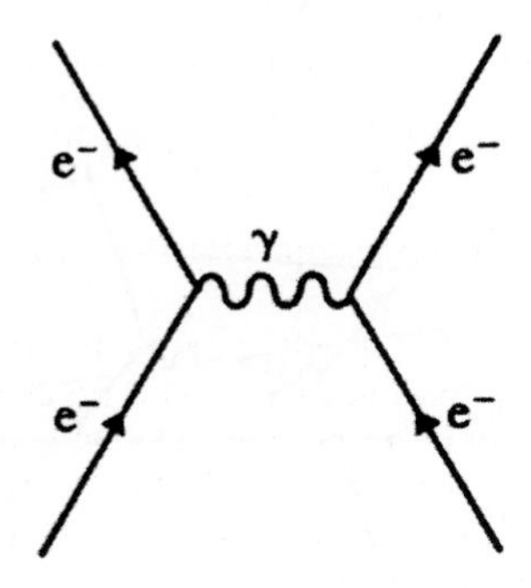

Figure 1
Electromagnetic interaction by photon exchange

2.3. The properties of virtual particles

The properties of the intermediate vector particles can be deduced directly from the conservation principles of the 'good' symmetries which are supposed to hold overall in each class of interactions. But in general it turns out that the rule

$$E^2 - p^2 = m^2$$

which links energy, momentum and mass for real observable particles is not obeyed by virtual particles. For example $E^2 - p^2$ is not zero for the virtual photon of electrodynamics.

So far I have discussed the use of conservation principles as bookkeeping rules in quite general terms. However, the methodologically prescriptive force of the adoption of a general principle of conservation does not come out fully until one looks in some detail at the physics of a particular class of interactions. The electromagnetic field appears in the form of an exchange of virtual particles in any coupling mediated by that field.

The naïve picture treats the coupling as the conjunction of the emission of a photon at one vertex and its absorption at the other, as in Figure 2. Suppose the energy of the initial state is E_A and of the final state E_B. Then

$$E_A = E_B.$$

But if E_N is the energy of the intermediate state of the system it can be shown that neither E_A nor E_B is equal to E_N. Book-keeping rules germane to the deduction of the properties of photons as IVPs of the electromagnetic field break down.

$$E_1 + E_2 - [E_1' + E_2 + \hbar_\omega] \neq 0$$

that is

$$E_A - E_B \neq 0.$$

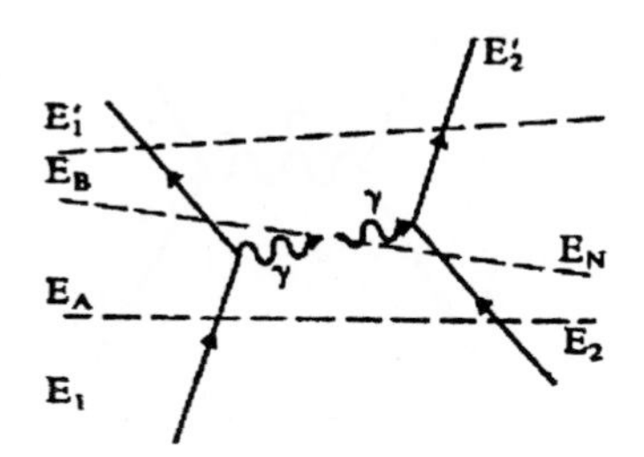

Figure 2
The naive emission/absorption picture

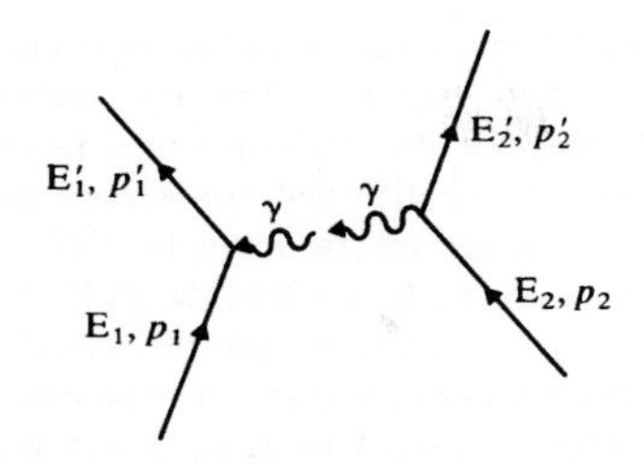

Figure 3
The alternative process

To get a more sophisticated picture we must incorporate the Feynman 'covariance'. The above picture includes the process direction A → B. But from the point of view of equivalent reference frames we must also consider the equally likely picture displayed in Figure 3. The total description will be a superposition of the two equally likely descriptions. From the analysis of the joint picture we get an expression for the amplitude which is Lorentz covariant. And in the combined picture both energy and momentum are conserved. According to the substantialist metaphysics to which we seem committed we are once again at least in touch with reality.

The properties of the IVP can now be calculated by using the restored conservation laws. 'Its' energy is $(E_1 - E_1') = (E_2 + E_2')$, and 'its' momentum is $(p_1 - p_1') = (p_2 - p_2')$. And for the virtual photon as IVP

$$q = (E_1 - E_1')^2 - (p_1 - p_1')^2 = (p_1 - p_1')^2$$

which is an invariant.

For the ordinary photon $p^2 = E^2 - c^2p^2 = 0$, but for the virtual photon $(p_1 - p_1')^2 \neq 0$, that is the virtual photon has mass. From a philosophical point of view the 'anomalous' property of the virtual photon seems to arise directly out of the Feynman restoration of the full bouquet of conservation principles typical of the substantialist metaphysics.

The choice of symmetry group and the achievement of gauge invariance relative to that group and the ascription of properties to IVPs through the use of strict conservation book-keeping rules on the basis of a conservative 'grammar' both seem to be mutually motivated pairs of operations. The empirical basis of this elaborate, internally supported structure is simple, namely the preservation of the observed initial and final states of an interaction or decay. Despite the complexity of the structure of quantum field theory the IVPs are apparently typical hypothetical entities of an advanced corpuscularian physical theory. The diagrams are, then, despite Feynman's disclaimers, treated as quasi-pictorial representations of the mechanisms of processes. We have already seen how the diagrams are *both* corpuscularian metaphors and guides to research designed to find the free species of being whose genus is defined by their 'hidden' constituents, that is those which do not have or cannot be assigned tracks in the real world on the basis of photography, computer reconstructions, and soon, but which can be associated in some way with particles which do leave tracks.

But should we call the electromagnetic IVPs 'photons'? After all, they have one startlingly anomalous property, 'mass'. The justification for the assimilation that the use of the common term implies lies, I think, in the natural history of quantum electrodynamics. Once the decision to separate the Hamiltonian for matter from that for radiation is taken, the theory of exchanges must develop from the conjunction of two 'pictures', one representing the emission of a free photon and the other its absorption. Thus the basic physics invokes standard photons as elementary quanta of excitation of the electromagnetic field considered as an infinite array of elementary oscillators. In this way the genus of the electromagnetic IVPs is fixed. It is only in the further development of the theory that we come to see that the IVPs of quantum electrodynamics are a different species from the photons of tradition. (Some would say that all photons are virtual. Those which are 'free' link distant vertices and are of near-zero mass.) Thus far the existence of virtual photons remains open. The framework of thought is still exactly that of the policy realism of part four. A free 'X' need not have exactly the same properties as a captive 'X' provided they belong to the same kind.

Suppose theory tells us of a necessary condition that must obtain for a certain, quite definite phenomenon to be possible. When the phenomenon turns up, physicists make a confident claim for the existence of the necessary condition, in just the same sense as the phenomenon exists. A cause, after all, must be at least as real as *its* effect. This reasoning is apparent in the following quotation:

> The researchers do not actually 'see' gluons in their apparatus [for instance they do not manage to photograph their trails]. At the highest energy at which [the apparatus] runs they find that a small fraction of electron-positron collisions produce three sprays or 'jets' of particles ... [of] three pronged appearance which all lie in the same plane. . . the physicists believe that the jets originate in a gluon, quark and anti-quark that materialize from the electron-positron annihilation (Quoted by Pickering 1981).

But the imperative to accept such particles is stronger than the force of a necessary physical condition. The 'intermediate vector boson' comes not only from the 'book-keeping' rules of the relevant conservation principles, but also from the conditions for the gauge invariance of the relevant Lagrangean. IVPs are also gauge particles, GPs.

A particular case of the use of this kind of reasoning is the Gargamelle experiment. On the basis of the properties assigned to the Z particle of the weak interaction a phenomenon called the 'neutral-current event' would be expected (see Figure 4). But neutrinos are involved in this episode. Since they are uncharged, they leave no ionization tracks, so a real-world neutral-current event would look something like the track in Figure 5. It is not hard to visualize the photograph of real tracks which would count as a picture of a neutral-current event.

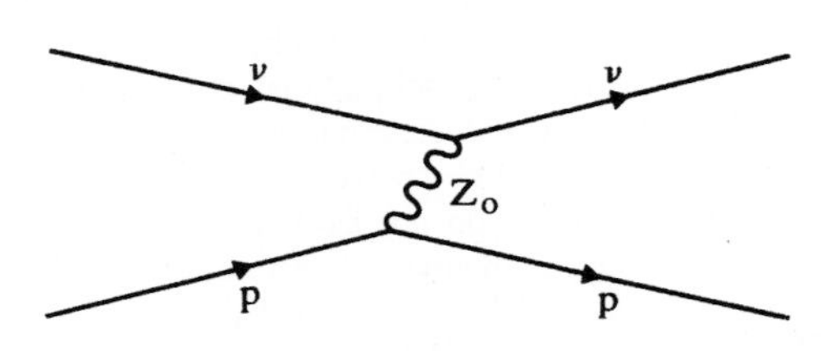

Figure 4
The neutral-current event

Figure 5
Schematic track of a neutral-current event

At first sight this looks like a simple case of hypothetico-deductive reasoning. It seems to make gauge invariance under whatever group of transformations is popular (in this case SU (3)) as ontologically pregnant as was covariance under the Lorentz group. But the matter is not so simple. In the relativity case the choice of the Lorentz group as the relevant 'base-line' is based upon considerations that are independent of the particular form that covariance takes in the harmonized laws. The 'contraction' rule can be deduced directly from the Michelson-Morley experiment. But, so far as I understand the reasoning that leads to the choice of symmetry groups in quantum field theory, it involves choosing a symmetry group just so that the favoured Lagrangean will turn out to be gauge-invariant. This makes the gauge invariance condition more potent than the group of transformations. However, there is another way of reasoning which leads to the community accepting the IVP in question.

The photograph which *we* read as a picture of a neutral-current event appears in the discussion *as interpreted.* I have already pointed out the importance of the balance between plausibility and revisability of theories. When the neutral-current event does turn up, that balance shifts slightly towards the left. But in a recent study of the history of the development of this approach to subatomic physics, Pickering (1984, pp. 188-95; 300-2) points out an interesting feature of the control of the experimentation. The 'observability' of neutral-current events is relative to the measures taken to eliminate from the tables of results another class of events, neutral background events. The latter are interactions between neutrons, which like neutrinos leave no tracks, and the circumambient environment, apparatus, etc. Pickering shows that the conventions of interpreting were so set in the 1960s as to strike out all events initiated by trackless particles as neutral-background events. Only when the convention was reset in the 1970s did neutral-current events, initiated by neutrinos, become observable. He argues, no doubt with justice, that the resetting of the convention had to do with growing confidence in the theory that said that neutral-current events should exist. Of course confidence in the theory will not make neutral current events come into existence. Neutral-current events might not have been detected within the reset convention. Resetting is a necessary but not a sufficient condition for their discovery. This point hardly seems worth emphasizing, but one must remember that there is a reading of the programme to sociologize all explanations of scientific decision-making which seems to suggest that conventions make existents, rather than making possible the setting up of experimental programmes for their detection.

But the history of the study of the weak interaction shows that it is actually much closer to the kind of example that I used to ground the case for policy realism for the beings of Realm 2. Experiments at CERN have produced sprays of particles, the immediate antecedent of which, though not itself leaving a trail, seems to have the physical properties assigned to the W particles by quantum field theory. If by 'free' particle is meant a being which leaves an ionization track that can be photographed or otherwise recorded, and which is the only particle required to explain the length, curvature and density of the track, then the CERN experiments have not quite identified a free W particle. But at least it is the next best thing. Quantum field theory then looks very like a theory-family in the sense outlined in Chapter Six. The difference lies in the mode of manifestation of the hypothesized beings when they are 'free'. Realm 2 beings appear to ordinary observation within the world of perception. Any problem about their existence is reducible to a philosophical problem about the status of Realm 1 beings. But even the most robust 'free' particles in high-energy physics are distanced from the naïve observer by the causal processes that intervene between the moving particle and the ionization track it leaves. These processes are epistemically ineliminable. It follows that when we, the scientific community, are using observational predicates in our description of these beings, they must appear in *dispositional attributions.* These are the characteristic attributions of Realm 3 discourse. [2]

3. A SUMMARY OF THE METHODOLOGY

From the point of view of quantum field *theory* there are two interlocking steps.

1. The mechanism of the fields (forces) is described in terms of exchanges of virtual particles. They are introduced as intermediate vector particles, IVPs, and their properties are determined by the 'bookkeeping' requirements of the conservation principles that control the attribution of quantum numbers for such properties as 'charge', 'spin', 'colour', etc. The properties that are conserved differ between the three interactions, strong, weak and electromagnetic. This aspect of the theorizing goes on within the framework of a corpuscularian metaphysical model.

2. Virtual particles also appear as gauge particles, GPs. Their properties are determined by the requirement that they compensate local field changes so that the Lagrangean for the whole interaction is locally gauge-invariant under some appropriate symmetry group. The choice of symmetry group is not independently motivated as is the choice of the Lorentz group for relativity. The symmetry groups which 'discipline' the invariances are derived from the rules for the conservation of quantum number properties, rules found necessary to keep the books in the descriptions of the results of experiment. Properties and symmetry groups 'grow up together' so to speak.

[2] A defence of a shadowy reality for IVPs as phenomenal "would be's" is attractive-the counterfactual might be explicated through an ontology based on vacuum physics. Dispositions, in general, can be grounded in some state, structure or condition of the substance to which they are ascribed. Photons can be treated as non-fundamental beings when they are taken as elementary quanta of excitation of the oscillator plenum of which the ground state is the vacuum. In the vacuum state only the average energy of the plenum is zero. Has the oscillator plenum any claim to be a representational model of reality? Can both 'real' and 'virtual' photons be interpreted within a common ontology based on the idea of the oscillator plenum? The metaphysics of the oscillator story is tricky, since the oscillators so invoked are mathematical abstractions of 'something else' (cf. Aitchison 1985, pp. 335-6).

Nevertheless the idea deserves exploration. The Lamb shift can be described in photonic talk, in terms of the emission and reabsorption of a virtual photon. It can also be thought of as the effect of an interaction between an electron and a random fluctuation in the vacuum field. Perhaps virtual photon talk is a photonic way of describing flutuations in the vacuum field, while real photon talk describes the behaviour of quanta when the average energy is non-zero. Approached in this way the apparently naive claim 'Virtual particles might be real' is a photonic way of claiming that the dispositions (affordances) of the basic stuff (the glub) can be notionally grounded in the oscillator picture. Since oscillator talk is also metaphorical the oscillator plenum is another icon, a Cartwright model of the glub.

If, by each route, we get concepts such that

$$IVP = GP$$

the community seems to be satisfied with the theory. It is plausible in just the sense of that term I developed for the assessment of theorizing for Realm 2.

The story of the W and Z particles also illustrates this procedure. By the use of the methodology sketched above the weak interaction is pictured as a process mediated by particle exchange, and the properties of that class of corpuscles worked out in the standard way. The Gargamelle experiment shares the weakness of all attempts to prove the plausibility of a theory conceived hypothetico-deductively, by the test of a prediction of phenomena of the same natural kind as the lowest-level 'facts' encompassed by the theory.

But if the parallel with Realm 2 methodology is taken more seriously the policy realist treatment of quantum field theory suggests a different kind of experiment, an existential search for examples of the beings referred to in the theory.

The quasi-pictorial reading of the Feynman diagrams, or if you like the taking of W *particle* talk fairly literally, motivates the 'search for the free W particle'. This amounts to the conceiving of and realizing experimentally a physical situation in which the vertices of the W process are well separated and the 'particle itself' can become manifest.[3] In fact the W particle does not manifest itself, *in propria persona,* but rather the phenomena are such that any alternative explanation of the 'cross-section' of the process is ruled out. Nevertheless the argument hinges on the general point about the difference between the status of 'virtual' and 'real' particles—that it is a matter of the genus encompassing two species, one of which can be tracked. But we are still in Realm 3. The W particle is not manifested by virtue of generating its own unique track. It has to be inferred from the tracks of a shower of particles to which it is antecedent. But that last and crucial condition can be satisfied at all only through the use of the appropriate bookkeeping rules—even when free, Ws are, in some sense, virtual!

The case against subsuming the virtual status as a species of physical reality turns on the unclarity of the particle concept in the context of exchanges. The analysis above, in which WPs seem to have most particulate a representation, is the lowest order of exchange. The 'true picture', so to speak, is a superposition of the higher-order modes of exchange as well-for instance, Figure 6. In the 'true picture' the number of virtual photons is not sharp, since the system fluctuates over an infinity of superposed states. On the reasonable principle that ontologically photons are entity-like and have some of the 'grammar' of individuals, the unsharpness of the number of virtual cousins counts against their reality. But the upset of the metaphysical clarity of the virtual particle picture goes deeper.

The metaphysics and methodology of quantum field theory in terms of particles meshes only with the simplest kinds of exchange. The more complex modes of exchange are conceivable within the discipline of the conservation rules which are represented by higher-order terms in the mathematical description of the interaction or decay. Integrating over these possibilities should

---

[3] In the reasoning that leads to a policy-realist interpretation of the W-particle there are two analogies. The photonic concept is first legitimated in the electromagnetic case through its traditional application to luminiferous phenomena (in Bohr's sense). Its success through explanatory power in 'parsing the amplitude' justifies the concept of the 'virtual' species. The whole scheme is applied by analogy to the weak interaction. But the analogy between real and virtual species is reversed. It is the concept of the virtual species that is legitimated via explanatory power in the tidy accounting of the weak interaction, and the search for the real species is dependent on an analogy in which the virtual species is the source, and the real species the subject.

give a gross description of the actual interaction. But the result of these operations are infinite values for momenta and other relevant physical properties, such as charge. It was shown by Feynman, Tomonaga and others that in general one could absorb all the infinities into a redefinition of the mass and charge of the beings adumbrated in the theory. When these quantities are replaced by the measured values of the same parameters, finite values are imposed. This technique is called 'renormalization'. The same way of cancelling infinities was extended from quantum electrodynamics to the combined theory of electromagnetic and weak interaction by 't Hooft, by adding scalar fields to the Lagrangean descriptions of the interactions.

Essentially what we are viewing here is a series of adjustments to the mathematics to make the ontology of the theory 'right'. But there are also adjustments to the ontology to save the mathematics. This comes out clearly in the history of the Higgs mechanism. In the gauge-invariant version of quantum electrodynamics the intermediate vector particles are not massless photons but the theory is renormalizable. However, this is not the case for the weak interaction. The conventional theory of massive charged vector particles is not renormalizable. Such a theory would be renormalizable *if* a gauge symmetry could be imposed. But the mass of the quanta are such that all gauge symmetries are violated. Suppose there were a 'hidden' gauge symmetry. This is the motivation behind the introduction of the Higgs mechanism.

Just as covariance depends on a commitment to conservation, so do the mathematical adjustments and ontological hypotheses that make up the story of the use of renormalization. But what are the properties whose conservation exerts such a powerful effect on the development of theory? We shall see that in general these properties are created so that there shall be conservation.

The unification programme conforms exactly to Aronson's analysis of theoretical science as directed to the construction of a common ontology for the phenomena in a domain. Suppose quarks are the basis of both electroweak theory and the theory of the strong interaction. They become a common ontology for all three theories. All the corpuscles invoked in the preceding theories are now thought of as complex and composed of quarks. Two independent sets of quantum numbers (new properties) were proposed for quarks. Electro-weak gauge theory corresponded to local invariance of quark flavours, while strong gauge theory corresponded to

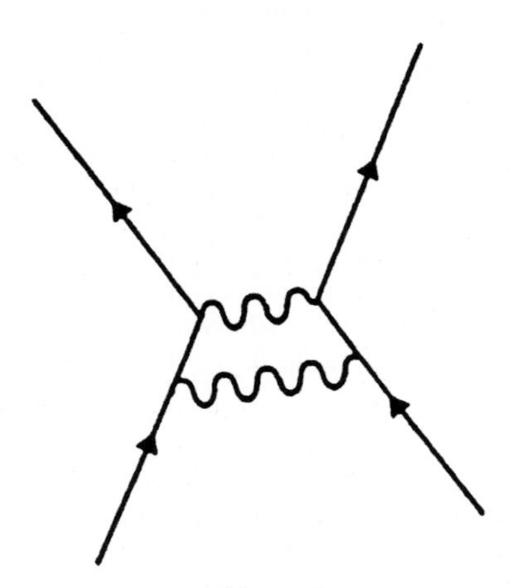

Figure 6
Higher-order exchange

local invariance of quark colours. The IVPs (which in a well-ordered theory are also the gauge particles) are coloured gluons. Looked at in this context colour, is a property that behaves 'grammatically' like charge. Later a further quantum number (property) 'charm' was added. In sum flavour, colour and charm, through conservation principles, maintain the bookkeeping, so to speak, of the initial and final stages of the subatomic event. Their bearers, quarks and gluons, provide the fields, whose adjustment to some suitable symmetry group results in a locally gauge-invariant Lagrangean description of the interaction. As I remarked above the choice of symmetry group does not seem to be animated by independent physical considerations, as was the choice of the Lorentz group in special relativity. Instead it seems to have been arrived at as the most elegant group of co-ordinate transformations under which a Lagrangean with the necessary gluon content is locally gauge-invariant.

Once again it is evident that the most fundamental metaphysical dogma, that which controls the way in which ever new ranges of hypothetical Realm 3 properties are invented, is the preservation of conservation principles. But that is only half the story. These uses of conservation do not lead physicists back to rooted properties. There is no hint of the a priori generalization of the primary qualities of bulk, figure, texture and motion to all existing beings, observable or unobservable, that set Newtonian science firmly in Realm 2. It was only our lack of microscopical eyes that prevented our experiencing those 'arrangements of [Newtonian] particles in the superficies of bodies' that were thought so to effect light as to cause our different colour sensations. Can anything further be said about the new properties? I shall argue in [the Introduction to section Five] that all these properties are affordances, a kind of dispositional property with which we became acquainted in the exposition of the Gibsonian version of the metaphysics of experience.

4. QUANTUM FIELD PHYSICS AS A THEORY-FAMILY WITH A TYPICAL SYSTEM STRUCTURE

. . . The corpuscularian metaphysics [of quantum field theory] seems to be a source analogue in just the way Newtonian metaphysics was a source analogue for classical physics. Disturbances to the internal balance of the theory-family structure lead to restorations of equilibrium by adjustments from within the structure. In this way it behaves like a system. These adjustments have characteristically taken the form of the invention of IVPs and gauge particles *under the discipline of the conservation principles* and the requirement that the Lagrangean descriptions of the interactions be locally gauge-invariant. Furthermore the introduction of new properties is controlled by the traditional menu. There are really only three kinds of properties. There are scalars like charge and mass-energy. There are vectors like momentum. And there are polar vectors like spin, isospin and parity.

But the conservation principles are yet more powerful. It is relative to them that new properties are introduced to permit standard bookkeeping exercises through which the physics of the events in question is developed. A theory-family in Realm 3 science could be described as an ordered pair consisting of a mathematical structure and a metaphysical scheme, which is itself composed of a categorical prescript (such as the corpuscularian scheme) and a set of appropriate conservation rules. The latter controls the readjustments we make in the former. For example, in the face of the non-conservation of parity in the weak interaction, physicists continue inventing properties, with the 'grammatical' constraints of existing physical theory, until they reach a set of quantum numbers which are conserved. And of course they may invent the particles to go along with them. This last step, as became clear, I hope, in the exposition of the way Feynman restored conservation to the lowest-order exchange picture of quantum electrodynamics, is constrained by a key feature of the mathematical structure, namely covariance. But why covariance?

The close link between covariance, symmetry and conservation, even in the context of the obscure affordances of quantum chromodynamics, suggests one obvious and commensensical explanation of why physicists use symmetry considerations. It is summed up in Redhead's introduction of the generic term 'stabilities' to describe the gamut of these principles. The commonsense principle, implicit too in Feynman's notion of the 'robust', runs something like this: Whatever persists unchanged through change is real. Call this the 'robustness principle'. It is an exceedingly powerful principle, since it would permit very direct inferences from Realm 1 phenomena, certain kinds of observable changes, through some bookkeeping calculations, to phenomena of Realm 2 or 3.

The robustness principle should be distinguished from the Kantian doctrine that whenever there is a change there must be a non-change against which it is picked out. Kant's doctrine of conservation licenses his transcendental deduction of the category of substance in B225 (Kant 1781/1978). Physics does not seem to need a principle quite as strong as this, to support modest realist attitudes. The robustness principle has the advantage of linking hypotheses concerning conserved quantities, substances, to empirically testable conditions, namely whether or not the accepted current bookkeeping rules can be applied. There is a measure of circularity in the genesis of these rules, that is the cluster of covariances, symmetries and conservations are mutually adjusted to one another. But that adjustment is not indefinitely flexible, as the history of particle physics amply demonstrates. The robustness principle is both metaphysical, controlling the formation of scientific concepts, and empirical, contingent in any specific application on the actual affordances. However, the robustness principle must be qualified if favourable consideration will be given to Bohr's warning that there can be no unequivocal *specific* application of any metaphysical principle to the ur-stuff of the physical universe. All principles are applied to that which Bohr called 'phenomena', those queer things which are half artefacts and half *objets trouvés,* and which we call experimental set-ups. Only the affordances for these phenomena are ascribable to the basic stuff of the universe. How remarkable it is that conservation principles can be applied to them!

The two stabilities upon which the edifice of Realm 3 science depends are both susceptible of commonsense explications. They both pick out what remains the same through changing points of view, that which is independent of the 'condition' of the observer. Stabilities differ from seventeenth-century primary qualities only in the non-subjective quality of what modern physics takes to be the 'condition of the observer.' In the seventeenth century the primary qualities were distinguished against the background of the subjective/objective distinction. 'Point of view' had as much to do with the qualities of independent consciousnesses as with differing location in space and time. (Galilean invariance and the principles of indifference associated with it were treated at the time as conditions governing the appropriateness of citing a cause in an explanation of change.)

Both of our stabilities have to do with independence of choice of a basis for measurement and physical description. Relativistic covariance is a stability which is defined just so as to be independent of the inertial frame which an observer uses as his or her bench-mark for the measurement of kinematic variables and the investigation of their relationships. Gauge invariance is a stability which is defined just so as to be independent of the particular calibrations which an observer chooses as his or her personal zero or bench-mark for the measurement of dynamic variables and the investigation of their relationships. I think that the reasoning which underlies the place these stabilities currently hold is essentially simple.

Whatever description (or measure) of a physical process whose form (value) depends on the arbitrarily chosen conditions of observation cannot be real. The robustness principle follows immediately as a necessary condition for the referent of a description (a process described by a law of nature; a quality whose value *is* given in a measurement) to be real. . . .

Finally we can see in a very simple and commonsensical way why renormalization has to be achieved for any quantum field theory to be taken seriously as a description of an unobserved reality.[4] The idea of the conservation of infinite quantities in a finite process, say beta decay, is incoherent. It is incoherent in just the way that the infinite forces required by Newtonian corpuscular theory during infinitesimal contact times were incoherent, against the background of Newtonian metaphysics. Unless renormalization can be achieved, a quantum field theory of a class of interactions would be at odds with the metaphysical basis of the overall enterprise. As a theory-family the system would not be in equilibrium. Renormalization restores the balance between the source concepts and the theory they inspire, and the experimental results which are taken up into the revised theory by the very act of renormalization.

Can the persistence of the same metaphysical basis from the proto-Newtonian physics of the seventeenth century to relativity and quantum field theory be explained in some fundamental way? We *could* say 'Nature has so far proved to be well disposed to open up its secrets to those who adopt the techniques which depend upon the robustness principle.' Or is it that by tacitly adopting the robustness principle our explorations of diverse phenomena have drifted towards techniques which promote the display of just those affordances whose manifestations are susceptible of that particular kind of bookkeeping? Perhaps that kind of bookkeeping is a reflection of the conditions for constructing an intelligible discourse about the phenomena our manipulations bring into being.

The robustness principle seems to be closely linked to an a priori commitment to a realist epistemology. The route to reality passes through symmetry and conservation. So the system structure of the quantum field theory-family is reality-preserving. But it became clear that at best the Feynman approach led only to necessary conditions for a theory to be susceptible of a realist construal. Particle physics, under the control of the general conservationist attitude, has been elaborated by the addition of a repertoire of particle properties that, despite their exotic appearance, are classical in the sense of obeying similar principles of 'deep grammar' to familiar dispositional attributes. Their general coherence with classical concepts is guaranteed by the overall control exercised by the corpuscularian metaphysics and the generic conservation principle. So we, as philosophers, on pain of circularity, cannot use the replication of classical 'grammar' in support of a general adherence to corpuscularian metaphysics and conserved quantities. It was an interest in the conditions of unambiguous communication that led Bohr to his 'correspondence principle.' Could we find an argument that would justify adherence to the generic conservation principle as a condition for a theory-family to engender ***intelligible*** theories?

If it could be shown that the conditions for our organic survival favoured creatures built to look for and make use of conserved quantities (stable entities) then that would be a reason why physicists, being just such creatures, find conservation principles so deeply attractive. . . . The successful evolution of organisms that detect and make use of stabilities in their environment would seem to entail that there are such stabilities at least among objects of daily use. But the deeper principle that the realm of the unobservable should also contain stabilities does not follow from the evolution argument. At best it would follow that physicists tend to imagine that realm in those terms. It will be intelligible to them only if it is thought of as containing the physical stabilities which appear as symmetries and conserved quantities. Even if it could be shown that only theories based upon conservation principles could be intelligible to human beings, it would not follow that such theories picked out the true architectonic principles of nature. At best this would

[4] For a very clear account of the inner 'logic' of renormalization see P. Teller (1987).

leave the basis of physical science exactly where Kant located it, in the synthetic a priori principles by which we construct a phenomenal world that would thereby be intelligible to us. It would come as no surprise to discover that we imaginatively extended that world in accordance with the same principles.

It has turned out that the force of both the 'mathematical' methodologies, covariance and gauge invariance with renormalization, is metaphysically conservative. At the end of the day the philosopher of science interested in the implicit ontological foundations of natural science is presented with the age-old problem: how does substance underlie phenomena? This is just the problem that Aronson's 'mappings' approach glosses over. In general there are two ways in which the problem can be tackled. In the one the part-whole principle is invoked, so that the tracts of substance presented in Realm 1 are treated as (structured) drifts of substantial individuals in Realms 2 or 3. But the analysis of quantum field theory has shown that this treatment is inadequate. A theory of properties is also needed. The right kind of ontological hierarchy to match the way in which hierarchies of theories develop in the physical sciences seems to be based on the principle that all properties are grounded dispositions. This idea will be confirmed by the analysis of the apparently 'new' properties invented to match the quantum numbers of the menagerie of particles called for by the corpuscularian trend in high-energy particle physics.

But the part-whole principle and the grounded disposition principle are in tension. If both principles are applied to natural science then an indefinite ontological regress is opened up, since there will be no properties which could serve as the attributes of the ultimate 'parts', because according to the grounded disposition principle it is always proper to ask how any property is grounded. . . .

CHAPTER ELEVEN

A DYNAMIC MODEL FOR SPACE AND TIME

Many years ago, I had the pleasure of attending John Lucas's lectures on Time and Space. They were delivered with his usual mastery in the splendour of Merton dining hall. While John, attired in an ancient gown, developed his themes and theses, a reel to reel tape recorder of almost equal antiquity spun quietly, making possible their transformation into *A Treatise of Time and Space*. Prominent among the principles enunciated in that book (Lucas 1973) were the Radio and Radar Rules, from which with the help of coloured inks and polyglot cast of widely distributed physicists, John extracted the basic principles of the Special Theory of Relativity. I was struck then, but have only recently realised the ontological significance of the way that this analysis established the viability of interpreting the Special Theory as a grammar for the constructing of an orderly and consistent cosmic discourse: But what was this discourse about? Surely it consisted of reports about where to find things and when this or that event occurred.

Subtle and powerful though the Lucasian derivation of Special Relativity proved to be, it incorporated an assumption that almost all philosophers of physics except Leibniz (and perhaps Kant) have subscribed to: that the perceptible world of phenomena and the imperceptible world from which those phenomena emerge are framed by the very same manifolds of locations and moments. John's more recent studies (Lucas 1985; Lucas & Hodgson 1990) seem to me to leave that assumption unchallenged. However, I believe that it is entirely in the spirit of a Lucasian challenge to the prevailing orthodoxy that I undertake the task of undermining that assumption. Abandoning it allows us to preserve the look of the mathematics of both the Newtonian and the quantum mechanical laws of particles in motion under the influence of non-uniform fields and the technical advantages that it allows, while advocating a dual ontology for the interpretation of the spatial and temporal variables that appear in both sets of equations.

In this paper I shall assume that there are good arguments, generally Leibnizian in character, to support the claim that the absolutist assumption that there are manifolds of places and moments or place—moments in addition to arrays of things and events is unjustified (Harré 1996). But arguments against the gratuitous doubling up of the manifolds of things and events with manifolds of places and moments do not necessarily go to the root of the substantivalist or absolutist thesis, that there exists a space and time independent of the existence of the material system of the known universe. At least one of the sources of absolutism is a pattern of mistakes about the grammar of spatial and temporal discourse, that in all contexts the concepts of space and of time and of space-time are subject to the same interpretation. As Wittgenstein remarked apropos of another case of pernicious reification: "We predicate of the thing what lies in the method of representation" (Wittgenstein 1953, Pt I, p. 104). Our familiar spatial and temporal grammars for everyday discourse are intimately related to the criteria we use for the individuation and identification of things and events, clarified by philosophical analyses of the criteria of numerical and qualitative identity. How does all this bear on the meanings of s and t, the ubiquitous variables that appear in all physical laws of motion, that is spatial and temporal change? Is the grammar of these concepts when used for the things and events we can observe unproblematically transferrable to the grammar of s and t when these expressions appear in the laws of quantum mechanics, for example? In the former context they are best interpreted as expressing certain relations between numerically distinct things and numerically distinct events. But if the everyday concepts of "thing" and "event" begin to wilt in the subatomic environment perhaps the ready transfer of the grammar of perceptible relations of the "space", "time" and "space—time" of the perceptible or manifest world, to that realm ought to be at least queried. In what follows I shall be assuming a generally realist stance to the interpretation of physics. It will be taken for granted that there is a viable distinction between regions of the world we can observe, the *manifest world,* and regions of the world we are not capable of observing, but which we must assume to exist as the causal substrate of the manifest world. I shall call this region the *nether world.* *S* and *t* have roles in the description of both the manifest and the nether world. The question to be addressed in this paper is whether it is proper to assume that the logical grammar of their employment is the same in both.

1. MOTIVATING THE QUESTION

Few would disagree with the claim that colour words are multivocal in their everyday uses, and in their role in the physical sciences. "Red" is used to describe how a thing currently looks, and it is also used to ascribe a disposition to a thing to look that in "normal" circumstances. This was famously institutionalised in Locke's account of secondary qualities as powers in a body to cause an idea of the corresponding perceptual quality in a conscious being attending to the body in question. Bodies had such powers because of the "bulk, figure, texture and motion of their insensible parts". In short, the word "red" serves both to describe a perceptual quality and to ascribe a grounded disposition. Misunderstanding Boyle (1666) and Galileo (1623/1957), Locke (1690/1947) proposed the disastrous thesis that ideas of primary qualities in the mind resembled primary qualities in the material world. He hoped that this principle would bridge the gap between the mental and the physical, while, at the same time, confining the contents of consciousness to nothing but ideas Boyle had used the distinction between the primary and secondary qualities to bridge the gap between the observed and the and the unobserved properties of material things, a distinction within an extramental realm.

Locke's version of the "resemblance" doctrine deservedly got a bad name. Berkeley's criticism is worth repeating in that it makes clear the complexity of the dispositional properties that Boyle (in the context of the observable/unobservable distinction) and Locke (in the context of the mental/material distinction) were ascribing to beings which existed independently of human sensibility. Berkeley (1710/1988) denied the viability of the distinction whether between primary and secondary qualities, or between ideas of primary and ideas of secondary qualities since, having denied the distinction between Lockean ideas and perceptible qualities, he held, rightly, that Locke's version of the resemblance doctrine was absurd. "An idea can only be like another idea." By identifying ideas with perceptible qualities in the *esse est percipi* principle, Berkeley came close to restoring Boyle's "classical" distinction amongst classes of qualities. His retreat from the full corpuscularian account of material things, in which matter had the power to act upon human sensibility, was marked by his denial of the causal efficacy of qualities-ideas. "Ideas [qualities] are inert." Power to act is reserved to spirits. Material things are passive whether they are observed or not. When a previously unobserved thing becomes observable, the ideas/qualities it presents must be from the same repertoire of perceptible qualities, since to become observable is just to become perceptible. From the discussions in *Sins* (see Moked, 1988) we know that Berkeley was favourable to corpuscularianism as a general philosophy of science, but though the corpuscles revealed by microscopes must have the Boylean primary qualities since they are now perceptible, they are for that reason inert. In Berkeley's system perceptual quality terms could not be doing the double duty that they do in Boyle's scheme and all its successors. To use them to ascribe causal powers to observed or unobserved material things is to violate the principle that ideas (perceptual qualities) are inert. Material things just are beings with those inert qualities. It follows that the perceptible world is not the manifestation of imperceptible material beings with causal powers. But, for Berkeley, it was the manifestation of the active powers of spirits. Sceptical of the citation of spirits, Hume's (1788/1978, bk. 1, sect. XIV) denial of the existence of active matter, left the fatal legacy of the regularity account of causation, in which an ontology of powerful particulars gave way to one of atomic events.

In the 18th century the consensus amongst physicists turned against Berkeley and Hume. Most philosophically minded physicists, for instance, Boscovich (1763/1966), and many philosophers with an interest in physics, for example Kant, (1786/1970), were dynamicists. They believed that the matter of the world was active, even that the primary qualities were best defined as powers, which could properly be cited in explanation of the dispositions that things manifested to act upon other beings in the appropriate circumstances. Active powers were manifested as forces, the potency of which could be expressed in felt thrusts upon the body.

I propose to follow Berkeley in his denial of Locke's notorious resemblance thesis, and to follow Kant (1786/1979) in his affirmation of the reality of forces and of the real existence of causal powers. We have then a general ontological duality that appears in the double meaning of predicates like "red". They are used to describe Berkelian perceptible qualities and also to ascribe Kantian causal powers to imperceptible material beings, endowing them with dispositions to manifest those qualities to normal human beings in ordinary circumstances. It is this ontological duality that I propose to extend to underpin the semantics of spatial and temporal concepts.

2. HOW MANY DUTIES DO *S* AND *T* PERFORM?

The Newtonian equations of motion and the quantum mechanical wave equation both contain spatial and temporal variables *s* or *x* and *t*. I believe that it is assumed, almost without question, that they mean the same in both contexts, whatever that meaning might be. Whether one is a substantivalist and believes in the existence of a matter independent space-time manifold, or whether one is a relationist and believes that spatial and temporal concepts are only meaningful as expressing certain relations among things and events, the *s* and *t* of the Newtonian physics of the manifest world and the *x* and *t* of the Schrödinger equation are variants on the same basic concepts. That this assumption carries with it the further ontological consequence that the individuals of the nether world are logically parallel to those of the manifest world, and are roughly thing-like and event-like, is not often overtly discussed. We must consider the ontology that seems to be forced on us by the Einstein-Podolsky-Rosen thought experiment, and Aspect's empirical replication of it, and admit the possibility of the dissolution of the necessary connection between the numerical diversity of events and the overt spatial separation of their occurrence. A more radical possibility springs to mind. What if the space and time of the mechanics of the beings we can observe, which is tied to their conditions of numerical and qualitative identity and diversity, and the space and time of the nether world of unobservable processes and beings which are manifested in observable beings and their qualities were related in something like the way the two uses of "red" are related? The powers of material things to look red are grounded in states of these beings that bear no resemblance whatever to colour. Why should it not equally be the case that spatial and temporal concepts, as employed in the discourses of the unobservable, might be best interpreted as ascribing powers to beings of that world to manifest themselves as numerically diverse things and events while the beings to which those powers are ascribed are not themselves subject to those criteria? Why should we transfer the "atomicity" of observable things and events, expressed in the grammar of spatial and temporal concepts for the manifest world, to the beings of the nether world? Why should we assume that a plurality of manifestations must be a consequence of a plurality of whatever it is that brings about those numerically distinct manifestations? Why should we adopt even the minimal correspondence of criteria of identity and difference suggested by my use of the plural in "beings"?

The standard answer would no doubt go something like this: when, in the past, this assumption has been made for a certain realm of unobserved beings, and those beings have been eventually revealed by some development of the technology of observation, lo and behold they do display the very modes of numerical diversity and identity that is expressed in the grammar of Newtonian space and time

My response to this is in two parts:

a. We already have good reason to believe that inductions from and analogies to the world as it is revealed to observation as the main conceptual resource for constructing a conception of the processes that deeply underlie what we can observe, even with sense-extending instruments, quickly prove fallible. In deve loping an ontology for subatomic physics we are obliged to invent a range of dispositional concepts, for few if any of which is there a plausible hypothesis concerning their grounding in entities, states and structures life any we can perceive. Furthermore, the very idea of a sense-extending instrument establishes an internal relation between the concepts of the everyday world and the concepts we use to deal with and to interpret the world as it is revealed with just these instruments. There are many instruments, specifically those involved in high-energy physics, which are not sense extending. Only to the revelations of the former do the Berkelian extensions of perceptible qualities to the realm of the contingency unobservable make sense.

b. The practice of ascribing powers to the world-stuff, realised in the dispositions of instruments to bring into being things and events to which perceptual qualities can be ascribed, has proved to be very successful. Coherent and powerful mathematical formulations of such ascriptions in terms of field theories have been constructed. If there is any place for a methodological induction in the metaphysics of physics it is to the general practice of ascribing powers to the unobservable "beings" which are the logical subjects of discourses of fundamental physics. Why should we assume that the criteria of identity of imperceptible beings with causal powers are anything like the criteria of identity of the things and events of the manifest world, that they give rise to?

Let us suppose then, as a working hypothesis, that spatial and temporal concepts have a dual role in the physical sciences, in much the same way as do colour concepts. However, unlike the case of colour there is no possibility of grounding the powers of the world to manifest itself as numerically distinct things and events in some occurrent, standing condition of the world-stuff. There is no such possibility because every such grounding makes use of spatial and temporal concepts. Colour is grounded in the spatial distribution of electric and magnetic fields, heat in the motion of molecules and so on. To ground the powers of the world to be realised in dispositions to manifest themselves spatially and temporally in yet more spatial and temporal concepts is to beg the question.

3. WHAT MUST A SPATIAL AND/OR TEMPORAL DISPOSITION AFFORD A HUMAN BEING OR A PIECE OF APPARATUS CONSTRUCTED BY SUCH A BEING?

Temporal durations, sequences of events which do not coexist with one another, are internally related to the conditions for the numerical diversity of qualitatively identical events, that is if they occur at the same place they do not coexist. I shall assume, for the purposes of this paper, that coexistence can be treated not as a covert temporal notion but as expressing reciprocal (coexistence) and asymmetrical (non-coexistence) causal efficacies. The temporalising powers of the complex devices formed by creating stable combinations of the indeterminate (in terms of concepts derived from the discourses established for describing the manifest world) stuff of the nether world with human perceptual systems and experimental equipment and other tools, are realised as dispositions to present properties that are coexistent but incompatible in the nether world were they to be treated as if they were characteristics of temporally diverse events in the manifest world. So, for example, the superposition of quantum states is described with a temporal parameter, expressing the fact that the quantum system is disposed to or has the capacity to display the corresponding observable phenomena arrayed as in time, for example two electrons cannot be at the same place at the same time. Complementarity is just a special case of the dispositional character of t as a variable in the equations describing those apparatus-world complexes we call quantum systems. But "moment" as a reading of a value of t is viable only in the phenomenal or manifest world.

Spatial extension, sets of things which coexist with one another, obey the conditions for qualitative and numerical identity and diversity for things. The spatialising powers of world-apparatus combinations (including human perceptual system) manifest themselves as the disposition to display things which are qualitatively identical but numerically diverse at different places, that is as spatially distant from one another. To be at a different place just is to be numerically diverse as a thing.

The tension between spatio-temporal concepts interpreted as appropriate to the manifest world, for instance the world of the laboratory, and the interpretation of the symbols x and t as they appear in the equations ascribing states to the nether world, emerges clearly in the Einstein-Podolsky-Rosen (EPR) experiment and its empirical realisation by Aspect. In the manifest world of the laboratory spatial and temporal concepts are internally related to criteria of qualitative and numerical identity through which individuals are recognised in that world. For example, if there are simultaneous events at distinct spatial locations then there are two events. It seems only natural that the x and t variables in a wave-equation and the x and t values in the solutions to such an equation express the same physical concepts. (Just as it seems natural that the word "red" when used to describe the hue of a surface and to ascribe a causal power to the object which endows the surface with the disposition to look that way in normal circumstances.) It is just this seemingly natural assumption that I wish to challenge.

Since it is clear that in the manifest world criteria of identity and spatio-temporal concepts are, so to say, two sides of the one ontological coin, in the nether world the same internal relation will obtain. Thus, if we revise our spatio-temporal concepts from sets of relations between numerically distinct things and events in favour of a dispositional interpretation in which they are used to ascribe certain powers to world-stuff which are realised in dispositions of apparatus in interaction with it, then the criteria of identity to be applied in the nether world change with them. So, for example, while the idea of space and time as manifolds of locations and of moments abstracted from the arrays of things and events to be found in the manifest world, it may be a serious mistake to project that structure on to the ontology of the nether world. There may be a place for individuality in the ontology of the nether world.

This analysis will prove its mettle in accounting for the seeming paradox presented by Einstein to Bohr in the famous Einstein-Podolsky-Rosen thought experiment, now empirically realised by Aspect. There are many excellent descriptions of EPR, but the one I shall draw upon for detailed examination is that offered by Sklar (1992). To set up the experiment we suppose that there is a source of paired particles, for example in the singlet spin state. This means that if one particle is experimentally shown to have spin up in a certain direction, then the other will be shown experimentally to have spin down. This follows from and is incorporated in the quantum mechanical description of the system containing the two particles. Equidistant from the source and on opposite sides of it are detectors. Suppose that the pair of particles are separated in such a way as to preserve the spin correlation. One particle, p_l, goes to the left-hand detector and the other, p_r to the detector on the right. In whatever direction we choose to measure the spin of particle p_l by adjusting the detection apparatus on the left after the particle has left the source, we find that the probabilities of measurements in any direction we choose for the right-hand measurement bear fixed probabilities given the result obtained on the left. The correlation of results cannot be explained by the conditions under which the particles are generated at their common source since the choice of a direction of measurement at the left-hand detector took place after the particles had separated. Nor can it have occurred by a causal influence propagated from A to B since the conditions of the experiment are such that any influence of the sort would require to have been propagated at a supraluminal velocity, a process inconsistent with Special Relativity.

Einstein thought that he had shown that quantum mechanics was incomplete, that the description of the whole system lacked a variable for a yet-to-be-observed property which would account for the correlation of results for the separated particles.

There are a great many ontological assumptions embedded, seemingly innocently, in this thought experiment and in particular in our way of describing it. The measurement events occur in the manifest world at two material sites individuated classically, and, since they occur simultaneously count as two events. But if we were to imagine a causal influence propagated from the left-hand detector to that on the right, it's not the result of any process in the manifest world. Even to talk of such an influence presupposes that there are nether world events corresponding to the measurement events, and subject to the same criteria of identity and individuation. But suppose that in the nether world no such criteria of identity apply. And suppose that there is no spatio-temporal manifold of things and events, such as constitute the beings of the manifest world. In the nether world there are powers endowing a suitably constructed Bohrian instrument/world ensemble with dispositions to display events of the kind imagined by Einstein and recorded by Aspect at distinct sites in the manifest world of the laboratory. There are no events or places in the nether world, only powers to manifest themselves as spatially separated phenomena, non-identical events in different places. In the wave equation that ascribes an evolving quantum state to the nether world s and t are to be interpreted via the ontology of powers and dispositions, while those expressions are interpreted in the *solutions* of the equation for the whole set up as the s and t of the familiar spatio-temporal manifold abstracted from the manifest arrays of things and events.

I have referred to Sklar's very clear account of EPR because it contains a very common misinterpretation of Bohr's position, a misinterpretation of which Einstein too was guilty. Sklar says: "... prior to measurement, the particles have only their disposition to come up with a given spin value in a given direction ... along with the disposition to have their spin values, if both are measured, correlated in the way described by a singlet-state wave function" (Sklar 1992, p. 216). And again on p. 217 Sklar, expressing what he takes to be Bohr's position . . . "After particle one has been measured and found to have a definite value in the *A* direction, the correct wave function to describe the world is the one that assigns to each particle a definite spin in the *A* direction." But are these particles the sorts of things that Bohr, in his deeper moments, would have countenanced as that which is described by a wave-function. Of course not! Bohrian "measurements" are not measurements in the classical sense. They are phenomena, the whole of which come into existence, *as such,* in the activity of a complex but unified object, the world apparatus ensemble. Not only does this entity bring into being such properties as "spin" but also the beings to which it is ascribed. Prior to the experiment neither property nor object exist. Using more modern terminology than Bohr had to hand, we would say that the apparatus that EPR imagined and Aspect built affords particles with correlated spins. What does the quantum mechanical description of the state of the world mean? It means that in conjunction with suitable apparatus, say EPR, the world has the power to manifest itself in the form of particles with certain detectable properties. All the variables used in the quantum mechanical description must be interpreted as ascribing powers, x and t included.

Why should one accept such an extravagant revision of our seemingly robust conceptual system? First of all, as I showed in an earlier section, there is a model for such an interpretation, the grammar of colour words. Second, there is the EPR paradox and its realisation in the Aspect experiment. EPR is no longer a mere thought experiment but a solid experimental result, posing an irresolvable paradox when it is couched with the traditional spatio-temporal ontology in which the nether world and the manifest world are assumed to be of identical ontological structure. If there are particles then there are criteria of identity for such particles implicit in the discourse, and they appear in the concepts of space and time. Without the assumption of ontological continuity between the manifest and the nether worlds there is no paradox. What better indirect argument could one hope for? The EPR "paradox" simply arises from projecting the classical conceptions of space and time and their intimately related notions of identity back from what can be observed and experimentally measured, on to whatever it is that engenders these phenomena. What we can say is that an Aspect-type experiment reveals that the world in conjunction with an Aspect-type apparatus has a power to display correlated spin states at two places, *in the manifest world.* But "place" as a reading of a value of s is viable only in the manifest world.

4. INDEXICALS IN THE NETHER WORLD

Suppose that we agree with the analysis so far. A very important part of the argument against substantivalists comes from the role of indexicals in the discourses of space and time. "Here" and "now" link acts of speaking to that which is spoken about, and so provide the basic relations of temporality and spatiality. If we abandon the idea of an array of nether things defining relationally a nether space, and an array of nether events defining relationally a nether time, does the dispositional interpretation of s and t for the nether world appearing as the s and t of the manifest world when it is in combination with people and their equipment, preclude the use of spatial and temporal indexicals for that world? Since indexicals link acts of speaking or writing with what is spoken or written about, if there were such indexicals they would bind the relational manifolds of manifest space and time with the dispositional possibility manifolds of the nether world. If, as I have argued elsewhere (Harré 1996) existence-causality and the unique present, the only "now", mutually define one another, then since we must suppose that whatever engenders the phenomenal or manifest world coexists with it, the same indexicals must serve for both. So whereas "here" and "now" are occurrent indexicals for both worlds "then" and "there" are indexical for the phenomenal world but dispositional for the nether world. For the nether world to have the power to engender an entity or event which is observed to be "there" rather than "here" when some equipment is working, that power must include the power to engender it incipiently "there" rather "here." But an incipient or possible "there" is not a place distinct from an incipient or possible "here" since it is not a place at all. A similar argument could be constructed for the incipient futurity of events. But what of incipient pastness? We shall have to allow, I think, that dispositions to manifest as events which are ordered as past, present and future, must, with respect to temporality alone, be indifferent with respect to past and future. The distinction of "already" and "not yet", which is so important a feature of temporality in the phenomenal world, will be analysed in terms of causal efficacy. Thus, a state in the nether world which has the power to manifest itself as an event must have the power to manifest that event as either a cause or as an effect with respect to some circumstances. The one will be past or future to the other, in terms of the relations that phenomenal events bear to one another, depending on which "sense" of agentive causality each is manifested with, cause to effect or effect to cause.

Unlike Leibniz's account in which God's propositions play the engendering role and Kant's where that role is played by the wholly unknowable noumenon(a), in this account dispositions and powers can be at least semi-ascribed to the primordial stuff, since it is in instrument-world or people-world complexes that things and events are manifest standing in those set of relations we abstract as the manifolds of space and time.

However, a smidgeon of doubt remains. If the nether world is to explain what happens in the manifest world, and there are changes in the manifest world, are we not obliged to suppose that the nether world too changes? A relationist must surely admit that if there are changes then, *a fortiori,* there is time in the relationist sense, namely, relations of identity and difference between changes. But manifest world "time" is constructed not only of relations of identity and difference, but also of successional concept intimately bound up with causality, in ways I shall simply take for granted in this chapter. Nether world temporal dispositions are characterised by identity and difference, but only dispositionally by succession.

5. ASCRIBING POWERS

By virtue of their powers material particulars are disposed to manifest this or that observable property in their actions upon other material particulars and by virtue of their liabilities in being acted upon. However, this simple subject-predicate logic with its associated metaphysics of individual substances and their attributes is inadequate to express the grammar of physics. By the end of the 16th century, Kepler's hypothesis of universal gravity and Gilbert's studies of magnetism had forced philosophers of physics to contemplate taking seriously the seemingly paradoxical concept of action at a distance. But how was it possible for a magnet or the earth, as a powerful particular, to act on some other particular without the intervention of a medium? Gilbert's concept of the *orbis virtutis* was the first full-blown field theory to emerge in physics. The magnet induced a distribution of active powers throughout the neighbouring space, such that at each spatial point a force would be exerted on a test body placed at that point. The logical subject of power attributions in field physics just is the field. It is spatially extended and temporally enduring. Just as a coloured surface can differ in hue from place to place, so a field can differ in its causal powers from place to place. Possessing these powers the field can act upon a body placed at a certain point in the field provided that the test body itself has the appropriate liabilities. A magnetic field cannot act upon chips of wood. By the mid-19th century we have a complete conceptual structure for expressing these institutions consisting of "charge", "field-potential", "force" and "acceleration". The two latter incorporating a further aspect of field physics, namely the direction in which the field can act on a suitable body. Fields merge according to quite specific laws. Two merged fields are a field with its own distinctive dispositional properties. Not surprisingly the conditions for identity and individuation of ordinary material things do not all hold for fields. Fields as beings in the nether world are *not* extended, but have powers to engender events in the manifest world that are spatially distributed. Our best reason for believing this must surely be the experimental confirmation by Aspect of EPR.

The *s* and *t* of field physics are used to locate first the observed action of the field and then the differentially distributed powers of the field itself, which is thereby endowed with extension, within the same spatio-temporal framework as the charged body that is the source of the field and the bodies it acts upon, or that are in mutual interaction with it, that is in the manifest world. Thus the *s* and *t* of the field inherit the structure of the manifold of Leibnizian relations that hold between observable material particulars.

Conceiving a field as a substance is facilitated by such models as the distribution of pressure within a fluid. At each point the pressure has a certain value, a measure of the power of the fluid at that point to act upon test bodies and upon the adjacent "element" of the fluid. If we imagine the fluid to be continuous the pressure at a point is not a property of any other material particular than the fluid itself. Unlike gravitational, electromagnetic or colour fields are brought into being by charges.

What of the siblings *s* and *t* of the nether world? The charge which engenders the field could be ascribed the power to manifest itself as a spatiotemporally distributed force field, so that the *s and t* in the description of the *charge* are dispositional concepts, while in the manifest world they are Leibnizian relational concepts expressing the principles of identity and individuation of the material particulars that are observed to serve as the source of fields and to be acted upon by them. According to this interpretation, the *s* and *t* in the field equations refer to dispositional properties of the charge, the *s* and t in their solutions refer to Leibnizian relations among observables. We have exactly the duality of observable states and unobservable dispositions to manifest those states that appears in the logical grammar of colour concepts. Red is an observable hue of a surface, and it is a disposition of the material being the surface of which we can observe to manifest itself that way to human beings, and in other ways in interacting with other beings, such as photons.

6. VELOCITY: A MORE COMPLEX CONCEPT THAN IT SEEMS

6.1. Velocity in the manifest world

Nothing would seem to be less problematic than the concept of velocity. Its everyday manifestations in such judgements as faster and slower, moving or stationary relative to some material framework, its everyday application in planning and executing journeys, seem entirely transparent. Yet in medieval physics speed was a property of a moving thing manifested in such phenomena as distance covered in a given time and the time to cover a certain distance. It was average speed that was measured, yet speed was not a property displayed over a period of time but a quality of body at an instance. Galileo was not the first to formulate the concept of instantaneous velocity. By his time the difference between the measure and the quality was clearly understood. It was to the problem of measuring the instantaneous velocity when at every moment the speed of a moving body was different that Galileo borrowed the general solution offered by Bradwardine and other Mertonians in the 14th century (Clagett 1959). The difference between the dispositional concept of velocity and the occurrent concept of average speed was not important for uniform velocities since what ground the moving body would cover in a given time, or how far it would go in a given time, was actually accomplished. But in uniformly difform or accelerated motion the instantaneous velocity of an accelerating body would never be actually realised. The concept was not only dispositional but counterfactual as well. The instantaneous velocity of an accelerating body is that distance it would have covered had it ceased to accelerate and moved uniformly from that moment. But all along speed had been dispositional. When direction of motion is added, the dispositions in each case become a little more complicated since we must add the direction of motion to each actual or counterfactual manifestation of the kinematic disposition The use of dispositional space and time concept was already well established in physics by the 14th century.

One should reflect too on the arbitrariness of the choice of manifestation as a measure of the dispositional property of fuel to propel a vehicle we measure how far the vehicle has travelled in consuming a fixed quantity of fuel, if we use the miles per gallon measure, but if we use the litres per hundred kilometres measure we measure how much fuel it has consumed for a fixed journey. Either will do as a measure of the dispositional property of the fuel. Equally we could measure the dispositional property velocity by how far a body would go in a certain time, so many metres in a second, or how long it would take to cover a certain distance such as how many seconds per meter. Either would do. In physics we use the former, in athletics the latter. We could have an Olympic game in which competitors were to run as far as they could in 10 seconds, 1 minute and so on. In both cases we could be measuring either an average speed (velocity) or an instantaneous velocity of disposition.

Whatever is said about velocity applies to acceleration as well. It too is dispositional. Both concepts have other applications in physics that point in a similar exegetical direction. Velocity appears in the definitions of kinetic energy as $1/2mv^2$ and of momentum as *mv*. In these cases too we need to go a step beyond what is offered in most but not all physics textbooks about these concepts. For example the admirable G.R.Noakes (1977, p. 72) says "The kinetic energy of a moving body is its capacity for doing work by virtue of its motion and its value is $4mv^2$.". So as properties of material particulars what are energy and momentum? Energy is the power to do work, *Fs,* and momentum is related to another power, "action", *Ft.* A body of mass *m* moving with velocity *v* has the power to displace a force a distance *s,* say in impact. So the seeming definitions are both specifications of how the powers of a moving body are to be measured. They are not definitions in the ordinary sense, that is some account of the meaning of the concept. Since velocity is itself a dispositional concept the formulae for the measures of kinetic energy and momentum are doubly dispositional. They provide for the quantitative assessment (measurement) of a disposition on the basis of the quantitative assessment of another disposition, velocity.

6.2. Velocity in the nether world

In the above analysis "velocity" has been treated as a disposition with respect to the possibility of changes in the Leibnizian spatio-temporal relations of the manifest world. In the setting up of measure for energy and momentum on either the *Fs/Ft* forms or the *m/v* forms velocity is firmly located among the properties of the material particulars of the manifest world. In the nether world, the corresponding powers are energy and momentum, but it is to the measure of their manifestations and not to their nether being that the concept of velocity applies. In the physics of the nether world the expression for momentum is usually *p* and for energy *e.* In the manifest world these powers are attributed to manifest particulars on the basis of the spatio-temporal formulae incorporating velocity. We must conclude then that there is no place for a concept of average velocity in the nether world, at least in its standard kinematic form as an instantaneous or as an average space/time quotient. We have seen that that form is the result of an arbitrary choice between that and a time/space quotient. Either would express in manifest terms the nether disposition to change Leibnizian spatial and temporal relations in the manifest world.

7. THE INTERPRETATION OF GENERAL RELATIVITY

7.1. Leibnizian space-time

The variables *s* and *t* have been interpreted as referring to relational structures amongst things and events by Leibnizians, and as referring to matter independent manifolds of locations and moments by substantivalists. The thrust of the arguments of my discussion of the dual signification of spatial and temporal concepts in physics has been towards the Leibnizian alternative. Matter-independent spatial and temporal and spatiotemporal manifolds are convenient fictions, characterised by Leibniz in the following "progression":

> When it happens that one of these coexistent things changes its relation to a multitude of others, which do not change their relation to the others, as the former had, we then say, it is come into the place of the former.

Then Leibniz introduces "abstract space" as a conceptual structure. The space of actual relations of situation among material things is detachable from actual coexistents only in thought. Place, as defined above, is a matter of the stability of local relations among coexistents. "But" says Leibniz (1717, pp. 15, 47) "the mind not contented with an agreement, looks for an identity, for something that should be truly the same, and conceives it as being extrinsic to the subjects; and this is what we call *place* and *space.*" The full enrichment of the abstract manifold that exists only in thought includes possible situations as well.

> One may always determine the relation of situation, which every coexistent acquires with respect to every other coexistent; and even that relation which any other coexistent *would have* to this, or which this *would have* to any other ... that which corresponds to all those places, is called *space.* (Leibniz 1717, pp. 5, 47).

Summing up these analyses we arrive at the following:

> Space is ... an order of situations ... abstract space is that order of situations, when they are conceived as being possible. (Leibniz 1717, pp. *5,* 104)

There is no space and time in Leibniz's nether world, a world populated by monads, the relations between which are logical. I doubt that many would want to go as far as Leibniz in so radically distinguishing the ontologies of the nether and the manifest worlds, but his line of thought points in the right direction. The illusion arises from the reification of an abstract set of mathematical relations.

Let us now turn to General Relativity and the interpretation of concepts such as "the curvature of space." This looks as if it might provide just the access we need to explore the spatiotemporal character of the nether world. But what exactly does Reimannian geometry refer to?

7.2. The thrill of "space-time curvature"

Einstein's resolution of the problem of how there can be action at a distance was to extend the concept of inertial motion by proposing that the joint Euclidean spatio-temporal manifold of Newton's cosmology should be abandoned in favour of a more complex geometry referring to a "space-time which was itself curved." Inertial motion along the curve replaces forced motion out of a Newtonian straight line as an explanation of motion under gravity. There are no mysterious force-fields, whose powers are engendered by distant masses. To take this image seriously as ontology takes us straight back to a conception of a universe in which there is a joint manifold of locations and moments *and* a manifold of material beings which in some mysterious way, can act upon that manifold. We are invited to consider what a matter-free space-time would be like. General Relativity is far from being devoid of mystery.

It trades on one unsolved problem, how can a massy body engender forces where it is not, for another; how can a massy body change the structure or the independent manifold of moment-places where it is not? Mesmerised by the mathematical relation established between the metric tensor, expressing the structure of space-time and the matter-energy tensor expressing the tendencies of material beings, the romantics among us ignore Leibniz's elegant ontogeny of the illusion of matter independent spatial and temporal manifolds to reify space-time.

It is instructive to read an ecstatic presentation of the reified space-time of General Relativity, equally as romantic as that of Minkowski's remarks about his space-time.

> What is it that pulls the apple to the ground, bends the circling moon to the earth, and makes the planets captive to the sun? It is no longer a force acting at a distance, but something far more primordial. It is intangible space and time themselves, acting in awesome concert as curved space-time and holding sway over all things in the universe (Hoffman 1983, p. 157).

This paragraph is full of romantic exaggerations and metaphysical extrapolations. Indeed, it is not forces that act at a distance, but charged bodies which act to engender spatially distributed fields of force. It is these forces which act where the test body is located since they act upon it. If there are no such manifolds as space and time how can they, even in "awesome combination" hold sway over anything?

None of this ecstasy is justified. The substantivalist conceptions of space, time and space-time are illegitimate projections from the abstract system of relations that express the criteria of identity and difference of things and events of the manifest world on to the world in general.

SECTION FIVE

A METAPHYSICS FOR EXPERIMENTS

SECTION FIVE

A METAPHYSICS FOR EXPERIMENTS

INTRODUCTION

There is no doubt that the philosophical study of experiments as such and of the apparatus and instruments with which they are conducted has been neglected. In recent years, the topic has been broached, but there is still much to say. The invisibility of the experiment as such in the period during which logicism dominated the philosophy of science will serve as a starting point for these investigations. The recent inroads that have been made into philosophy of science from the sociology of science have brought experimental activity to the fore and have led to some attention being paid to laboratory equipment. However, there is still a tendency to see science wholly in terms of the discourse of scientific communities. In that light concrete experimental procedures and the *instruments* and *apparatus* used to perform them are still almost invisible. If they are attended to, they are given a constructionist interpretation as allies in the argumentative discourses of scientific communities. This paper is about laboratory equipment, in all its materiality as pall of the material world.

1. SOME PHILOSOPHICAL VIEWS OF EXPERIMENTS

An experiment is the manipulation of apparatus which is an arrangement of material stuff integrated into the material world in a number of different ways. I shall refer to such integrated wholes as apparatus-world complexes. In the course of the manipulation some process of interest is made to occur in an apparatus-world complex. The process more often than not results in discernible transformations of the apparatus. A close analysis of the kinds of equipment in use in the sciences will lead us see that there are two very different ways of interpreting these transformations. In those cases in which an apparatus is serving as a working model of some natural system, the changes brought about by experimental manipulations must be interpreted as *analogous* to states of the natural system being modeled. In those cases in which an instrument is causally affected by some natural process, the changes in the instrument are *effects* of the relevant state of the material world. These effects must be interpreted in terms of the causal relation presumed to hold between the process in nature and the state of the instrument. It will be convenient to use the word 'instrument' for that species of equipment which registers an effect of some state of the material environment such as a thermometer, and the word 'apparatus' for that species of equipment which is a model of some naturally occurring structure or process, such as the use of a calorimeter to study the effect of salt on the freezing point of water, or in vitro fertilization. Science has advanced so far now that there are pieces of apparatus in which processes are made to occur that have no analogues in nature, for instance the equipment used for cloning large animals.

The manipulation of material stuff extends beyond the experiment as such, into the past and into the future. The apparatus has to be created, designed and built by technicians from available material. As Toulmin (1953) pointed out, the materials used must be subjected to processes of purification. This demand covers not only the reagents used in chemistry but also the material of which an apparatus is made. There are all sorts of presumptions then about the past of an instrument. There are also presumptions about the future of an instrument. Though a specific piece of equipment need not survive in order for an experiment to be replicated, it must be possible to reproduce versions of the material structures of the apparatus in the future. The replicability of experiments depends on that condition. Even something like a space probe that is burnt up in the Jovian atmosphere should, in principle, be able to be duplicated, and the experiment done again. If an instrument does survive to be used again it must be presumed that no radical changes have occurred during the course of the original experimental procedure.

Whether the equipment is an apparatus built to be a working model of some feature of the world, or an instrument causally transformed by processes in the world, the result of an experiment is a reading of the end state of a process. Much can happen to equipment after the moment at which the experiment ends. However, the moment at which the experiment is taken to have ended, and so which state is taken as the end state of the experiment, is not an arbitrary cut in a continuous causal process. It is determined by the project for which the experiment was undertaken.

1.1. Experiments in Logicist Philosophy of Science

The long-standing idea that philosophy of science is a branch of logic has two aspects. There is, or might be, an inductive logic that legitimizes generalizing the results of observation and experiment. These generalizations are often presented as laws of nature. Scientific knowledge, so obtained, is presented in writings organized according to the principles of deductive logic. The laws of nature and other generalizations are used as axioms in the deductive process. The result of working within this framework has been the privileging of the *proposition* as the entity around which discussion and debate as to the acceptability of putative contributions to scientific knowledge must centre.

For the most part the discourse of 'experiments' was shaped by a standard format. According to logicism, the relation between the material world and the discourses of science was taken to be entirely captured by the relation between the propositional forms of an updated version of the Aristotelian Square of Opposition. Experiments and observations were reported in the 'Some A are B' or 'Some A are not B' forms, and general laws or law-like propositions, of the 'All A are B' or 'If x is A then x is B' forms. Occasionally the 'No A are B' propositional form was required. Philosophy of science was thought to start where the propositional universe began. All this seemed entirely natural during the domination of logicism in philosophy of science.

Experimental results were just there to be described, brute facts, so to say. Observations were usually mentioned in the same breath as experiments. Even when the first intimations of a more subtle account of how states of the world were captured in propositions appeared, in the thesis of the theory-ladenness of descriptions, the role of the apparatus as such, as the locus of the genesis of phenomena, was ignored.

Logicist philosophy of science is the legacy of two philosophers above all, Mill and Mach. Mill's methods (Mill 1872) were the result of his attempt to identify the logical forms of patterns of reasoning that led from particular propositions to general propositions. His general propositions were causal laws in the Humean sense, that is they reported exceptionless correlations between types of observable states of the world. Mach's contribution was two fold. His 'sensationalist' metaphysics confirmed a phenomenalist strand in philosophy of science, having its source in British empiricism. Science was concerned with discovering the relations between sensations. His account of laws of nature was Humean. They were nothing but summaries of instances of correlations between types of sensations, serving as mnemonics to recover any instance at will. The Mach-Mill analysis of scientific knowledge confirmed the project of the search for scientific rationality as the development of an inductive logic (Mach 1914).

The once popular Popperian fallibilism involved an inversion of the general schema of inductive logic—'From true statements describing the results of experiments or of observations one can infer a true general statement which can be judged to be worthy to be accepted as knowledge.' The logic of scientific discovery for Popper could be expressed as a general schema from deductive logic — 'From a true statement describing the result of an experiment or observation that contradicts a prediction deductively drawn from a general hypothesis, one can infer the falsity of the hypothesis, which must be judged worthy to be rejected as knowledge'. Both the inductivist and the fallibilist principles are offered as rules governing scientific discourse. Popper's only discussion of the sources of descriptive propositions is confined to a few comments on the conventionalist origins of descriptive predicates and the need to take certain basic statements to be true by convention (Popper 1967).

From a logicist point of view, experiments are simply the unproblematic sources of descriptive propositions in the Aristotelian L and 0 forms, 'Some A are B' or 'Some A are not B'. The only feature of experiments that matters in logicism is the restriction of their spatial and temporal span to the here and now. This means that descriptions of empirical studies in science can exhibit only the I and 0 forms. It is also worth noting that experiments have traditionally been lumped in with observations, in the frequently recurring phrase, 'observations and experiments.' Both are cited as sources of I and 0 propositions.

The upshot is a complete neglect of the apparatus itself. Since it was never attended to the variety of kinds of apparatus was never discussed.

1. 2 Experiments in Constructionist Philosophy of Science

Latour (1987) and others have discussed apparatus and experimental manipulations. I shall use his account as a stalking horse.

We can use technically crafted things to *do science.* The relation between apparatus and Nature needs to be analyzed. Its value to science as the best means for obtaining representations of aspects of Nature needs to be justified. Latour's move, which, as we shall see, sidelines this issue, is to delete the idea of Nature as an independent being from the account of science. In his treatment, apparatus is wholly a humanly created artifact that brings other artifacts into being. These determine what Nature is for the scientific community.

When we use experimental equipment of either kind, apparatus or instruments, according to Latour, we are taking up material things into discourse. Here is Latour's interpretation of an experiment.

> The guinea pig alone would not have been able to tell us anything about the similarity of endorphin to morphine; it was not mobilizable into a text and would not help convince us. Only a part of the gut, tied up in a glass chamber and hooked up to a physiograph, can be mobilized in the text and add to our conviction (Latour 1987, p. 67).

The observation that Latour uses to push one towards the textual interpretation of experimental activity is his account of an instrument. It is deliberately severed from its place as a material thing integrated into a material system that includes the world. The material entity, though ultimately of natural origin, is detached from its place in nature. According to Latour, it is 'mobilized in the text'. The laboratory, he says is 'a set of new [rhetorical] resources devised in such a way as to provide the literature with its most powerful tool: the visual display (Latour 1987, p. 68).

> I will call an instrument (or inscription device) any set-up, no matter what its size. nature or cost, that provides a visual display of any sort in a scientific text (Latour 1987, p. 68).

> What is behind a scientific text? Inscriptions. How are these inscriptions obtained? By setting up instruments (Latour 1987, p. 69).

Just as in the logicist neglect of the instrument that makes it invisible, in this rendering of its role it becomes relevant only in so as it produces inscriptions. Only these are relevant to and define its importance for the enterprise of science. The triad 'world/apparatus/inscription' is replaced by a dyad, 'apparatus/inscription.' An apparatus is something that produces inscriptions, like a ticker-tape machine.

There are two obvious difficulties with this account. It does not seem to me to differ in any substantial way from the logicist account. In that too experiments are relevant only in so far as they are the source of inscriptions to be added to the text. Secondly, if inscription producing were all that apparatus is good for, then surely any material set-up that produced inscriptions would do! By what criteria do we reject De La Warr's boxes[1] and accept Wilson's cloud chamber as genuine scientific equipment?

At the beginning of its definition the "thing" is a *score list* for a series of trials. . . the 'things' behind the scientific texts are all defined by their performances [and are similar to the heroes of folk tales]. Thus the thing is *not identified or picked out* by its performance: that is a cloud chamber and it makes droplets, and that is a Geiger counter and it makes clicks' (Latour 1987, p. 89). The thing is *defined* by what it does.

That is not yet all. As an ally in a scientific dispute, the cloud chamber of metal and glass dissolves not into lines of droplets, but into a visual display of subatomic particles, in short, a kind of statement. Similarly, the Geiger counter, having dissolved into clicks, dissolves even further into part of an auditory display of a wave-like distribution of particles. This raises a nice point, since the cloud chamber favours a particle interpretation and an array of Geiger counters favours the wave picture. What is the relationship between the displays? It cannot just be that one is an auditory display and the other visual. Why should they have anything to do with one another?

That this is a constructionist account is confirmed by Latour's third rule of method (Latour 1987, p. 99): 'since the settlement of a controversy is the *cause* of Nature's representation not the consequence, we *can never use the outcome - Nature - to explain how and why a controversy has been settled.*' 'Nature's representation', something propositional and abstract, is equated with 'Nature' something concrete and material. This follows inexorably if we are willing to accept that it is the use of apparatus that creates *Nature* by displaying readable representations of *nature.*

The final step in assimilating the material and the propositional is Latour's special notion of the 'black box.' The story of Herr Diesel and his engine (Latour 1987, p. 105) provides the setting. For the use of the engine to spread through the world, it must be possible to treat it wholly in terms of its performance as a source of motive power. Those who use the engine need not be forever worrying about the fine details of how it works. However, to get these fine details right required the recruitment of a great many 'allies' to the project. At times these allies fell away

---

[1] De La Wan produced diagnostic 'machines' that were made of odds and ends. They were shown to be worthless.

and the engine's fate was uncertain. Only as a ***unified whole*** with all the components working properly did it make its way into the practical world. According to Latour a 'black box' exists when many elements are brought together to act as one.

This step having been taken the reconstitution the formerly material thing, the apparatus, as text seems to be a natural step. As Latour says, 'no distinction has been made [by him] between what is called a "scientific fact" [a proposition] and what is called a "technical object" or "artifact"' (Latour 1987, p. 131).

There is only one difference between colleagues as allies and machines as allies in settling scientific controversies, according to Latour. It is easier to see that the gathered resources are made to act as one unified whole in the case of machines and pieces of hardware than in the case of colleagues and communities and runs of learned journals.

Latour has attended to the importance of practical skills, in such matters as 'making the equipment work.' However, these appear as bargaining encounters in social competition for community hegemony. Such skills are defined in a quite complex way with respect to the material nature of experimental equipment and the tasks of technicians in using it to display what it is expected to display, and sometimes does not. Failures are not always due to incompetent manipulations but to the intransigence of Nature which, as integrated with *this* apparatus, is other than the scientific community thought it to be. In Latour's treatment, it is hard to see that any distinction between these *kinds* of failures can be made.

No doubt we get the *idea* of what Nature is like from experiments. However, it is not that *idea* that is integrated with a bit of itself in a laboratory apparatus. Despite attention to the real world of the laboratory, Latour (and others) do not clearly distinguish the role of people talking and writing and so producing 'science', in the sense of a community discourse from the role of people making and manipulating material things and using them. The former are manipulating an idea or representation of Nature, the latter are manipulating Nature. It is evident that Latour does not treat these as different. He is led by the illuminating power of his metaphors, especially 'ally', to assume an identity. Nevertheless, the fact that a metaphor 'sticks' does not justify taking the similarity of source and subject that the metaphor makes visible to be an identity.

2. WHAT IS APPARATUS?

2.1. The Instrumentarium

I owe this useful phrase to Robert Ackermann (1985). Each laboratory has its characteristic *instrumentarium,* the actual equipment available to an experimenter. Depending on the generosity of the budget the *instrumentarium* will consist not only of what is in the storeroom, but also what is to be found for purchase in the catalogues of instrument makers.

There seem to me to be two main groups of philosophical questions that the use of apparatus raises, once logicism and constructionism are set aside.

1. What is the ontological status of a piece of laboratory equipment? Is it part of the material world, or can it be treated as if it were detached from or outside the material world, a probe that is affected by but does not affect that which it samples?

2. What is the epistemological status of the induced states of an apparatus? In particular, what can we 'back infer' about Nature from those states?

I will show that the answers to (2) depend on the category of the equipment, apparatus or instrument. To that end, I will develop a classification scheme in some detail. It is unlikely that a comprehensive survey and analysis of all that is called 'apparatus', 'instruments' or 'equipment' would reveal a common essence. What follows is a preliminary effort at a taxonomy setting up some generic categories, based on an analysis of laboratory equipment world relationships.

The broadest distinction, which gives us distinct families of laboratory equipment, is between *instruments* some relevant states of which are causally related to some feature of the world in a reliable way, and *apparatus* that is not so related because it is serving as a working model of some part of the world. Causal relations relevant to apparatus are within the model system. Causal relations relevant to instruments link the equipment to the world.

Among the most important genera of apparatus are working, bench-top models of natural processes and the material systems in which they occur. Allied to these are computer generated models of natural systems.

However, there is another genus of apparatus of great interest in contemporary physics. Apparatus, as conceived by Niels Bohr, does not model the production of naturally occurring phenomena as a discharge tube might. It creates phenomena that do not occur in nature in the absence of the apparatus. I will offer a detailed analysis of Bohrian apparatus below.

2. 2. Apparatus as Models of the Systems in the World

2 . 2. 1. Material Models as Domesticated Versions of Natural Systems

An apparatus of this genus is a domesticated and simplified version of a material set-up, which has two main features of interest to us.

a. It is found in the wild, that is occurs in Nature in the absence of human beings and their constructions and interventions. For example, a model of a ferro-concrete structure would not fall into this genus, however useful it might prove in architecture.

b. The feral set-up is such that certain phenomena can be perceived, seen, heard, tasted and so on.

The apparatus is a material model of the naturally occurring material set-up. I shall use the metaphor of 'domestication' to explore the relation between apparatus as model and that of which it is a model. The history of experimental science offers us a rich catalogue of apparatus as domesticated material systems.

Let me illustrate this genus of models with some examples.

Example i: Theodoric of Friborg set up a rack of water-filled spherical flasks as an apparatus to study the formation and geometry of the rainbow. It is not too fanciful to think of his rack as a domesticated version of the curtain of raindrops implicated in the coming to be of a rainbow. The drops replace each other sufficiently quickly in the falling shower that they can be considered as if they were a fixed array. If certain conditions on manipulability of the domestic version are met, for example finding a moveable light source to simulate the sun, the whole set-up makes possible an experimental laboratory study of the rainbow. The rack of water filled flasks *is* a curtain of spherical drops, which is *like* the curtain of spherical drops in the naturally occurring shower of rain.

Example ii: A drosophila colony is a domesticated version of an orchard replete with a breeding population of fruit flies, which display variation by selection. If certain conditions on manipulability of the laboratory colony are met, it makes possible the experimental, laboratory study of inheritance.

Example iii. An Atwood's machine is a domesticated version of a cliff down which a stone falls or a Leaning Tower from which objects can be dropped. The machine consists of a graduated vertical column with various moveable attachments, allowing for the releasing of standard weights from different heights, and the determination of the locations of the falling masses at different times.

Example iv. A tokomac is a domesticated version of a star. A powerful magnetic field confines hydrogen atoms in small volume, fusion to helium being ignited by an external energy source. The process so set going is a domesticated version of stellar fusion.

A useful image with which to explore the metaphor of apparatus as domesticated versions of feral originals could be farmyard creatures. A cow is a domesticated version of the auroc, which was found in the wild and which did give milk. However, cows are more tractable than aurocs. One notes that life on the farm is simpler than in the wild. There is abundant fodder, and there are no predators. Such a life is lived with more regular and less extreme mental and bodily states than life in the wild. Hunger and fear are rare in the farmyard. Not only is a cow spared the anxieties of feral living but it has been bred to be docile. Furthermore, unlike its feral relatives, it is therefore easily manipulable, for example, it will stand patiently to be milked.

Domesticated versions of material set-ups and processes that occur in feral form in Nature, versions that we know as experimental apparatus and procedures, are, relative to their feral ancestors, simpler, more regular and more manipulable. The drosophila colony in the laboratory is a simpler bio-system, with more regular life patterns and is more manipulable than the swarms of flies in the apple orchard.

Things can happen in the domesticated model world that do not happen in the region of nature that is the source of the model. For example strange variants of the insect appear and can even be maintained as living examples of mutations that would either not occur at all in the wild or be immediately eliminated.

2. 2. 2. Back Inference from 'Domesticated' Models to the World

Domestication permits strong back inference to the wild, since the same kind of material systems and phenomena occur in the wild and in domestication. An apparatus, of this sort, is a piece of Nature in the laboratory. Of course, the richness of back inference will depend on how relations of similarity and difference are weighted by the interests of the researcher in performing the experimental manipulations.

There is no ontological disparity between apparatus and the natural set-up. The choice of apparatus and procedure guarantees this identity, since the apparatus is a version of the naturally occurring phenomenon and the material set-up in which it occurs. Theodoric's apparatus brings a rainbow to be by refraction and internal reflection in spherical volumes of water, just as the curtain of raindrops does. So, whatever can be learned about the paths of rays of light in the laboratory can be back inferred to the wild, to Nature.

A weight falls in an Atwood's machine just as a stone falls from a cliff or an iron ball falls from the Tower of Pisa. What we study in the laboratory by the use of this apparatus is not the effect of a causal relation between a state of Nature and the corresponding state of an apparatus. It is a simplified version of the phenomenon itself.

Galileo's inclined plane experiment is, in some respects, an intermediate ease. In order to use the results of times and distance of descent as a test of his hypothesis of free fall, he could not treat the ball rolling down the plane as just a domesticated version of free fall. He had to perform a mental operation on his results, effectively resolving the inclined motion into a vertical and a horizontal component. The back inference required this intermediate step.

2. 2. 3. Apparatus-world Complexes and the Production of Phenomena

The analysis offered above does not seem to fit the case of the Wilson Cloud Chamber, the Stern-Gerlach apparatus and many other well-known pieces of laboratory equipment. At first sight, it might seem that these belong in the causal family. We might be tempted to say that the lines of droplets in the cloud chamber are the effects of the ionization produced by the passage of electrons. Yet, are there *electrons as moving particles* in the absence of this kind of apparatus? Are there *electrons as interfering wave fronts* in the absence of double slits and photosensitive screens? There is molecular motion in the absence of thermometers and the roughly global earth exists in the absence of cartographic surveys.

An apparatus is a piece of junk until it is integrated into a unitary entity by fusion with Nature. A retort exhibited in a museum is not an apparatus. Let us call the apparatus/world complexes which scientists, engineers, gardeners and cooks bring into being Bohrian artifacts. Properly manipulated they bring into existence phenomena that do not exist as such in the wild, that is in nature. In general, there is no material structure in nature like the apparatus. Ice cream does not occur in nature, only in kitchens with refrigerators.

In the famous Bohr-Einstein debate around the EPR paradox, it is possible to see the outlines of Bohr's account of experimental physics. While Einstein is insisting that for every distinct symbol in a theoretical discourse there must be a corresponding state in the world (logical atomism under another name). Bohr (1958) is concerned with the concrete apparatus, and its relation to the world, *as part of the world.* An apparatus is not something transcendent to the world, outside it, interacting causally with Nature. That is the role of the instrument. The apparatus and the neighbouring part of the world in which it is embedded constitute *one thing.*

Bohr realized that the seeming ontological paradoxes of subatomic physics could be resolved by taking a right view of experimental apparatus. It is possible to see, for example, how Nature can yield both particles and waves, by treating particle *phenomena* and wave *phenomena* as products of the running of different apparatus world complexes. Particles and waves are phenomena that occur in such complexes. They do not occur in nature.

Bohr's philosophy of experimentation was misconstrued as some kind of positivism. It was never Bohr's intention to argue, as Mach had argued that science was the study of the properties of apparatus. In an experiment, what was 'run' was not just the apparatus. Nor was Bohr advocating a straight-forward realist interpretation. Physicists could not treat the apparatus in this class of experiments as a transparent window through which to see the world as it would have existed had the apparatus never been constructed and switched on. Science was the study of the apparatus/world complexes. Neither component could be detached from the reality which produced phenomena.

The laboratory is full of equipment, apparatus, drawn from the local instrumentarium. People are setting it up and making it work, and so bringing phenomena into being. In some cases, the apparatus is a materially independent entity, with all relevant causal processes entirely internal to it, for instance the experimental drosophila colony. Its relation to the world is analogy. In Bohrian experiments, the apparatus is indissolubly melded with the world. In that case, the phenomena are properties of the apparatus/world complex. It is materially part of the world. The question of whether there are natural set-ups like the Bohrian complexes we have constructed, such as tokomacs, is germane to this class of apparatus as it is to the simpler, materially independent models, such as Theodoric's flasks. The flasks contain spherical masses of water. That is what rain-drops are.

Therefore, we have two matters to examine. There is the issue of the nature of the apparatus as a constructed material object in relation to a naturally occurring material system. We must also examine the nature of the phenomena created by using it. These may be states of the apparatus that conic into being as effects of causal processes in the world. On the other hand, they may be phenomena that are brought into being by running the apparatus as a model of some material system. If the apparatus is a model of something in the world, we can ask what is the relation between the phenomena we produce in the apparatus and those which occur naturally. It is well to bear in mind that apparatus has properties before it is switched on, heated up or otherwise manipulated.

In thinking through the meaning of the products of experimental activity it is important to keep in mind that the phenomena generated by experimenting with Bohr-type apparatus are properties of a complex unity, the apparatus/world entity. Bohr was driven to this insight by the duality of types of quantum phenomena, but the point is quite general. In this chapter, I shall use Humphrey Davy's isolation of sodium in the metallic state by electrolysis as my prime exemplar of Bohrian experimentation. As far as I know, there is no set-up similar to Davy's equipment anywhere in the universe. Free metallic sodium exists only by virtue of the apparatus/world complex Davy built. Humphrey Davy used electrolysis on molten common salt in a crucible to bring metallic sodium to light. Think of how much is presupposed in describing this experiment as the 'discovery of sodium' or as the 'extraction of sodium.' There is no metallic sodium in the universe to the best of my belief Sodium-as-a metal is a Bohrian phenomenon.

This experiment contrasts sharply with Faraday's use of a tube of rarefied gas to study discharge phenomena. A similar set-up to the apparatus existed 'in nature', in the electron wind in the rarefied upper atmosphere. Therefore, we can understand the glow in the laboratory tube as an analogue, in a domesticated version of the upper atmosphere of the *aurora borealis.*

Just as the cow and the auroc can serve as a metaphor for the relation of apparatus to the world, so the homely image of a loaf of bread can serve as a metaphor for the Bohrian apparatus/world complex. A loaf is brought into existence from wheat and other ingredients by the use of material structures that do not exist in the wild, such as flour mills and ovens. Loaves do not appear spontaneously in nature.

At the Cern Laboratories a huge apparatus/world complex brings certain tracks into existence, which simplifying, we could imagine are recorded in photosensitive plates. One such set of tracks was lauded as the 'discovery of the W particle'. There are probably no free W particles in the universe now. They are exchange particles, intermediate vector bosons, postulated in quantum field theory. They are wrenched into momentary isolation at Cern. The pattern of reasoning that lies behind the 'discovery of the W particle' seems to have been something like this: photons can be studied in the propagation of light, and they also play a role as exchange or virtual particles in quantum field theory. So we have the idea of the free version of the virtual exchange particle. The W particle was introduced to physics as the virtual exchange particle for a certain class of interactions. By parity of reasoning there should be a free W particle, analogous to the free photon.

Is there an analogy between the discovery of the W particle and the isolation of metallic sodium? I think that few chemists would interpret the Na atom as a virtual constituent of common salt. So, Davy's experiment brought metallic sodium to light by aggregating enough pre-existing Na atoms. Reflecting on the possible analogy of this ease to that of the W particle we can see that what would be at issue of the analogy were to be taken seriously is the ontological status of virtual particles in quantum field theory. This is a deep issue which cannot be gone into in this discussion, except to point out that the virtual W particle is a representation of just one of the exchange modes possible in any relevant particle interaction. It is hard to make a case for a pre-existing particle in any exchange process (Brown and Harré 1990).

2. 2. 4. Back inference from Phenomena to Nature

Back inference from phenomena created in Bohrian artifacts is problematic, since there is an ontological question to be solved. What is the standing of the apparatus/world complex in relation to the world, to Nature?

The general form of this question seems to be whether it is legitimate to analyze the situation in Aristotelian terms (cf. Wallace 1996), that is in the principle: An actual phenomenon produced in an apparatus is the manifestation of a potentiality in the world. This would allow a back inference that would simply ascribe a natural propensity or potentiality for whatever occupies some region of Nature in contact or casual connection with the apparatus to appear as the experimental phenomenon. This sounds as if we could say 'an apparatus makes actual in the laboratory that which is potential in nature'. This still treats apparatus as a kind of 'window on the world'.

It ignores the contribution of the apparatus to the form and qualities of the phenomenon. Reflecting on this issue takes us deeper into the Bohr interpretation. The Bohrian phenomena are neither properties of the apparatus nor properties of the world that are elicited *by* the apparatus. They are properties of a novel kind of entity, an indissoluble union of apparatus and world, the apparatus/world complex.

This makes the question 'In what form does metallic sodium exist before the electrolysis begins?' illegitimate. Nature, in conjunction with Davy's apparatus, affords metallic sodium, just as Nature, in conjunction with Wilson's apparatus, affords tracks, and thereby affords electrons as particles. By parity of reasoning, the question 'in what form do electrons as particles exist before the cloud chamber is activated?' is equally illegitimate.

To follow this line of analysis further would take us into the metaphysics of powers, dispositions and affordances, the neo-Aristotelian metaphysics of physics implicit in the writings of Nancy Cartwright (1989) and explicit in the recent work of William Wallace (1996). I will return to this issue in the final section.

2. 3. Instruments in causal relation to the world

The distinction between apparatus and instruments is vital to an understanding of how knowledge is acquired in laboratories, and what sort of knowledge it is. In many discussions of the nature of experiments it is simply assumed that the state of an instrument is an effect of an independently existing state of the world. In the ideal experiment the producing of the effect in equipment, the instrument, does not change the state of the world of which the state of the instrument is an effect. Sometimes the thermometer requires so much heat to expand the mercury that the liquid being studied cools down substantially. Skilled experimenters know how to compensate for these exceptions. Usually the pressure in the car tires is not significantly reduced by the amount needed to activate the tire gauge.

2. 3. 1. Kinds of Instruments

There are two main genera of 'causal' instruments. They can be differentiated by the use of an extended version of the old but useful distinction between primary and secondary qualities.

Ideas of primary qualities are those, which, as apprehended by a sensitive organism, resemble the state of the material entity that caused the experience. For example, 'shape' is a property of a material thing whether or not it is being observed. It is experienced by an observer as a shape, according to some rule of projection.

Ideas of secondary qualities are those conscious states of an organism that do not resemble the states of the material entity that induce them. Rapidity of molecular motion is experienced not as motion but as heat.

A distinction somewhat similar to that between primary and secondary qualities and the corresponding experiences or 'ideas' can be used to identify the two main relations that the states of an instrument can bear to the states of the world which caused them. Thus, the image on a photograph of a spiral nebula is a spiral shape, a projection of the shape of the nebula. Shape is a primary quality, since the representation in the photograph is analogous to shape as experienced by a conscious being and both are of the same nature as the spatial structure of the nebula itself. The energy of molecular motion in a material entity results in a change in the length of a mercury column in a thermometer. The length of a mercury column in thermometers is something like an 'idea of a secondary quality', representing the energy of motion that caused it but not resembling that motion.

Some instruments are calibrated so that they yield numerical measures of qualities that can occur in different degrees. Other instruments simply detect the presence of some natural state, process or entity. For example litmus paper detects the presence of acidity, free hydrogen ions, without indicating the pH or strength of the acidity.

2.3.2. Back Inferences from States of Causally Based Instruments

It may seem at first sight that inferences from 'primary' qualities of instruments to the qualities that cause them via various manipulations must be based on a different principle from the back inferences from 'secondary' quality detectors and measurers. I would contend, however, that the principles are essentially the same. Both kinds of instruments depend on the reliability and verisimilitude of the relevant causal relations, the evidence for which can be found *somewhere else in physics and chemistry*. Change in internal energy of a sample of gas is a change in the root mean square velocity of the constituent molecules, which causes a change in the motion of the molecules in the mercury column. This motion causes the column to expand. This is a fact in physics, if it is a fact at all. A similar but longer chain of causes links an infection conceived as the invasion of an organism by viruses, to the presence of antibodies, which can be chemically identified. This is a fact of biochemistry, if it is a fact at all.

2. 4. A Taxonomy for Laboratory Equipment

2.4.1. The Basic Classification

Laid out as a tree diagram the classification of the items in an instrumentarium might look something like this:

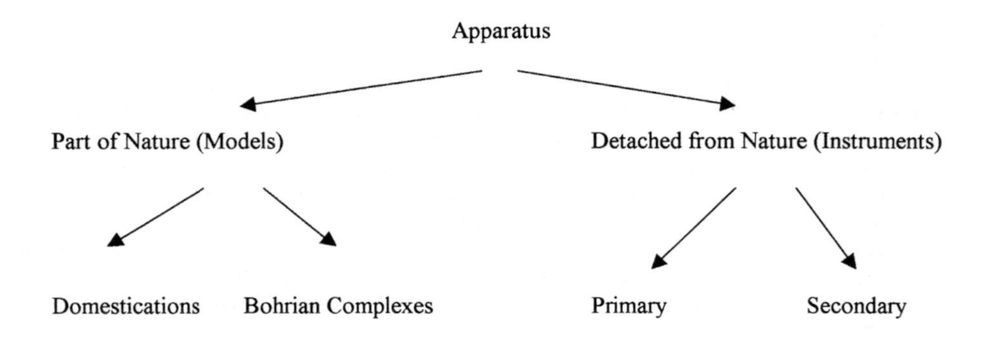

Figure 1
An Instrumentarium

2.4.2. Two Kinds of Back Inference Schemata

The schemata by means of which legitimate back inferences can be made from the state of laboratory equipment to the relevant state of Nature are quite different in the two families of equipment, apparatus as models, be they domestications or Bohrian, and instruments.

For apparatus as models, the back inference schema is not causal. It is ontological. The relevant state of the model is not caused by some state of Nature. The relation of model to Nature is analogy within the framework of a qualified ontological identity. Showers of rain and racks of flasks are both subtypes of the ontological supertype curtains of spherical water drops'. A Bohrian apparatus is a part of the material world with its characteristic products. So there is no question of seeking a ground for back inferences elsewhere in physics or chemistry.

For instruments, as I have outlined above, back inferences are based simply and directly on some causal laws already established with whatever degree of certitude the relevant branches of physics and chemistry admit. These may include quite theoretical matters with their own foundations in models. For instance, there is a causal law linking molecular motion in a sample to the extension or expansion of the substance used in the detector. There is a law linking the vibrations of electrons in hot atoms to the spectral lines detected in a spectrometer.

3. TOWARDS AN ONTOLOGY OF A BOHRIAN THEORY OF EXPERIMENTATION

A superficial glance at the situation would suggest that attributions based on experimenting with apparatus/world complexes would be conditional: If such and such a material set-up is created and activated, then it will display certain phenomena. The Bohrian interpretation of experimental physics suggests an ontology of dispositions. There is something right about this intuition, but it is too simplistic to do duty to the subtlety of Bohrian experimentation. To prepare the way for an understanding of what Bohrian experimentation can disclose I will begin by taking a general look at the metaphysical foundations of physics and chemistry.

3. 1. The Metaphysics of Physics

The success of the corpuscularian metaphysics in the modern period, particularly in inorganic chemistry and the physics of gases, great triumphs of the nineteenth century, supported heroic efforts to generalize this ontology and the grammar in which it was immanent, across the board. There were held to be a myriad of material things, strictly located in space, with continuous temporal existence, and having clusters of essential properties. In different versions, Descartes and Newton shared this metaphysical vision. This metaphysics was naturally realized in the use of subject/predicate grammar to describe the natural world.

However, there were important dissidents, especially Alibis, Bucolic, Kant and Faraday. They espoused versions of an ontology of forces and active powers, distributed throughout space, and structured around charges and poles taken as the sources of the field. This was the metaphysical basis of the field ontology. Trying to work with the presuppositions of the substance-attribute conceptual framework in setting up the science of field physics around strains in a substantival ether, came to seem unnatural. Instead, we have intimations of new ontology, of potentials at space-time points related to charges and poles. These are not some strange new kind of entity. There are no material entities in this metaphysics. Fields are distributions of dispositions through which the powers of charges and poles are manifested.

Which ontology shall we choose?

There are the familiar problems with justifying claims about imperceptible substances with imperceptible properties that are characteristic of the ontology of the traditional mode of theorizing by invoking hidden mechanisms constituted of localizable and enduring or quasi-enduring particles. How are alternative accounts of the hidden realm to be decided among? There are difficulties in understanding the role of apparatus in Bohrian apparatus/world complexes in the genesis of knowledge if we persist in thinking in terms of substances and their occurrent properties. It seems quite natural to think in this way about the metaphysics implicit in the use of instruments and for many of the model worlds realized in laboratory apparatus.

3. 2. The Structure of Field Thinking

Starting with the metaphysics common to field theories from Gilbert's *De Magnete* to Faraday's *Experimental Researches* the story has been something like this:

Task	Ontology
Descriptions of phenomena	Eclectic
Explanations of observable phenomena	Dispositions
Explanations of dispositions, etc.	Causal powers

Table 1

Philosophers and logicians have put a huge effort into developing only of one of the possible options available in the eclectic grammars of ordinary life, namely substances and their occurrent properties. This privileges subject/predicate sentence structures. After Descartes, the attributes in focus became more or less restricted to extensive magnitudes. Physics, however, continued to develop its own version of the dispositions and powers scheme, implicit in the mathematical analyses of the field concept. I contend that the adoption of that scheme will help reveal the role of apparatus in a much clearer way than trying to do re-jig the substance/occurrent property scheme yet again. The exigencies of field physics demands conditional sentence forms.

It is worth remarking that the substantive content of a dispositional attribution, expressed in the usual conditional form consists of observables only. There are various truth-functions of observables, for example 'p & q', 'p v q' and so on. The conditional form by means of which we express simple dispositional attributions, p $\rightarrow$ *q*. is just another one. Dispositional ascriptions have no explanatory value. However, it is here that the distinction between occurrent and non-occurrent properties gets its first outing. A disposition is not an occurrent property. A substance can be properly said to have a certain dispositions when it is not manifesting it. Thus we say aspirin is an analgesic when the tablet is still in the bottle.

3. 2. 1. Powers and the Concept of Energy

Dynamicist ontologies, that is metaphysical schemes based on activity concepts such as 'force', 'energy,' and so on, as used by Leibniz, Boscovich, Kant, and Faraday and modern field theorists, offer explanations of the same general form as corpuscularian explanations. They invoke unobservables to account for observable phenomena. However, unlike the corpuscularian unobservables, which are potential perceptibles, the unobservables of the dynamicist ontology are imperceptibles. Forces are manifested in their effects. They are not observable as such. The visualisability constraints in theoretical thinking that seem so natural within the corpuscularian tradition are simply not relevant for dynamicists.

The simplest dynamicist layout uses imperceptible energy transformations to explain perceptible accelerations, via the relevant field potentials, that is as described dispositionally. The structure is tightly bound together conceptually since the dispositional attributions ascribe tendencies to accelerate to test bodies located at specific points and moments in the field in question. The content of a dispositional attribution is wholly observable, indeed perceptible, but as properties, dispositions are not occurrent. For example, a gravitational field is described at the level of observables by the spatio-temporal distribution of dispositions of test bodies to accelerate, while it is described at the level of unobservables in terms of the distribution of potential energy. In mechanics, 'energy' is neither a substance nor a disposition, but a power which accounts for the existence of dispositions. Energy is the power to do work.

I have shown elsewhere how the dynamicist scheme can be generalized to quantum field theory (Harré 1990). However, to do that a new kind of dispositional concept must be introduced. This is the affordance, which will prove indispensable in making sense of the role of Bohrian apparatus.

3. 2. 2. Affordances

Gibson's concept of 'affordance', which he deployed to such effect in his account of the dynamics of perception (Gibson 1979), can serve a similar purpose in the analysis of experiments. An experiment shows people what a particular apparatus/world complex can afford people who are skilled enough to use it. Of course, any piece of matter can afford all sorts of things. Ice affords walking, skating, cooling drinks, and so on. Which affordances are invoked depends on what human beings are proposing to do.

Gibson introduced the concept of an affordance to distinguish a certain class of dispositions. The general form of the ascription of a dispositional property to something is conditional: 'If certain conditions obtain then a certain phenomenon will (probably) occur'. In many cases the outcome of activating a disposition does not depend on any particular human situation, interest or construction. However, in some cases, the phenomenon has a specifically human aspect. Compare the generic outcome that ice of a certain thickness can bear a certain weight per unit area, expressed in a generic disposition, with the claim that ice of that thickness affords walking for a person.

Generalizing the notion of an affordances we can say that an apparatus/world complex can afford *things*. For instance, wheat, yeast and a stove can afford loaves of bread. A lathe can afford chair legs, and a discharge tube can afford gamma rays.

An apparatus/world complex can also afford activities. For instance, some rapids can afford white water rafting. A reamer can afford boring and a chemistry laboratory can afford gravimetric analyses.

In both groups of cases, what is afforded would not have existed without human action to bring it into being. Thus, ice affords walking only if there are actual and potential walkers. The heterogeneity of the first group shows up clearly, in that while there are no loaves of bread or perfectly regular cylindrical lengths of wood in Nature, there are gamma rays. However, even that is not quite right, since there is fermentation and there are fairly regular cylindrical branches on some trees. Perhaps the gamma rays in the Cavendish laboratory were not quite the same in type as those from a sunburst.

Back inference from Bohrian artifacts to the attribution of affordances never escapes wholly from the human domain of material constructions. However, it is not confined to that domain. The Bohrian artifact is a hybrid being, part construction, and part Nature. The phenomena that are produced in activated Bohrian material systems are the manifestations of affordances. These are dispositions that bring together two sets of causal powers that cannot be disentangled. There are the powers of the material stuff organized as an apparatus, and the powers of the world realized in the phenomena. What can we say? Bohrian apparatus displays the affordances of the world, relative to that apparatus. At the deeper layers of scientific work there are no transparent windows on the world, as it would he, did the apparatus not exist.

4. CONCLUSION

It has been important to distinguish two broad families of laboratory equipment. The first family consists of domesticated versions of material set-ups in Nature and Bohrian artifacts. The second family consists of instruments that yield knowledge of Nature by virtue of causal relations with states of the world.

The first family of equipment, apparatus as part of Nature but isolated in the laboratory, can serve as a springboard for back inferences from the laboratory to the world outside the laboratory. Simple modelling depends on and requires only simple ontological identity, of subtypes under a common supertype. We are still in an extension of the everyday world of substances and their occurrent properties. In Bohrian modeling, the apparatus is also part of Nature. However, unlike simple models, a Bohrian apparatus cannot be isolated from Nature, since it is indissolubly melded with it. The interpretation of the states of Bohrian apparatus, phenomena, requires a shift in ontology from Newtonian occurrent properties to dispositions and powers. These dispositions are affordances and permit limited inferences from what is displayed in Bohrian artifacts to the causal powers of Nature.

The various pieces of equipment that one finds in the second family, instruments in causal interaction with Nature may throw up problems of the justification of back inferences from the state of the instrument to the presumed state of the world. However, these are in principle resolvable by doing more physics and chemistry, taking for granted the verisimilitude of the reactions of the instruments used for that subsidiary work.

CHAPTER TWELVE

APPARATUS AS MODELS IN THE PHYSICAL SCIENCES

The current popularity of anti-science has many roots and takes many forms. But there is one for which we ourselves, as philosophers of science, have been in part responsible: the dethroning of the experiment as the touchstone of truth. Again there are several ways in which this has happened, with different putative effects. I want to focus on those that could roughly be grouped together as 'post-modern'.

It seems to me that the authors of the post-modernist account of science, from Fleck to Latour, have been animated, in part, by the need to account for the disparity between the weak *logical* power of empirical propositions reporting the results of experiments and observations with respect to the assessment of the belief-worthiness of hypotheses and theories and the actual effectiveness of descriptions and observations in forming and changing people's beliefs, *their persuasive power*. How is this gap to be filled? In order to assess the post-modernist attempts to close it, and perhaps to supersede them, it will be necessary to set out the classical account of the experiment in some detail. I shall set out this account as simply as possible, using the Aristotelian square of opposition as a format.

1. PROBLEMS FOR SCIENCE AS DISCOURSE

Traditionally the problems that have attracted philosophical attention have been centered around the relation between general and/or theoretical hypotheses and descriptions of experiments and reports of observations, rather than around the relation between experimenting as a material practice and the world. The traditional scepticism that infects all discursive accounts of the empirical aspects of science could be summed up as follows:

The internal logical relations of the discourse of science are not strong enough to support the knowledge claims made on its behalf. Confirmation alone no more rationally enforces acceptance of a hypothesis than falsification alone rationally enforces rejection. So, by the standards of the logicism of the Enlightenment, scientific methods, as the privileging of the results of experiments and observations, are flawed. How is the gap between the 'logic' of the square of opposition and the role of experimental reports in supporting or undermining the beliefs of scientists to be closed?

The external semantic relations of the discourse of science to the world are equally problematic. Notoriously, there is a need for a correspondence account of truth which has never been met, while 'degrees of truth' remains a stubbornly intractable notion.

Post-modernism offered resolutions to both problems. The explanation of the way beliefs are strengthened or weakened is to be found in features of discourse other than its logical structure, for instance its rhetorical tropes and the social positions of those who speak and write it. The second, and even more thrilling post-modern critique of the Enlightenment project, was the assertion that Reality, the World, does not enforce belief via scientific method, since it is not an essential component in experiments. This point was not so much that experiments and theories are internally related, a claim derived from a strong generalization of the theory ladenness of the observations and readings of the states of experimental equipment, but that the World, 'Reality' is not an essential component of the working of the apparatus or the making of observations.

Both of the problems of the traditional account are resolved. The gap between logic and belief is filled with literary theory and sociology. The failure to find an acceptable account of the discourse to World relation is by-passed since the World has no part to play in science.

1.1. Debatable Assumptions behind the Neglect of the Role of Experimental Apparatus

Why is the material practice of experimenting and the needful apparatus 'bracketed' in the classical account? For the purposes of the discussion to follow the answer can be found in tacit adherence to two principles:

THE PRINCIPLE OF TRANSPARENCY: experiments reveal some aspect of the natural World, as it is, *ceteris paribus.* It follows that the analysis of experimental procedure itself can be omitted from the epistemology of natural science.
THE PRINCIPLE OF ACTUALISM: experiments reveal occurrent aspects of the natural World. It follows that the results of experiments can be described in indicative propositions.

Though both of these assumptions have been queried in discussions that seem to me to have been independent of the advent of post-modernism in philosophy of science, as seen from the perspective of post-modernism, they represent the Achilles heel of the classical account.[1]

1.2. The Principle of Transparency

The Principle of Transparency entails a Detachment Corollary, that phenomena can be described by what, at first sight, appears to be a description independent of the apparatus that gives us access to them. Take a very simple experiment: to use a mercury thermometer to measure the boiling point of a liquid, say alcohol. The State of the liquid causes the thermometer to come into a certain state, which is then taken as the measure of the state of the liquid. To make this move we must assume a tight and tidy causal relation, from World state to apparatus state. But the apparatus is itself a material thing and part of the World. It is assumed that the tight and tidy causal relation can be investigated like any other causal relation. If it 'holds up' the apparatus drops out of the story. Does the World state pre-exist the thermometer state and is it independent of it? The tight and tidy relation is causal, and causes are materially and conceptually independent of their effects. The apparatus, in this case the thermometer, is transparent, and provided care is taken to allow for the differential expansion of the glass, and that parallax errors are eliminated from the taking of the reading, and so on, the state of the thermometer represents the state of the liquid. Note that it is assumed that there might be, and students of calorimetry worked to provide, other ways of representing the same state of the liquid. This makes sense only under the Detachment Corollary. If criteria of identity for states of the liquid were derived from the nature of the measuring apparatus then they would not be measuring the same state.

---

[1] The Principle of Transparency has been brought into question in popular discussions. However the Principle of Actualism came under attack as early as R. Bhaskar (1978), an analysis of experimental activity in terms of a distinction between the actual and the real, the latter consisting of natural powers and tendencies. More recently N. Cartwright (1989) develops an argument in favour of natural tendencies and can also be read as opposed to actualism, though it is not exorcised as a critique of experiment. Ironically it will emerge that one of the taken-for-granted assumptions that 'powers' post-modernist critiques of the experiment is adherence to a form of actualism. This was perhaps the most serious problem that operationalism encountered.

1.3. The Principle of Actualism

Traditionally, science has been taken to be about actual states of the World. For both Hume and Mach laws of nature are summaries of repeated sequences of actual states.

The Principle of Actualism expresses a choice of ontology. The main alternative realist ontology of tendencies, dispositions and powers has an equally long history. If we take experiments to be attempts to find out the powers of material things, then experimental results are not adequately analysed in an actualist framework. But the classical account of the experiment seems to be strongly actualist. General ontological considerations are germane to the interpretation of the practice of experimenting. However, as we shall see, post-modernism involves a rejection of the Principle of Transparency while preserving a commitment to an actualist ontology.

The correspondence theory of truth fits in to the traditional point of view as the alleged link between I and 0 propositions and the relevant state of an apparatus, *and* between the aforesaid state of an apparatus and some state of the world. There are two correspondences in the traditional account. Both have been interpreted actualistically. The basic statements of a scientific discipline describe occurrent states of apparatus, say the length of a column of mercury, while the states of the apparatus are in one to one correspondence with states of the World, namely the state of motion of the molecules of the substance the temperature of which is being measured. This is in keeping with the Detachment Corollary above, and indeed could have offered as a premise from which to derive it.

1. 4. The upshot

I believe that both the Principle of Transparency and the Principle of Actualism are false as having general application to experimental natural science. But it will be my task in what follows to try to show that the anti-science position, associated with post-modernism, is equally unsatisfactory. Experiments play all sorts of roles in the natural sciences, some of which I catalogued and illustrated in my study (Harré 1981) of apparatus and its uses, and we all agree that some account of the apparatus cannot be eliminated from the description of experiments. I intend to show that neither can Nature. To show this I must set out, in some detail, the most recent line of argument in support of the post-modernist case. The recovery of the experiment will not be a return to either the Transparency Principle or to the Principle of Actualism.

The argument to be developed is intended to show that while post-modernists reject the Principle of Transparency, they still adhere to the Principle of Actualism, a strong and disputable ontological thesis. The 'third way' is based on a rejection of the metaphysics implicit in the Principle of Actualism. An alternative ontology will permit the recovery of the traditional disciplinary role of the experiment without a return to the Principle of Transparency.

2. POST-MODERNISM

I will first present some characteristic but general post-modernist theses, and then show how they lead to an equally characteristic philosophy of science.

2.1. General principles

The catalogue of theses that I will present is not exhaustive, but highlights those insights that seem to me to have been of the greatest importance in the post-modern movement as a philosophy of science. Let me set out the post-modern position first of all as it has appeared in the philosophy of the human sciences, and then modulate to the treatment of the natural sciences within this framework. Summarized rather brutally the position can be expressed in four main theses:

> THESIS ONE: CONTEXTUALITY: Every intentional act, be it linguistic or accomplished in some other symbolic medium, derives at least part of its meaning from the context in which it is produced. The boundaries of contexts are variable and negotiable. There is no universal context, which would endow intentional acts with trans-situational meanings.
>
> THESIS TWO: MULTIPLICITY: In any context there is an indefinite multiplicity of readings available for any sequence of intensional actions. No reading has precedence over any other.
>
> THESIS THREE: POSITIONALITY: Actors occupy 'positions' in relation to one another in episodes of meaningful engagement, that is, each assumes a set of rights and obligations as actor, which may or may not be conceded by the interactor. Positions are essentially contestable. Multiple contexts and multiple readings derive in part from multiple positions.
>
> THESIS FOUR: REFLEXIVITY: Any explanatory or descriptive account of any aspect of human life is itself a reading and so a fit subject for a discursive analysis, and subject to the above principles, in particular there is no universal and/or privileged account of human activities including giving accounts and readings.

It is not too difficult to reformulate these principles to deny any privilege to scientific discourse as an account of our experience of Nature. Experiments are context bound and their interpretations are legion, depending only on the relative positionings of the interpreters.

2.2. Post-modernism as a philosophy of science

At the core of the post-modernist resolution of both the internal and external sources of scepticism in the traditional situation there are two alleged insights, remedying what is evidently missing from the traditional account. Neither microsocial nor macrosocial relations and structures, nor experimental equipment and using it are mentioned in the traditional account. The post-modernist move is to introduce both. From that standpoint the traditional account is no more than an obscure way of analysing the conventions of a certain kind of agonistic discourse, of the rhetorical devices used in offensive and defensive moves in controversies.

This sounds interesting, perhaps at the level of superior gossip, but then something quite thrilling is proposed: that Nature has nothing or almost nothing to do with the assessment of the value of scientific discourse as a repository of knowledge. The experiment, which, we prided ourselves, gave us access to Nature, and which we took for granted, in the transparency of experimental *apparatus,* is reconstrued either as a rhetorical device, an ally in some local controversy, or, though less frequently, as a pilot plant to be realized, perhaps, in some large-scale industrial process.

Detailed discussions and analyses of the relation of the experimental apparatus to the material world were rare in classical philosophy of science. What is the relation between the members of one class of material things, experimental setups, and the members of another class of material thing, physical structures and processes in the (or 'an') independent world?

According to Latour (1987) and Hacking (1992), there is no such relation, or if there is, it is of marginal importance in understanding the production of scientific knowledge. The laboratory is a closed system, and outside it is the wild, forever *terra incognita*. Unlike the classical philosophers of science Latour and Hacking focus in interesting ways on the role of experimental apparatus in the generating of scientific discourse. If these insights could be sustained the outcome would be thrilling. In the absence of an apparatus to world relation, there would be alternative but a wholesale rejection of any correspondence theory of truth in all contexts, including that of the experimental results themselves. What seemed to be a transparent material relation turns out to be intratextual, social and discursive.

According to Latour, Pickering (1984) and others, even the technical standards upon which the power of an experiment to convince rests, are a function of the social power of the group performing them. Therefore, it is argued, the explanation of why some accounts are believed and become authoritative, some rejected and others are not even formulated, will have a wholly sociological-historical explanation, probably in terms of social and institutional 'power'. There is no role for the experiment as a reflection of Nature or the World in explanation of why one account is preferred to another. That certain experiments are regarded as authoritative is a status created by the members of the relevant community.

2.3. The post-modernist critique of the experiment

The post-modern critique of the experiment amounts to the seemingly thrilling claim that Nature or Reality has no place (or a very limited place[2]) in the work of the natural sciences. It amounts to an exclusion of Nature/World/Reality from philosophy of science. For example Gergen and Gergen make the position very clear in the following quote:

> [T]he confirmations (or disconfirmations) of hypotheses through research findings are achieved through social consensus, not through observation of the 'facts'. The 'empirical test' is possible because the conventions of linguistic indexing are so fully shared ('so commonsensical') that they appear to 'reflect reality' (Gergen and Gergen 1991, p. 81).

There are three versions of this position. There is the 'textuality' version, in which all is discourse. There is thus a closed cycle between empirical and theoretical discourses and apparatus and the World are themselves texts to be read with no privileges as permanent bases of the assessment of discourse in general. The second version admits a categorical difference between discourses and material setups, say in laboratories, but excludes Nature, non-human Reality, from a role either in the creation of the laboratory or in the way the apparatus works in that laboratory, though 'society', perhaps as the client for the scientific entrepreneur is not excluded. In the third version, advocated by Latour neither Nature nor Society penetrate the walls of the laboratory nor the community within. In all three versions the Principle of Transparency is implicitly rejected. But all three are strongly actualist.

2.3.1. Goodman: the Closed Cycle of Discourse

For an example of a strong actualist and post-modernist philosophy of science we can go to Goodman's several presentations of his Ways of World making argument. The 'multiple

[2] How limited is 'limited'? Hacking's account, to be discussed later, sounds thrilling because of the implication that 'limited' is vanishingly small.

versions' argument, is implicit in Goodman's earlier and well known study (1978). Here is Goodman's latest and perhaps strongest exposition of his view:

> Worlds are version dependent (that is vary with different versions) with respect not only to *what* they are but also to *that* they are. The answer to the question, 'Version and what else?' is 'Nothing.' The non-versional auxiliary, thought to be needed, is all encompassed within the versions (1996).

The difficulty of grasping Goodman's thesis is exacerbated by the immediately following qualification: 'Such version dependence does not imply that versions make their Worlds but only that they have Worlds answering to them'. But in the illustrative example we are back to making rather than answering to. If each version reveals an aspect of the World, I have no quarrel with Goodman, but as his argument unfolds it discloses a much more radical thesis. Goodman asks us to consider two versions, D and T, descriptions of the sky at night. In D we describe a section of sky with a star B at its centre. In T we assert that B is million light years away and what we see is how B looked a million years ago. T 'puts the life of B in the far past; while for D, B still sparkles.' 'T', says Goodman, is typical of versions that distribute things over time and space, and in doing so make time and space'. Neither version has priority nor subsumes the other. However, all depends on Goodman's implicit actualism. According to the alternative dynamicist ontology, B, the real star, has the power to cause us to see a light in the sky a million years after it was in a certain state. The appearances of the night sky are actual. The World is a complex structure of powers and dispositions. How the sky looks now enables us to ascribe powers to distant stars as material things. It took a lot of hard work to reach a version that conformed to the requirements of a science, capable of accounting for the widest range of appearances. The science of astronomy clearly distinguishes between the powers of things that could be displayed in observation and experiment and the actual appearances. Versions in cosmology no more 'make time and space' than do versions that one uses to explain to a child why the woodcutters axe on the other side of the Mississippi is seen to strike the log before it is heard to do so.

If the project is explanatory, that is, is a part of natural science, then version T subsumes version D. But notice that version D is necessary for version T since version D is nothing other than the observational/experimental basis for version T. The grammar of 'D-type versions' includes dispositional concepts, while the grammar of 'T-type versions' includes causal power concepts. A 'T/D' science is created by a supergrammar in which the explanatory relation is created by the generic ontological (grammatical) relation 'powers/dispositions'. Goodman's refusal to privilege the 'scientific version' is surely a consequence of prior adherence to an actualist ontology. In the dynamicist framework T subsumes D, while in the actualist framework D is robust and T scarcely more than a sci-fi speculation.

In his recent article, cited above, Goodman tries to make time version dependent, and thus defend and strengthen the 'Versions make Worlds' thesis. In his example there are three stellar explosions, say K, L and M. According to one version, V_1 the cosmologist's, they occur in the order L > M > K. According to an observer on the earth, V_2, they are seen in the order M > K > L. According to Goodman V_1 and V_2 are both true and so they are of different Worlds if of any. But this is absurd. There is one World such that stellar explosions ordered as L > M > K are observed on Earth, which is in that one World, in the order M > K > L. The versions can be ranked in that while one explains the other, the latter includes the description of supporting observations.

The moment we shift our metaphysical basis from actualism (occurrent states) to dynamism (natural powers and dispositions) the thrill of Goodman's argument is dispelled.

There is a very simple rebuttal of the 'equal standing of versions' argument if we think in less intellectualist terms and pay attention to the role of discourse in the management of human projects, some of which will be practical, others political, others social and so on. Relative to a well defined project, versions can be ranked. If the project is to land a robot on one of the moons of Jupiter, the Newtonian version of the Solar System Story is better than the astrological. However if the project is to give some sense of a place in the universe to people not deeply versed in the physical sciences, the astrological version of the cosmos might do a better job. Versions are competitors only relative to projects, and in the absence of projects they are neither of equal standing nor ranked in some preference order. In some project environments there are uses for 'true', while in others different forms of assessment are appropriate. So the post-modernist claim that all versions are equal can only be made in a project vacuum. But in that context the contrary claim that some versions are better than others can make no sense. If a counter-claim makes no sense in a context, then neither does the original claim.

Since each version 'creates' its own world there is no place for experiments, as transparent windows giving on to one World, enabling us to rank versions with respect to their quality of representing aspects of that World. Goodman's position seems to lead to a denial of the Principle of Transparency. However, returning to his account of the world making power of versions it seems to me that that claim rests in part on a tacit and prior rejection of the Principle, together with a strong actualism. So his argument cannot be taken as a disproof of the Principle of Transparency since among its grounding premises is a tacit rejection of that very principle.

2.3.2. Rorty: The Closure of the Laboratory of Nature

In both his major essays on the topic, Rorty bases his philosophy of science on three principles:

1. 'A repudiation of the very idea of anything [social or material] having an intrinsic nature to be expressed or represented' (Rorty 1989, p. 4). So, 'Water is an oxide of hydrogen' would not express the nature of water, since it seems, according to Rorty, it has none.
2. Truth is confined to sentences which 'cannot exist independently of the human mind', though 'the world contains the causes of our being justified in holding a belief' (Rorty 1989, p. 5). But this does not support a correspondence theory of truth since there are no facts in the world.
3. We should not look for criteria for adopting a certain vocabulary, for instance that of Newtonian physics, even though adopting this vocabulary lets us predict the world more easily.

Rorty's extrusion of Nature from its traditional role in the epistemology of the natural sciences seems to depend on the thesis that concepts like 'knowledge', 'essence' and so on are wholly explicable within a discourse genre. If facts are propositional and it is facts that are juxtaposed to hypotheses then natural science does not reach into the World. From the point of view of this analysis Rorty's position can be seen as a rejection of the Principle of Transparency, but in discursive form. The three-step structure, hypothesis-fact-World is not uniformly linked. The second pair are linked causally, while the first pair are linked semantically.

Presumably, as human constructions, experiments are sited within the discourse of science, not in the World. World-to-experiment relations are causal, but experiments, as interpreted phenomena, Rortian facts, must be discursive. One can see how this is a tacit denial of the

Principle of Transparency. The Rorty scheme is really four-step. Hypothesis-experimental result as fact-experimental result as material state-World.

The treatment of these issues is more or less the same in Rorty's earlier and well known work (1980).

At the same time as he admits that the world causes us to entertain certain beliefs Rorty denies that it is the natures of world stuffs that does this useful work. The experiment, for example a manipulation to decompose water into hydrogen and oxygen, and to reconstitute it from them, seems to be forgotten. It is not Nature that causes our beliefs, if we can use that notion for a moment, but Nature as domesticated in the experimental apparatus. Only having forgotten the materiality of experimental apparatus could he say that 'physics fits the world' is an empty compliment, not an explanation of its success.

To restore the role of the World, while steering clear of the Principle of Transparency, will require a thorough analysis of the nature of experiments and the status of apparatus. My final commentary on Rorty's pragmatist account of science must await the detailed discussions of experiments. The point to be established is that though the relation of apparatus to World is a material correspondence it is not analysable in simple cause/effect terms. In a certain sense apparatus is Nature domesticated. In so far as Rorty's position depends on a causal analysis of the Nature/Apparatus relation it is flawed.

2.3.3. Latour: The Isolation of the Laboratory and its Inhabitants from both Nature and Society

Latour's post-modernism, like Rorty's, is based upon the intuition that science is a wholly discursive activity. But his position is stronger than Rorty's in that he pays close attention to the importance of experimental apparatus and its skilful manipulation in the genesis of bodies of knowledge. But he goes further than Rorty in denying the World or Nature any role in the genesis of science either as discourse or as practice. This step beyond Rorty's modest position is driven by his account of experimental apparatus. People make it and people use it, adjusting it to get the results that will serve to settle some controversy in accordance with their beliefs and in their favour. In taking part in a controversy, scientist needs allies. He or she finds these both among the human inhabitants of the laboratory and the non-human, the apparatus. Fact builders have to "enlist and interest the human actors ... and ... the nonhuman actors too so as to hold the first" (1987, p. 172). Once the controversy is settled the winners redescribe the results of their work by deleting the role of the apparatus and the local character of their manipulations, and presenting their results as the face of Nature itself. But this is a discursive or rhetorical transformation, achieved by deleting all indexical markers from the text. Seen clearly Nature does not penetrate the walls of the laboratory.

Sokol, following up his famous hoax with a philosophical criticism of post-modernist interpretations of the natural sciences, is intent on getting rid of the thrill of the post-modernism take on the natural sciences. In a recent paper (1997), Sokal examines the structure of a key principle in Latour's argument. Latour condenses the post-modernism aspect of this analysis of laboratory work into a list of rules. The most thrilling is Rule 3, which runs as follows:

> Since the settlement of a controversy is the *cause* of Nature's representation, not its consequence, we can never use this consequence, Nature, to explain how and why a controversy has been settled.

The thrill is, of course, the implication that the explanation must be wholly sociological-political.

But as Sokal points out, in the quotation there are two 'nature' expressions, namely Nature's representation' and 'Nature.' They are not synonyms, and in the course of the paragraph the latter is substituted for the former, and that is what generates the thrill. Suppose one insisted on semantic uniformity. Using 'Nature' in both places we get an absurdity, 'the settlement is the cause of Nature. . . ' But if we substitute 'Nature's representation' for 'Nature', we get the banal truism that if the settlement is a cause of Nature's representation we can never use that to explain how the controversy was settled. The thrill is the consequence of semantic slack, readily dissolved. If we take 'Nature's representations' to be propositional, stories, narratives, accounts or versions, then, if we bought into the general post-modernist account of discourses about any subject matter whatever, Latour's cross substitution would make sense. But one cannot justify the general post-modern account by claiming support from an analysis of natural science that presupposes that account.

It is by his construal of pieces of apparatus as allies that Latour tries to close the laboratory to Nature. So it is to the role of apparatus that our analytical efforts must be directed.

3. THE ANALYSIS OF THE ROLE OF APPARATUS IN EXPERIMENTAL PRACTICE

The deep point behind the whole treatment I am proposing is that the important forms of Nature's representation in natural science are material, namely experimental set-ups, that by isolating fragments of nature, we domesticate it, making it available for material manipulation. The discursive style of Nature's representation flourished *prescientifically*. The significance of the invention of the experiment in the late Middle Ages, by such as Theodoric of Freiborg, was not so much a check on discourse, the logicist tradition that has persisted to the present day, but as *an alternative form of representation*. For example it is not hard to treat Gilbert's Terrela, or model earth, a 10 cm. sphere turned from loadstone, as a material representation of the earth as Gilbert believed it to be. Instead of saying, 'the earth is a sphere' and drawing inferences from it, we make a microworld spherical mini-earth, and see what it will do, and how things behave on its surface.

We can accomplish our escape from the locked laboratory by looking more carefully at the nature of experimental apparatus. I shall try to show that the most general characterization of a large number of pieces of equipment, an 'instrumentarium', to pick up Ackermann's elegant phrase (1985), in use in laboratories would be as bench-top models of the World or parts of the World. As such come under the general logic of models be they cognitive or material. But since experimental apparatus is material an instrumentarium is also part of the World. Let me illustrate the point with two examples:

3.1. Faraday, Crooks and the discharge of electricity in attenuated gases

With the ready availability of a source of high voltage electricity, the Ruhmkorff coil, and developments in the technique of implanting electrodes in glass, the discharge tube became a popular 'toy' in mid-Victorian London. With after dinner demonstrations of the 'electric light' one had one's own aurora borealis in the drawing room. The discharge tube is a model (a domesticated version) of the unreachable but visible spaces of the upper atmosphere, and the glow a domesticated version of what can be seen in the wild. The drawing room shades into the laboratory. Faraday records visiting J. P Gassiot (Vice President of the Royal Society) to see a demonstration of the 'electric' light. He notices the 'dark space' near one the electrodes, spotting something not perceptible in the wild. Later he manipulated the glow by moving a magnet along the tube. He and others ask themselves 'How is the electricity transmitted in a near vacuum?'. Eventually Crookes begins to pay attention to what might be happening outside the tube. And so the familiar research programme leading to the discovery of alpha, beta and gamma rays gathered momentum. According to the logic of model building, the aurora and the glowing discharge tube are sub-types of the same super-type. The discharge tube in the laboratory does contain an attenuated gas and a miniature 'solar wind' passes through it. A little bit of Nature is reproduced in the laboratory in a domesticated version, and so represented in material and manipulable form.

3.2. Theodoric of Freiborg and the Rainbow

Theodoric used an illuminated rack of water filled urine flasks as a model of a curtain of water droplets. Thus he recreated in the laboratory the familiar pattern of differing refractive colours in a domesticated and familiar form, wonderfully well described by Wallace (1959). The whole set-up is a material analogue in the laboratory of the rain and the rainbow in Nature. One could say that the flasks are a domesticated version of the rainbow in the wild. And by-the-same-logic the apparatus/colours is a sub-type of the same super-type as the rain-drops/rainbow. Water filled urine flasks *are* spherical masses of water.

The distinction I wish to make is not so much between the artificial and the natural, in which experimental apparatus as artefact is detached from any but an accidental relation to the World, rather that between the domesticated and the wild. The laboratory is like a farm. It is like neither an art-gallery nor a zoo—neither wholly artifactual nor wholly wild. The material setup has been tamed, rather than represented or caged.

3.3. A refinement in the 'logic' of models

Let us look again at the experimental equipment and what happens when a Stern-Gerlach apparatus is switched on; or when a reagent is poured into a test tube; or a bunsen burner is ignited under a retort, or a population of drosophila are left to breed in a suitable enclosure. We have certainly created models of certain aspects or portions of the World. But we have also done something else, since the apparatus is made of material stuff and fruit flies are flies. We have domesticated, and so brought partly under our control, certain aspects of the wild. Experiments are not just discursive representations of Nature in a material medium. They are natural phenomena.

In so far as models are treated as representations of reality they are in no better case against the post-modernist critic of science than propositions, but if they are taken as miniworlds, they are part of nature, bench-top apparatus, that belongs within the same type-hierarchies as some wild/naturally occurring mechanisms. Models, bench top, not cognitive, are domesticated portions of the wild World. Just like cows are domesticated versions of the aurochs, primeval wild oxen.

Treating ontologies as type-hierarchies makes clear several aspects of natural science. For the purposes of this discussion the authentication of a type-hierarchy is achieved through the demonstration that among the sub-types are one or more that are discernible in Nature. In this way the representational power of a laboratory model is established, if it is sub-type of a type-hierarchy some of the sub-types of which are known in Nature. Domestication permits representation. This is not the simple causal relation that Rorty takes it to be.

Pieces of experimental apparatus are artifacts and the stuffs they are used to manipulate are selected and often reprocessed from raw material. However the types instantiated in apparatus and the samples are material, and as such must have a place in some type-hierarchy along with other material sub-types, of some of which they could be models. Pieces of apparatus are made, not discovered. But the 'logic' of their placement in the natural sciences must be the same as the classification of new sub-types, members of which have been stumbled across in empirical investigations. For example Thagard's analysis of the discovery of the bacterium *helico pyloris,* as the cause of stomach ulcers, required the adjustment of two type-hierarchies (1997). A new sub-type was added to the bacterial type-hierarchy, authenticated over centuries, and, in this case, importantly established by comparison (in relevant respects) with the syphilis bacterium, already having a place as a sub-type. At the same time the type-hierarchy of diseases was revamped, so that the sub-type "gastric ulcer" was relocated under the type "infections" and deleted from the type "sychosomatic conditions".

The performances of experimental apparatus are natural phenomena, though domesticated. What can we ascribe to Nature on the basis of these phenomena? The Principle of Actualism would counsel the ascription of these actual phenomena. This ties in with the traditional logicist move of using the I and 0 forms as the raw material of scientific knowledge. The post-modern critic is quite right to point out that these phenomena exist only in the laboratory settings, created and interpreted by a scientific community. However, the performances of experimental apparatus are open to another non-actualist interpretation. They display what Nature is capable of in conjunction with apparatus. In so far as the apparatus is a bench-top model of some Natural state of affairs, it displays real possibilities.

The argument shows that despite the falsity of the Principle of Transparency a proper understanding of the status of experimental apparatus and so of the process of experimenting coupled with a non-actualist ontology, scientists do have access to Nature. The crucial step is the abandonment of the Principle of Actualism. In consequence the Principle of Transparency can be rewritten. Apparatus is not transparent to the actual states of the World, but to its powers and tendencies.

4. TESTS OF THIS INTERPRETATION

4.1 Hacking's disjunction of apparatus and world

One must be struck by the relative stability of certain branches of the physical sciences. Geometrical optics, Newtonian mechanics, the 'nineteenth century' parts of inorganic chemistry, and so on have been stable, in some cases for four centuries. Within certain limits, defining appropriate ranges of environments, processes and substances, there appears to be not the slightly prospect of their being upset. How is this stability to be explained? The traditional explanation is that the laws of nature expressing what is known about the relevant phenomena are true *ceteris paribus.* This simple explanation collapses in the face of the philosophical assault on propositional truth, together with the problems of all logicist accounts of science.[3] According to Hacking, there is a quite different explanation, essentially a post-modernist one. Experiments are not simple displays of correlations of observables. They are many-way manipulations. The aim is to achieve a stable mutual adjustment of theory, discourse conventions, data collection and analysis and practical techniques. When these are nicely adjusted this bit of science is stable. The aim is to achieve a stable mutual adjustment of theory, discourse conventions, data

[3] For instance the underdetermination of theories by data (Clavius paradox), the paradoxes of confirmation, and so on.

collection and analysis and practical techniques. When these are nicely adjusted this bit of science is stable. There is thus a near perfect discourse/apparatus circle.

Hacking claims (1992, p. 30) that '... [as] a laboratory science matures, it develops a body of types of theory and types of apparatus and types of theory and types of apparatus and types of analysis that are mutually adjusted to one another. . . . They are self-vindicating in the sense that any test of a theory is against apparatus that has evolved in conjunction with it . . . and in conjunction with modes of data analysis.' And he goes on to say 'our preserved theories and the World fit together so snugly less because we have found out how the World is than because we have tailored each to the other.'[4] Hacking points out, quite rightly, that experimenting produces phenomena.

But Hacking draws a strong post-modernist conclusion.

> When [we have brought matters into consilience] ... we have not read the truth of the World. There usually were not some preexisting phenomena that experiment reported. It made them. There was not some previously organized correspondence between theory and reality that was confirmed (1992, p. 58).

By opting, tacitly, for an actualist account of the science/World relation Hacking misses the crucial point about experimentation, namely that a large class of experiments are directed to making natural affordances available for human contemplation and use. In some cases there were preexisting phenomena, such as moulds producing a secretion that killed hostile microorganisms, the Aurora Borealis, rocks falling from cliffs, and so on. We create material models of these setups in the laboratory and call them apparatus.

We are now set up for the post-modern. We thought that Nature must have a place, however humble, somewhere in this symbiotic relationship but according to Hacking (1992, p. 56) 'in referring to nature I do not imply that nature causes or contributes to such symbiosis in some active way. I do not invoke nature as an explanation of the possibility of science' and more of the same.

But there is a third explanation of the stability of well established science that is neither traditional nor post-modern. It is not at all thrilling. It does involve a quite different emphasis on the goings on in the laboratory from its rivals. The third explanation is that the Apparatus, the instrumentarium, consists of a set of material models of certain parts of the material World. The science derived from the use of an instrumentarium is stable, not because the laws of nature describing the behaviour of the apparatus are true of the World, but because the apparatus materially resembles that aspect of the World of which it is a model. An Atwood's Machine is like a cliff from which a loose rock can fall. A sealed tube of attenuated gas is like a layer of the upper atmosphere. A Petri dish is like an open wound. (Of course, in the relevant respects!). If the bits of apparatus in an instrumentarium are domesticated chunks of nature, then the general outline of Hacking's account of the stability of some parts of science is a consequence. The phenomena are in the laboratory, but the powers to produce them, in these circumstances are the powers of Nature, and they are very much present in the laboratory. In many cases the phenomena of the laboratory are domesticated manifestations of affordances. Sometimes, however, we may suspect that some apparatus/world complexes permit the display of

---

[4] It is difficult to grasp what 'tailoring the world' could be. It sounds rather like another example of the thrill-inducing slip Sokal discerns in Latour's Rule 3.

affordances that are never realized in nature.

4.2. Apparatus not modelled directly on Nature

The discourse-experimental set-up that manifests 'Hackerian closure' does so because the apparatus is a good material model of something in the World. But not all instrumentaria are so simple. Some other cases need to be looked at to defend the thesis of this paper in full generality.

Many of the ways cows behave are pretty much like the ways aurochs behave, but not all. Cows are a reasonable model for an ethology of the auroch, but there are some things they do that are not very auroch like, for instance standing around while someone other than a calf milks them with a machine which raps all four teats. at once. Working with material model requires good judgement and sometimes meta-experimentation to get the similarities and differences between apparatus as model and its natural analogue clear.' But there are instruments that are not simple models in the sense I have been discussing so far.

David Gooding has described a nice case, Faraday's demonstration of the mechanical effect of electromagnetism, that requires an apparatus that is highly unlikely to find an analogue in nature. The apparatus of mercury bath, rotating wire and so on is created for a somewhat different purpose. It is a demonstration of a natural affordance, but not by the construction of a material model of a naturally occurring setup. More about that below. This needs careful qualification. Compare the analysis of an experiment to make electrons available to people as phenomena and one which is designed to make bacteria available to people as phenomena. How much of each is a product of the technique? In the electron case we can choose to make 'them' available as particles (using a cloud chamber to create tracks) or we can choose to make 'them' available as waves (using a double slit to create diffraction patterns). Whatever 'they' are, even if there is a 'they', is indeterminate given just these experiments. Nature has these powers relative to these material setups. But making bacteria available optically, with different stains, and making them available electronically by exploiting the wave-like properties of electrons, leaves robust criteria of identity in place. We have no conceptual difficulty with the idea of 'aspects of' or 'properties of' or 'internal fine structure of' bacteria.

Chemistry is rich in examples of Goodingesque 'makings available' to people of yet another kind. I became hooked to science the night my father and I, using a simple retort, isolated bromine. He assured me that this 'stuff' never appeared in Nature as the dark brown liquid we had brought to light. Faraday's electro-mechanical phenomena do not exist in Nature, even as analogues. The apparatus does, however, reveal a natural tendency. Bromine atoms nowhere exist in liquid form but even when locked into molecular combinations they do possess that capacity.

There is more to the requirements for 'making something available'. We need apparatus but also stuff for it to work on. Daniel Rothbart has discussed the question of the preparation of pure samples (1997). Thus all chemistry, biochemistry, metallurgy etc. is based on the idea of the pure/impure distinction. Once again we have a version of the domesticated/wild distinction. The specimen is domesticated in that it is designed to enable the experimenter to access some its natural dispositions. But there are complications. The experimenter prepares samples for manipulation. At the very least the dispositions of raw samples are rearranged in the processes of purification, unmasking some that impurities obscured. In breeding cats and dogs, cows and horses, farmers and their like enhance, suppress and indeed sometimes create dispositions. Natural powers can be enhanced and deleted, while new powers are created.

Part of the art of experimental science is to devise apparatus that enables us to manipulate something perceptible, and so to manipulate something imperceptible. For instance the Stern-

Gerlach experiment and many experiments in thermodynamics, purport to achieve this. A Stern-Gerlach apparatus makes the nuclei of silver atoms manipulable, in that we can use it to sort them into two classes, depending on a certain quantum number. But they are not rendered perceptible in that apparatus. The tie between the two kinds of availability is loose. Sometimes perceptibility facilitates manipulability, for instance cloning by 'surgery' on the blastosphere. Sometimes techniques that make something perceptually available do not facilitate manipulability (for instance what the Hubble reveals) as in astronomy (Bohr 1958). But to justify this claim the above argument based on the creation of bench-top models by domesticating part of the wild World needs supplementation. Nor are these cases so simple as the use of a novel experimental setup to display a natural power for a human audience, as in the wonderful demonstrations put on in the Christmas Lectures at the Royal Institution. Like the Stern-Gerlach, the Joule experiment on the mechanical equivalent of heat trades on a huge amount of theory, stories about reaches of nature unavailable to anyone however well equipped. Rotating the paddles by a falling weight (twenty descents of 63 inches) increases the velocity of water molecules, which in turn, increases the velocity of molecules in the thermometer which in turn caused the level of the mercury to rise. Assuming proportionality among all these processes, we reach the mechanical equivalent of heat, via manipulation of the imperceptible. Joule's apparatus has a pedigree, starting from Rumford's studies of the heat produced in solids by mechanical action, through Meyer's qualitative demonstration that mechanical action on liquids also warms them. That the claim to have speeded up the water molecules is legitimate can, I believe, be justified. But it requires a long historical/inductive argument which would not be germane here. In Joule's paper of 1843 the molecular story is not mentioned. Still, the link with Nature, though tenuous, is not broken. The water at the foot of Niagara is warmer than the water above the falls.

For the adherent of a dynamicist ontology the actual phenomena of the laboratory, created in apparatus built by human beings are *not* the building blocks of scientific knowledge. Rather they enable us to display the powers and tendencies of the World in a domesticated form, the final step in the argument will be to offer a general account of the dynamicist interpretation of experiments.

5. BOHR'S PHILOSOPHY OF EXPERIMENTATION

For the full development of the dynamicist interpretation of a large class of experiments we must turn to the philosophy of science of Niels Bohr (1958). Philosophers and physicists took a long time to digest Bohr's point of view, perhaps because it was so very radical that it did not fit into the positivist/realist dichotomy, within which philosophy of science was then conducted.[5] The point of Bohr's interpretation was to acknowledge the ineliminability of the instrument/apparatus, the constructed character of laboratory phenomena, and at the same time to avoid eliminating the World. We can read Bohr as pointing out that the Principle of Transparency is false, that therefore the Detachment Corollary fails. The World is made manifest only in interaction with apparatus (and people). What range of powers and capacities can be

[5] Not surprisingly, many commentators interpreted Bohr's views as some kind of Berkelian idealism. The subtlety of his position, and the debt it owed to Kantian influences, did not begin to emerge until the commentary by J. Honner (1987). It might even be said that the so-called 'Copenhagen Interpretation' was not true to the deep aspects of Bohr's insights.

made available for human beings is relative to the method of exploration. But this does not mean that the World plays no role in the outcome of experiments. To bring out Bohr's insights we need a non-actualist ontology. The basic distinction at the heart of dynamicist ontologies is that between dispositions and powers. Expressions for perceptibles can be related by truth functional connectives in a variety of ways, such as conjunction, disjunction and conditionality. A disposition is an observable because in the 'if A then B' form with which it is ascribed to a material set-up both 'A' and 'B' are expressions for perceptibles. The notion of an affordance drawn from J. J. Gibson's psychology (1968) of perception is just that of a disposition the second term of which involves some human activity. Thick ice affords walking. The concept of 'causal power' is reserved for the unobservable property of at least some of the material entities in the setup that accounts for its observable dispositions.[6] The Bohrian line, in Gibsonian dress, goes as follows:

> Apparatus 1 x the World manifests phenomenon P1 to a human observer, and this shows that the apparatus/World complex affords P1.

> Apparatus 2 x the World manifests phenomenon P2 to a human observer, and this shows that the apparatus/World complex 2 affords P2.

The World appears in both the above condensed sets of instructions for bringing some of the powers of Nature to light. Move the Atwood's machine to the moon, say in the form of a golf ball and golf club and the machine/World complex differs by Worlds. The new set-up affords a different motion from the earth-bound one.

Experiments cannot be interpreted without paying attention to a metaphysical scheme. The apparatus alone would not have afforded either P1 or P2 or anything else. Nor would the World. One or the other is made manifest by the use of the apparatus to explore the World. However, what the World affords is relative to apparatus. The Bohrian answer, I believe, would have been that the World/apparatus complex has certain dispositions to display certain states thus affording human beings such phenomena as tracks and interference patterns. Of course if one were a committed actualist, it could look very well as if Bohr had been some kind of positivist, since what we read off is 'on the surface' of the apparatus. Apparatus is not transparent. But these dispositions are explained by reference to the causal powers of the world. How the causal powers of the World are manifested is relative to the apparatus or technique employed. Nevertheless its use allows us to make the World display its powers in so far as this apparatus and not that calls them forth. What has the power, the World or the apparatus/World complex? In dynamicist terms the World has the power, but while it is the apparatus/World complex that affords the display, that is to which dispositions are ascribed.

Plainly there is an ontology immanent in Bohr's philosophy of experimentation. It is certainly not actualism. According to the dynamicist point of view, from which this paper is written, observed dispositions are explained by the causal powers of the beings or clusters of beings to which they are ascribed. In the particular case of sub-atomic physics, in which apparatus manifests certain phenomena, the apparatus/World complex displays the appropriate

---

[6] If someone were to press for an analysis of the concept of 'causal power' in actualist terms, to what state of a material substrate does the term 'power' refer?, they would have missed the point of dynamicist ontologies. These have been offered from Kant to Cartwright as alternatives to actualism. To explicate a dispositions/powers ontology one can only offer examples.

surface appearances, just in so far as the World has the power to afford them when conjoined with apparatus. According to the Honner-Brock interpretation of Bohr's point of view, natural science, and in particular physics, is the attempt to explore the powers of Nature through the exploration of the dispositions of a special class of material things, apparatus locked into a system with Nature. It is apparatus/World complexes that afford perceptible phenomena.

To defend the possibility of using experiments and experimental equipment as a source of reliable knowledge I need to defend the ontology within which and only within which experimentation makes sense, namely the depth ontology of real causal powers as opposed to the surface ontology of actual appearances and manifestations. According to the ontology of dispositions and powers the Principle of Transparency is false. Since this was also claimed by post-modernism, a defence of the experiment that does not depend on that principle is not vulnerable to criticism on the basis of the accusation of adhering to a naive correspondence theory of truth. However post-modernism does seem to make use of the Principle of Actualism, which is at odds with a dynamicist ontology. Post-modernism, at least as it is exemplified in the writings of Gergen, Latour and now Hacking, seems to be a close relative of the positivism to which it seemed at first to be opposed. It is certainly deeply at odds with any account of Nature in terms of observable dispositions and deep causal powers.

The basic particulars of classical physics are not so much Newtonian mechanical atoms, but charges and their spatio-temporally distributed fields of potential. Experiments in the physical sciences are concerned with isolating and making manifest in phenomena the powers of basic particulars as observable dispositions.[7] The simple charge/field physical scheme, mapping so nicely on to the dynamicist ontology of powers/dispositions that has persisted from the eighteenth century, has been further refined in quantum field theory, and general relativity, in which charges have dissolved ontologically into fields. These developments do not, I believe, affect the argument, since they are even less actualist than was the classical field theory. It is not my purpose in this paper to defend a dynamicist ontology in detail, but to show how Bohr's philosophy of experimental physics shows us how to escape from Latour's laboratory to an engagement, of a sort, with Nature.

---

[7] Hume's actualism should be seen against the background of a general adherence among leading philosophers of the time to dynamicist ontologies. These have never been better summed up than by Thomas Reid in his *Essays on the Active Powers of the Human Mind* (1788/1969, Chapter One).

1. 'Power is not an object of any of our external senses, nor even an object of consciousness.

2. 'There are some things of which we have a direct, and others of which we have only a relative conception, power belongs to the latter class.

3. ' ... power is a quality, and cannot exist without a subject to which it belongs.

4. 'We cannot conclude the want of a power from its not being exerted; nor from the exertion of a less degree of power, can we conclude that there is no greater degree of power.

5. 'There are some qualities that have a contrary, others that have not; power is a quality of the latter kind.'

CHAPTER THIRTEEN

REINTERPRETING PSYCHOLOGICAL EXPERIMENTS

1. INSTRUMENTS, EXPERIMENTS AND MEASUREMENTS IN PHYSICS

Physicists make use of two broad kinds of laboratory apparatus. Psychologists have borrowed the terminology of the physics laboratory to describe the devices they use in empirical studies. We need to decide to which of the two main categories the so-called instruments' of psychology belong. This is of the greatest importance since the ways they are each related to the domain they are used to investigate are wry different.

There are devices that change their state under the causal influence of some changing property of the environment in a way, which varies, systematically with changes in the environment. For example, a thermometer measures the degree of heat in its immediate environment because the length of the mercury column is causally related to the level of molecular energy in its surroundings. The same principle governs the use of the barometer, the hygrometer, the vo1tmeter, and many other instruments. Let us call a piece of equipment of this kind 'an instrument.'

There are also devices that are material analogues or models of some real physical system. For example, a gas discharge tube is a model of the upper atmosphere and the current in *it* is an analogue of the solar wind. The glow in the tube is an analogue of the Aurora Borealis. A calorimeter with a mixture of ice and salt can be treated as an analogue of the sea in winter, and soon. Let us call this kind of equipment 'apparatus'.

Only the first kind of device can yield measurements. They are simply read off the changing state of the instrument. The height of the mercury is a consistent effect of the molecular state of the hot liquid. So we say the temperature *of the air* is 22° C, though what we observe *is* the length of the mercury column. The relation of the measured to the measurer is causal and deterministic.

Devices of the second kind have all sorts of important uses. However, no measurement can be derived from a property of the model. The relation between device and reality is analogy, not causality. We can say that the freezing point of the sea is -4° C on the basis of the behavior of the saline solution in the calorimeter. It might be a good estimate but it is not a measurement of a property of the sea. It is an inherence by analogy from a property of the model to a property of the subject of the model. We can use the calorimeter model to explain why ice floats on the surface of the sea, leaving liquid water below.

2. INSTRUMENTS, EXPERIMENTS AND MEASUREMENTS IN PSYCHOLOGY

Psychologists use expressions like 'instrument', 'experiment', and 'measurement', which are almost certainly borrowed from the physical sciences. Could they mean the same thing as physicists mean by the use of these words? Since there are at least two kinds of experimental equipment used in physics, instruments, apparatus, with very different logics, we might find that while psychologists could not be using something that conforms to the logic of the other. That is indeed just what we find. Unfortunately, most psychologists are seriously confused about these matters. They tend to interpret the study of set-ups corresponding to apparatus, which are actually models or analogues of that which they help us to investigate, as if they were measuring instruments, the properties of which are effect of causally efficacious states of that which is measured.

Psychologists use the word 'instrument' for such devices as questionnaires and checklists. "Subjects" answer questions or check off items. The experimenter forms statistical analyses of the answers. If this were an instrument of the same type as those used in physics, then the answers should vary systematically with some varying property of the subject. The relevant property that varies should be varying in that subject, and causing a variation in the properties of the instrument.

Are questionnaires really instruments like thermometers? To answer this one must pay close attention to what is going on when someone provides written or spoken answers, marks a checklist, or indicates a point on a Lickert multi-point scale. The participant is answering questions posed by the psychologist. This joint activity is a kind of formal conversation. As such, a questionnaire is a *model* of an informal conversation. The answers to the questionnaire are not caused by some mysterious unobservable property of the person answering it. The results of using a questionnaire as an 'instrument' in a psychological investigation are not measurements at all. They are logically parallel to the results of using an apparatus as a model or simulation of a real world set up and reasoning analogically from the one to the other.

The point of the experiment can be understood only by examining the analogy between the questionnaire as formal conversation and an ordinary discussion of the same topic. The results of the whole procedure, particularly the bringing out of correlations between types of questions and types of answers, are neither more nor less than expressions of narrative conventions and semantic rules governing the kind of conversation modeled.

Interpreting empirical studies in terms of concepts like 'instrument' and 'measurement' presupposes a causal metaphysics. Reinterpreting such studies in terms of concepts like 'model' and 'simulation' presupposes a meanings/rules metaphysics. Let us now look closely at an example in which the reinterpretation enables us to recover interesting and cognitively significant results from what otherwise would have to be dismissed as nonsense.

SECTION SIX

THE ETHOGENIC POINT OF VIEW IN SOCIAL SCIENCE

SECTION SIX

THE ETHOGENIC POINT OF VIEW IN SOCIAL SCIENCE

INTRODUCTION

The ethogenic point of view in the social sciences conceives of human beings, not just as passive responders to the contingencies of their social world, but as agents deploying in their social lives a theory about people and their situation, and a related social technology. Ethogenists believe that this point of view has been kept alive in anthropology, and the symbolic interactionist tradition in sociology, while it was denigrated or denied in those branches of social science that were nearer psychology. Nevertheless, one is not merely preaching to the converted in laying out the principles involved in this theory, since I think it survived as a largely unexamined assumption and is still in need of explicit formulation.

Any individual person's theory about himself and others, and their social world, is like the kind of theory we have identified in sciences like chemistry, in which the pattern of events and the mechanism productive of the pattern are capable of separate description, and frequently of separate empirical investigation. Theories of social action, whether they are those held by an ordinary social actor in terms of which he develops his own action and construes the actions of others, or whether they are those special versions of that kind of theory developed by social scientists, have a similar structure. On the ethogenic view, a person's social capacities are held to be related to the relevant cognitive equipment he possesses, and this equipment includes items most naturally treated as models. These may range from very specific representations of the natures of his near and dear, to a typography of schemata for managing the identities of those who are more remote. And any given person's social theory will also contain representations or models of the social order, one or more of which may be models of that order which have diffused through from social science. 'The class system' is a case in point.

All this cognitive equipment is remote from inspection even perhaps from the actor him or herself and any other person, be it social scientist or layman, must perforce conceive a model of relevant fragments of the cognitive structure of the person be interacts with. In this way, the actor forms a model of cognitive elements which are themselves models of varying degrees of abstractness.

The fact that some kind of social order exists shows that our models of each others' models and of the social order have some degree of homology, i.e. formal correspondence. This seems to be ensured amongst humans by the existence of common language and other forms of symbolic interaction by which a psychic community of shared meanings is created, ensuring a sharing of symbolic forms.

This is one of the points at which structuralist ideas assume great importance. In *The Savage Mind,* and in other places, Lévi-Strauss (1966) can be understood to be offering hypotheses about the shared sources of models for representing certain abstract structures, an understanding of which plays an essential part in the management of social life, and representations of which must be supposed to be present among the cognitive resources of each member of the community. This is the sense I give to his famous concept of *bricolage,* i.e., it is the common source of each individual's models, that ensures that they will be sufficiently homologous to create a common social world.

Not only are representative models shared, but so too are models in the sense of ideal types. The stylistic features of performance by which a person projects a chosen self-image in public, suitable for what he or she takes that interaction to be, must be adjusted to the range of types recognized as authentic by the audience (Goffman 1959). Otherwise the actor will fail to be understood and the efforts at self-preservation are liable to be written off as weird and idiosyncratic aberrations. The aetiology of such models as these is extremely complex, but it is clear that they may involve both imitative or iconic modelling as well as elements which could more properly be taken to be arbitrary. Contrast the loose, hip-swinging walk of the young men entering Harold's Club in Reno, Nevada, by which a culturally standardized image of 'big spender' is projected, derived from well-known human exemplars, and the head-toss of a young woman, an equine metaphor by which 'unbridled autonomy' is indicated, with the open and shut palm with which an Italian waves goodbye. The young man may surround his walk with further iconic accoutrements, such as a string tie, big hat, and so on, derived from his exemplar, though for the young woman to elaborate her point with a whinny would be regarded in most circles as extravagant.

1. OCCASIONS FOR MODEL USE IN THE SOCIAL SCIENCES

An ethogenist sees a social world existing primarily in the episodes of individual encounter, and the constraints exercised on previous and subsequent conduct by the possibility and realization of culturally sanctioned forms of microsocial interaction. In a sense, we know such episodes well and our techniques for their successful management are really remarkably good. But as social scientists, we must join with Machiavellians, con-men, and theatrical producers in both occupying roles and taking distance from our social performance. By this means, we become conscious of a complex reality, in need of analysis. Following Burke (1969) and Goffman (1959; 1963), I will assume that we must take a generally dramaturgical perspective, and look for the broad principles of construction of homoeomorphs of the episodes of microsocial life in such concepts as Burke's dichotomy between scene and action. By seeing the social world as a self-directed production on the stages available in home and factory, in office and airliner, one possible, skeletal structure emerges from the complexity of the action, a skeleton capable of enormously fruitful development.

What must people be like for them to be capable both of the action as conceived in the dramaturgical perspective and of the distancing by which controlled social action, and social science (one of many possible forms of commentary upon the action) become possible?

The natures, psychological 'mechanisms', and cognitive resources of the actors are by and large unknowns, and for these an interrelated system of paramorphs will be outlined, using as a common source the resources and intellectual skills of the most self-aware Machiavellian. Finally, the macrostructures of the social world, existing themselves only as iconic models within people's minds, as part of their cognitive resources for interaction management, will be looked at very briefly. The fact that social scientists describe such structures as if they had an independent reality serves merely to illustrate their instrumental role as images, since in the minds of both social scientists and ordinary men such images have the very same function. To take this view of the representation of the social macrostructure is already to adopt a radical stance to the problem of the form of its reality (Berger and Luckmann 1966). Indeed, in each problematic area the model bears a different relation to reality. Homoeomorphs of microsocial episodes are abstractions from an independent reality; paramorphs of the natures of people are representative anticipations of a possible reality, while the totality of individuals' models of the larger social order exhaust the reality of that order itself.

2. MODELS OF MICROSOCIAL EPISODES

Whether we are looking at the matter from the point of view of a layman or social scientist, it makes sense to begin the task of social analysis with an analytic of social episodes. With that begun, we can then turn to the prerequisites required of people that they have the necessary powers to operate in the way we find them operating. In this way, the model of the social order and a close analysis of the forms of microsocial interaction logically precedes the development of a model of man, though as a good ethogenist I take it for granted that if there is any priority in causality between the two essential elements it is human powers that are ontologically basic and the social order is one of their epiphenomena.

Social action can be examined from a great many points of view, but I shall begin, somewhat arbitrarily it may seem at first sight, from the aspect of the performance of those overt and conscious social acts by which the social order is created and maintained. I do not wish to imply by this choice of priority that I am not aware of a vast covert (i.e. non self-conscious) stream of interaction which is also functional in this way. I mean the stream of paralinguistic interaction, and other mutual influences of a non-verbal kind by which a kind of atmosphere, social aura or tone is created and maintained (Argyle 1972).

I begin then with such overt social acts as the interaction among a group of people by which a certain status hierarchy is maintained and reaffirmed. Such interaction episodes can be arranged along a spectrum by means of which we insert a source for fine-structure models within the general model of the staged performance. At one pole are the wholly formal, in which every move is laid down in advance and the actors do what they do by explicit reference to the rules. In a state banquet, status is affirmed and confirmed by explicit acts according to a strict protocol. We commonly identify formal episodes in which formal social acts are achieved as rituals. At the other pole are wholly informal interactions where a pattern of action directed to the performance of the social act is manifested but where there is little or no explicit attention to an antecedent specification of a procedure, and where the actors may not be conscious of the ceremonial character of the episode in which they are engaged. The ceremonial character may be capable of being seen only from the standpoint of a detached observer of the scene, though of course it will affect the perceptions and expectations of the actors who have taken part in it (Goffman 1959, pp. 114-16). As an example of an episode at this end of the spectrum I have in mind a family quarrel, which seems to the participants to be open, that is, various outcomes seem to be possible. But if an outsider could observe it, the various phases of a sequence, sufficiently strongly patterned as to lead to the humiliation of one of the participants and the triumph of the other, would seem inevitable to him. One might, from the point of view of that perception, follow Levi-Strauss in glossing the quarrel as a ceremonial reaffirmation of a status differential which was once the result of a genuinely problematic quarrel (Levi-Strauss 1966, p. 30). At the formal end, both actors and outsiders, privy to the forms of the society, agree on the social act brought about by the action, while at the informal end though social acts *are* brought about, the lack of antecedent specification of the form of the action may lead to disagreement between social commentator, be he scientist or family friend, and the actors. It is the possibility of such disagreement that allows a social therapist room to manage the perceptions of the actors in such a way that they come to see the episode as falling within a class of episodes having a formal structure which, though they acted it through, they were too involved to perceive. As I see it, this is the use of a process of modelling.

Let me make this somewhat clearer with the help of a schema. For any given social act we can order the episodes by which the act is performed along a continuum, from the formal, to what I shall now more properly describe as the enigmatic, that is those for whom the underlying productive processes are unknown and for which there are, *a fortiori,* no explicit rules. People who think they are quarrelling nevertheless follow, day by day and quarrel by quarrel, a highly stylized and very stable form. But they are not aware of the sources of this ritualistic feature of their quarrels nor even perhaps that their quarrels have taken on a ceremonial character. To describe the quarrel as an 'informal ritual' as I have done in introducing the case is to subsume it under a model, that is to conceive of it as produced according to some analogue of the process by which the structure and actions of the formal episode is produced and to have an analogous structure and social meaning to an explicit ritual directed at achieving the same or a similar end. This is to model the enigmatic on the formal in two different dimensions. It is to offer an analytic for the *episode,* under which the various things that happen will have a different construal from that given to them by the actors, and to offer a model of the generative process which would produce an episode like that which actually occurs, were the episode formal. And this last step may lead to empirical hypotheses about the cognitive equipment of the actors, about the development of their social skills and even, by a further process of modelling, about the structure of their nervous system, as I shall explain in the next section. Thus, the structure of an enigmatic episode is conceived of on the model of a formal episode comparable as to social act achieved, while the mechanism by which the episode is produced is conceived of on the model of the mechanism by which the formal episode is produced, namely explicit rule-following. In this way we generate both a homoeomorph of the episode, and a paramorph of the processes of its production.

This kind of double use of the formal model is part of the methodology of Goffman (Harré and Secord 1972, pp. 205-26) in his studies of the forms of everyday life, and it is clearly an important part of the technique of field anthropology. But the anthropologist's use of the

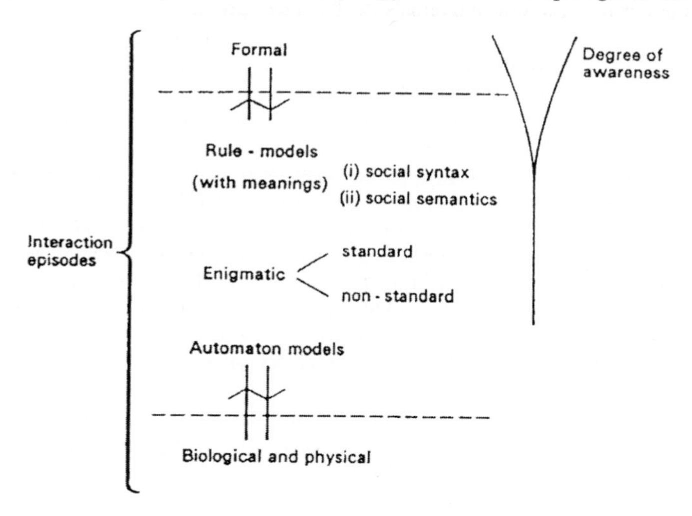

Figure 1
Structure of analytical modelling for episode analysis

technique is problematic in a way which Goffman's is not. There are etiquette books and manuals of instructions about in our culture, so that Goffman knows which part of his subject matter is genuinely enigmatic and which part is the formal product of explicit rule-following, or of residual habits having their source in formal prescriptions. In many cases, an anthropologist's notes provided the very first etiquette book that a society may have known. Of course, the fact that he is *told* that there is an oral tradition specifying the forms of tribal ritual may mean no more than that the oral traditions start from here. It must be immensely difficult for an anthropologist to distinguish the formal from the enigmatic, and once having distinguished it, prevent his perception feeding back into the society converting the enigmatic to the formal.

Ritual (or the liturgical model) is not the only source of, i.e. model *for,* analytical concepts and paramorphic models of the generating processes of the action. Of equal importance is the game or agonistic model (Scott and Lyman 1970). Both fit within the generalized dramaturgical perspective.

But the idea of using the formal as source for concepts for analysing and explaining the origin of the enigmatic has recently led ethogenists towards a complementary perspective. The game-ritual source of concepts provides a device for analysing the content of the elements, moves, actions, and sayings of social interactions. One might naturally be inclined to say that the application of this conceptual scheme enabled one to understand the social meaning of gestures, expressions, sayings (and non-sayings), and so on. It would be natural to call the products of this analysis a social *semantics.* From this point, it is a simple step to the *linguistic analogy.* Perhaps there is a social *syntax* that determines the structure of interaction rituals and games. Perhaps the order of kinds of actions in an interaction (say a greeting 'ceremonial') is as rule-governed and fraught with significant propriety as the order of the component elements in a saying. The pursuit of the study of social syntax is at a very rudimentary level (Lehner 1969). It has been suggested (Tiger and Fox 1971, pp. 12-13) that the linguistic analogy might be taken so seriously as to encourage the pursuit of structural universals on the lines proposed by Chomsky for language (see Katz 1966). At this stage, one can only say that the linguistic analogy opens up the possibility of intriguing lines of research.

Having found out what people are doing, and this, as I have argued involves identifying the social meaning of what is done, i.e. what social *act* has been performed, and may involve the discovery of the rules governing the structuring of the component actions by which the act is performed, the next phase of modelling is to conceive of a model of man such that people are capable of doing what we know quite well they do in fact do. I believe that in our everyday interactions with people we are pretty thoroughly behaviourist, identifying people by their appearance, and thinking about them only in so far as we form opinions as to what they will be likely to say and do. The best we usually hope for is some kind of untroubled passage through an encounter. But social *scientists* must do better, they must do at least as well as moderately sympathetic and intelligent laymen do from time to time. They must try to get at what it is to be that very person. They must explore the phenomenology of the social world.[1]

[1] The close parallel between ethogenic social psychology and phenomenological sociology can be seen in Schutz (1972).

3. MODELS OF PEOPLE

When we turn our attention to the problem of modelling people we find a situation fraught with complexity and heavy with the debris of history, exploded myth, and of sustained prejudice. It can hardly be disputed that human beings are physico-chemical mechanisms *and* conscious, self-monitoring, rule-following, intention-pursuing, meaning-endowing agents. I propose to turn my back on the siren simplicities of reductionism, and offer a set of model schemata, complex enough to capture our presently perceived reality, but open enough to be capable of absorbing a certain range of future discoveries, conceptual or biochemical.

To do justice to what we know about people, *four* kinds of person-model are required, each individual being capable of being minimally represented in his social capacity by a set containing one of each. They are the ethogenic, the cybernetic, the system theoretic, and the physiological. What is required is in fact something more than just a set of models. I shall be outlining something more in the nature of a set of skeletal theories, each with an embedded model. The ethogenic theory, with its embedded models, is required to do justice to the intentional agency of human action, exemplified in such activities as conceiving, critically reviewing, and realizing a plan. The physiological theory is required, not only because one can hardly deny that people are ultimately physiological mechanisms, but also because some straightforward physiological responses, such as the trembling hands of the nervous, are an intimate part of social life, in that they are both indications of a state, and are absorbed into the ethogenic model as the bearers of meaning.

3.1 The Ethogenic Model

An ethogenic theory takes as the source of its models of the generative mechanisms of enigmatic human action the known modes of acting of human beings when they follow a rule, or conceive and realize a plan, in full awareness of what they are doing. In the course of such action the fully self-aware human being monitors those actions, and monitors the monitorings, making himself thereby capable both of stylistic control and commentary upon (or accounts of) what he or she is doing, thus justifying his or her actions and explaining them at the same time (Harré and Secord 1972, pp. 84-100). One kind of such account is social science.

How does the most self-aware person plan social actions? He imagines a situation and his own and others' actions therein, adjudicating the propriety of what he does by reference to the imagined judgment of a few significant others; the provincial try-out before the opening night, one might say. He may run over several possible scenarios in the theatre of his imagination. I claim that this is what happens, as a matter of fact, in the most self-conscious form of social planning. I acknowledge that most social action is habitual action, and there is little self-consciousness about the actions and mind of the confident and experienced social actor. But I am advocating the adoption of the state of mind of the adolescent and the consciousness of self-management of a Machiavellian, as sources of models for the cognitive processes of the mature social actor whose self is lost in the interest of the action. This closely corresponds to the phenomenologist move from the 'natural attitude' to retrospective and reflective attention.

I am inclined to think that the analysis of the inner lives of adolescents and Machiavellians yields four main components relevant to the genesis of appropriate social action. They are 'situations', 'selves', 'arbiters', and 'rules for the development of the action'. There is a certain number of standard social situations which we must be able to recognize. In each situation, there are appropriate social selves to be presented in a recognizable way (see Goffman 1959, pp. 47-

95). There are usually a few people whose opinion of our social skills and performance we would be shamed to forfeit. Finally, each situation and its proper self constitute a scene in which action may develop in a limited number of permissible ways. One can imagine a set of rules, like the rules of a ceremony, which prescribe the proper course of the action.

This complex of definitions and rules, images and imagined actions, constitutes not only a resource for an individual, but knowledge of it is the central stuff of social science. Knowing the cognitive resources of the people of a society we know its possible social forms, and thus, the only ways action can develop which can be recognized as such. These resources could be laid out diagrammatically as a matrix, each element of which is a conceptual complex, to the unravelling of which the methods of Kelly might be appropriate (see Bannister and Mair 1968). Social action is intelligent action utilizing these resources, or it is the habitual residue of what was once intelligent action of that sort. The ethogenic point of view repudiates the idea that the human social world comes into being as a kind of automatic, multiple reproduced response to the signals of one's conspecifics.

3.2. The cybernetic model

In the cybernetic model the rules and plans which are perceived or imagined as guiding the conduct of the self-managing man of the ethogenic image are expressed formally, and 'following as a conscious act' is replaced by the formal 'following' of the operation of an algorithm. The cybernetic model must be functionally equivalent to the ethogenic, for any given person in any given culture, i.e. the algorithms must generate prescriptions of patterns of action etc. which are sufficiently similar to those generated by the workings of the ethogenic model, which are themselves sufficiently similar to what people actually do. This model is a formal expression of the ethogenic, and could be treated as not independent at all, but as a homoeomorph, ideal and abstract, of the ethogenic. Its formal instructions would be related to the naturalistic rules of the ethogenic model as their direct formalization.

3.3. The system-theoretic model

There is no place in the cybernetic model as such for a representation of the hardware capable of functioning in the way the formal algorithms of the model do. This deficiency is made up by the conception of an abstract structure capable of performances formally described in the cybernetic model. The systems so conceived must be functionally equivalent to the algorithms of the cybernetic model. The system-theoretic model can be conceived of as an abstract homoeomorph of the real physiological system, and as a paramorph of the mechanism responsible for the abstract operations of the cybernetic model; and through the functional equivalence of that model to the ethogenic, as a paramorph of the mechanisms responsible for the actual, concrete operations of conscious, human agents.

3.4. The physiological model

The abstract structure of the system-theoretic models can take concrete form in iconic models of putative, real structures in the nervous system, thus generating physiological hypotheses. This form of model closes the circle, since physiological models are models of the nature of humanity, that man or woman who in real life, with real resources and real cognitive and imaginative powers, can think and act in the manner represented in the ethogenic models. Physiological models will then be homologous to certain favoured system-theoretic models, the homology being a reflection of the fact that we can conceive of the system-theoretic model as an abstract and idealized form (a homoeomorph) of the concretely realized physiological model.

3.5 Models of the social order

I have claimed that the social order is nothing but a series of homologies between models of an imaginary structure each of which are part of the cognitive resources of people involved in the management of social interaction. It is just because we can form models of reality that we have the power to create a reality by conceiving of a structure which has the status of a model but whose subject we create on the basis of the model. As natural scientists and protoscientists we create a model of an existing world: as social actors we create a world for an existing model. For example, social power is an endowment to an individual from us, as we give him deference. It derives from the place in our image of social reality reserved for somebody of his kind. Of course, he and his kind may be part authors of our image, but his social power lasts no longer than the image with our place for him persists.

In the metaphysics of any science, we set up a kind of general vision of reality in which the kind of things that can exist and the kinds of events that can occur are represented. Of course, it is an empirical question as to which possibilities are realized, and as to what exact form that realization takes. In an analysis of the theories about people and their social actions that we deploy in the management of our self and our interaction with others, images of society have the place that metaphysical visions of the world have in natural science. The image of society gives us our ideas of what roles there can be, and then we discover empirically if they are filled, how and by whom—the only difference from natural science being the crucial fact that we can and do construct worlds to fulfill images. Finally, it is important to notice that in social theories a viable set of interactions may exist between people whose images of society are very widely different, provided only that some minimal homologies exist in terms of which some structure of meaning can be given to an interaction. An important area of empirical study here awaits a bold and unprejudiced observer. The cognitive structures which we ascribe to other people are given form and content by an imaginative process similar to that by which develop our own cognitive resources. They are invented but not wholly freely. People have ways of representing abstract social realities in which they employ analogues having similar formal relations to the structure they wish to represent, but which may reflect only the logical or formal relations of that structure. This, as I have suggested, is how we should interpret the homologies of Levi-Strauss.

Each person has a mode of representing to himself the structure of the social world, and he manages his social action, when rules and habits run out, by reference to that representation. One device by which the social order might be stabilized and the necessary homologies between each actor's societal image be guaranteed would be a ritual reaffirmation of the legitimacy of the local mode of representation. It is very important to remember that the modes of representation may be realistic, and have an immediate relation to real differences in the world, such as a world conceived of as male and female; or the mode of representation may be wholly fictitious in form, such as a social world pictured as hierarchically structured by class, or ordered as in the feudal pyramid.

The world of male and female is a really real world, while the world of man and wife, though superficially resembling the world of male and female, gains its reality through the relative and potential universality of the homologies that exist between conceptualizations of an image of order. A marriage ceremony creates a set of homologous images, it does not manufacture bonds. The ceremonial element in social life, whether formal or informal, is the one ineliminable and universal prerequisite for a social world since it is the means by which homologies are continuously created, and the social world brought into being and sustained.

The only place I know where the physical layout of the people is an exact homologue of their status hierarchy is at the wives' end of the room during an American middle-western faculty party. The florid extravagance of this form of analogue can be grasped if we reflect that even at the court of the Sun King the metaphor would have been deemed too banal. Despite the fictitious form of the *representation* the method of concretely visualizable metaphor may serve as a useful model for preparing solutions to the microsocial problems with which one is likely to be beset, and those images are an important part of the cognitive resources of competent social actors. *We* play with Durkheimian ideas of social facts at our peril. But no doubt the great man was aware in his heart of the metaphysical points just made.

4. TESTING MODELS BY SIMPLE PREDICTION

All this talk of the role of the imagination in science must not be allowed to obscure the necessity for bringing its products into some kind of contact with reality. Models must be checked for authenticity. Once again, we find ourselves faced with a very complex situation, by contrast with the elegant economy of the older view of science. The older view saw a theory as a logically organized structure of hypotheses, from which predictions were made by deducing the consequences of supposing that certain boundary conditions held. If the prediction turned out to be correct when those boundary conditions were realized, inductivists held that this added a modicum of weight to the theory, while if it turned out to be mistaken, fallibilists held that this showed that the theory was worthy to be rejected. Something of this structure survives into the era of models. If the adoption of an iconic model leads to the incorporation of statements in a theory from which testable predictions can be drawn, then the outcome of those tests will bear upon the acceptability of that model. But the source of a model of the unknown inner structure of some entity is usually sufficiently rich to allow ameliorating additions to be made to the original model should it fail hypothetico-deductively, additions whose pedigree ensures they escape the stricture of being merely *ad hoc*. The most striking case of this is the continuous modification of the gas molecule, as it becomes more and more like its source, namely a solid thing, under the pressure of the discovery of the breakdown of the original forms of the gas laws in extreme conditions.

But iconic paramorphs can be the source of other kinds of hypotheses, particularly existential. I have already pointed out that an iconic model is often a model of the unknown inner structure of a thing or material, and often has the form of a system of elements, organized in some way. The commonest form of such organization in the natural sciences is spatial. For instance, the von Laue explanation of X-ray diffraction describes a model of the inner structure of crystals in which the crystal is envisaged as a lattice, at the vertices of which are ions. From this, an existential hypothesis can be derived, since the lattice vertices specify places at which a definitive search can be made for specified kinds of things. Should ions be found at the specified places the model has been authenticated. Its status changes from a simulation of the unknown inner nature of crystals to that of a realistic picture.

But there are very obvious limits to the existential checking of iconic models. These are the limits of observation. From an existential point of view, it seems proper to draw the line at the limit of what might be called the 'extended senses'. Just where this line is to be drawn is a nice point, and could be the subject of argument. But it is also obvious that much of science transcends the most permissive drawing of such a line. How then do we judge the authenticity of the imaginative simulation of reality? Jacques Monod (1972) argues that science is made possible only by the acceptance of the possibility of such simulation. Our trust in it, he argues, can be rationally grounded in the fact that our powers of imaginative simulation are the product of an evolutionary process. But their adaptiveness might as much derive from systematic self-deception as from genuine authenticity, it seems to me. I am inclined to think that we must seek yet another method for checking the authenticity of our creations.

The final criterion is that of plausibility according to the prevailing metaphysics of the period. This criterion, while being perhaps the most powerful of all checks upon scientific theorizing, is both vague and evanescent. It is vague because the metaphysics of an era determines only in very broad outline what is assumed to be obviously real. Consider the variety of forms the material atom took, each of which was acceptable at some time and to some members of the scientific community. It is evanescent since powerful new models of reality may lead to shifts in the metaphysics of a cultural community, while changes in the groundwork of metaphysics have often enlarged imaginative possibilities in one direction while closing them in others. Both directions of influence can be discerned in the history of the field as a model of the fundamental physical reality. *Naturphilosophie* and the Romantic movement prepared the metaphysical ground, while the scientific development of the concept by Faraday and Maxwell has encouraged the growth of a metaphysics of potentiality.

Science is pursued under the intellectual discipline of plausibility, and the empirical discipline of fact.

5. THE REPLICATION OF REALITY

I have already argued that the humdrum methodology of chemistry provides a better scientific exemplar for social studies to follow than does the heady but atypical methodology of fundamental physics. In addition to the processes of testing I have outlined above, chemists use a technique of checking their models of reality, the development of the analogue of which by Mixon (1972, pp. 145-77) for the social sciences must be regarded as a methodological breakthrough of considerable importance.

Organic and biochemists not only try to discover the structure of the compounds that come their way, and to check their hypotheses as to that structure by seeing that the predicted products of decomposition actually appear, but they regard the ultimate triumph of their science as consisting in the synthesis of the very compounds they have analysed. The crowning achievement of chemistry is the replication of reality.

The analysis of accounts both from actors and the ethogenically aware participating spectators of social action yields three main kinds of interlocking material: images of the self and others, definitions of situations, and rules for the proper development of the action. Mixon has shown how, by ferreting out this material and constructing a sufficiently detailed scenario, a particular range of social realities can be replicated by amateur actors, complete with appropriate feelings! The method is in its beginnings, but I believe that in its further development lies much the most promising line for future empirical research in social science. On the analytical side the phenomenological tradition has much to contribute, while the pursuit of the linguistic analogy, and the further identification of non-verbal components can complement and complete the dramaturgical perspective. By the exploitation of homoeomorphic models of episodes, paramorphic models of the human resources necessary for action in those episodes, we can create model social worlds, complete, as they say in the model railway catalogues, 'in every detail'. Our capacity to do this is the measure of our achievement as social scientists.

6. TESTING PEOPLE MODELS

Criteria of authenticity, including the replication of reality, provide the most pertinent method for the checking of those models in use in the social sciences which simulate cognitive features of individual people. I have argued that the development of such models is essential to the ethogenic approach in social science, which emphasizes the construction and management of the social world by individuals. Such models represent the cognitive equipment with which one tackles the problematic character of the social world, and is, of course, one of the items put to the test in a Mixonian replication of reality.

But developing a model of the unknown, cognitive structure of a human being not only passes limits analogous to those of the extended senses, in that it may involve elements and structures of which we are unaware (and perhaps never could become aware), but it may also involve modes of organization that are not found in ordinary experience. It will certainly involve modes of organization that are not found in the structures that the traditional natural sciences study. For instance, it may be necessary to explain the succession of one thought by another by the principle that the latter is a reason, in a context of justification for the other; rather than that the former is the cause of the latter. How can we check the authenticity of such models?

Corresponding to the use of the microscope is the reportage of participants in a social encounter. But it is clear that cognitive modelling in the service of explanation goes far beyond what can be identified in accounts. To provide a justification of modelling in such terms I shall elaborate the concept of functional equivalence already introduced above. In discussing the necessity for introducing models of unknown mechanisms I argued that the model would have to behave in such a way that it could be thought to produce patterns of phenomena analogous to those patterns produced by whatever was the real mechanism. If the model produces a pretty good simulation of the known patterns it could be said to be functionally equivalent to whatever was really producing the patterns. In the case of natural science the check on functional equivalence is confined to the effectiveness of simulation of the observed patterns.

But in the human sciences the situation is more complex and for once the complexity is to be welcomed. In the very long run, one wants to say that the mechanisms producing the patterns of human action are physiological. A cognitive model capable of simulating known patterns of social behaviour (or of language-using for that matter) must, in the last analysis, be functionally equivalent to some physiological mechanism. Suppose, for instance, that the cognitive models required to explain the genesis of human social action are all of at least the degree of complexity of a system that can not only monitor its behaviour but monitor that monitoring, and in the light of what it learns, controls not only what it is doing but the manner or style of that doing, then the physiological mechanism by which our sort of organism brings this off must be at least of that degree of complexity and have at least that measure of structure as a system. Thus, functional equivalence between cognitive model and physiological structure requires that certain homologies hold between them, mediated by the intermediate levels of modelling that I have christened the cybernetic and the system-theoretic.

Thus, we have a third check upon our modelling, though in the present state of the art the realization of this check may be very distant. Only if there are identifiable physiological mechanisms capable of performing the control and monitoring functions of the cognitive models are those cognitive models authentic. These checks operate with no regard whatever to the deliverances of introspective awareness.

CHAPTER FOURTEEN

SOCIAL ACTION AS DRAMA

The theatrical performance is among the oldest sources of models for social episodes. It was much favoured in the sixteenth and seventeenth centuries. In Ben Johnson's *New Inn* the psychological complexities of taking a dramaturgical stance to the events of everyday life are explored not just by locating a play within a play, a common Elizabethan device, but by imagining that the innkeeper has asked his guests (who are themselves a company of players) to perform the play. They must take up the roles of performers in amateur theatricals as a professional act, and that twice over.

The revival of the theatrical model for the analysis of social episodes must be credited to Goffman (1969). However his innovations derive, in part, from the works of Burke (1969). I shall begin this chapter with a brief sketch of Burke's 'Pentad'. This is a scheme of five inter-determining theatrical concepts, which in the manner argued throughout this work, have dual application. They are analytical tools for the description of complex social phenomena, and they are discursive tools for the accounting procedures by which actors can render their actions intelligible and warrantable. In this way and by the use of these concepts actors can present their actions as acts.

Social psychologists of the old paradigm school, have and indeed still do neglect or ignore the settings of the social episodes they analyze. To remedy this I have devoted the bulk of this chapter to a detailed presentation of scene analysis. As Burke makes very clear neither of the complementary concepts, 'actor' or 'action', can be adequately specified for any actual occasions without attending to the scenes in which they have their place.

Can the dramaturgical model be exploited in the explanatory mode as well as the analytic-descriptive? There is no simple answer to this question. Once the myth of the information processing engine is set aside the question still remains of how the social actor's everyday skills are to be compared with those of the stage actor. Again Goffman (1974) addressed the question. As a preliminary to further studies of the matter, I draw on his metaphor of 'key-change' to approach the question of how a pair of qualitatively identical actions are experienced one in a stage play and the other in an episode of everyday life each of which fall under the same episode-type. We shall see that the ideas of seriousness and of commitment loom large in any account of the distinction.

1. ACTION ANALYSIS

Of course, a great deal of action in a great many scenes from kitchens to carpenters' shops, from space-modules to cow-byres, is practical. The actions undertaken by the folk legitimately on the scene are explicable in terms of the practical aims of the undertaking and the practically or scientifically sanctioned means of bringing these about. But it is the argument of this work that for most people, in most historical conditions, expressive motivations dominate practical aims in the energy and even in the time expended on co-ordinated social activities. In many cases too the prime motivation of the practical activities is to be found in their expressive value — space flight, scientific research and cooking are obvious examples. I shall devote no space to a discussion of the practical activities of mankind since I believe that they bear only tangentially on social life during most of human history. With the exception of the nineteenth century in Western Europe and in certain countries whose 'nineteenth century' is still to come, practical motivations are and will be secondary.

Paying most attention, then, to expressive aspects of action, I propose a threefold division of dramatic scenarios — remedies, resolutions and monodramas. Each has to do with the management of a situation of tension, in such a way that the social order, though it may be changed, is maintained in some form or other. Remedies are action-sequences which serve to restore lost dignity or honour; resolutions are action-sequences which resolve a growing tension between expressive and practical activity by formally or ritually redefining relationships on another plane, for example friendship stages, marriages and so on; while monodramas are action-sequences in which an actor achieves his or her personal expressive projects while continuing to have the good will and respect of those who have to subordinate their aims and wills. Doubtless there are very many more kinds of scenarios played out against the background of well defined social scenes. These are offered as illustrations.

1.1. Remedies

If the major human preoccupation in the complex interweaving of practical and expressive activities is the presentation of an acceptable persona, appropriate to the scene and the part in the action (the social collective component) associated with a sense of worth and dignity (the psychological/individual component), then since the possibility of loss of dignity, of humiliation and expressive failure exists, we would expect an elaboration of remedial activities for their restoration.

The existence of boundaries creates the possibility of their violation, and violations require remedies. The general form of remedial exchanges has been analysed by Goffman (1972) and I shall follow his treatment closely. The first point to notice is that for a remedial interchange, say an apology, to be required, there must be someone who has proprietorial rights on that space or time. It must in some sense be *their* space or *their* time. For instance, a lesson is a teacher's time, and a party is the time of the hostess, just as my office is my space, and the kitchen, the cook's. If the space or time is 'owned' by no one, there can be no occasion for remedy, so if I miss the train and thus exclude myself from that period of time, that is train-journey time, I cannot apologize for my lateness, for there is nobody whose time it is, except of course mine. It is the guard's train but not his journey-time.

Goffman's analysis depends upon an underlying distinction between virtual and actual offence. To arrive late is to commit an actual offence, and the person whose time it is must be apologized to in the proper ritual form. But the generality of Goffman's analysis is made possible by the extension of the notion of offence to virtual violations, which are remedied in advance, so to speak. In order to get the water-jug I must violate your table territory which I remedy in advance by asking politely, that is in proper ritual form: 'Would you mind passing the water-jug please?' which allows, but never admits, the response, 'Yes, I would mind'.

The general form, then, of remedial exchange is as follows:
A: Remedy: I'm terribly sorry I'm late.
B: Relief: That's O.K.

There are two further elaborations of this basic form. Frequently the Remedy-Relief interchange, whose referent is the actual or virtual violation of someone's space or time, is supplemented by a second interchange whose referent is the first interchange. Thus

Al: Remedy: I'm sorry I'm late.

Bi: Relief: That's O.K.
A2: Appreciation: Gee, I'm glad I didn't upset things too much.
B2: No, no, it was O.K.

where A2 in the second bracket expresses appreciation for B's granting of relief, and in B2, B minimizes the extent of his condescension, thus restoring to A his status as a person in equal moral standing with B.

But particularly where time is concerned there is another form of remedial interchange, the counter-apology. So far as I can see the final product, that is maintenance of the boundary and equilibration of the moral standing of the people involved is just the same as in the Goffman ritual. Consider the following:

Al: Remedy: I'm awfully sorry I'm late.
Bi: Relief: That's O.K.
B2: Counter-apology: I'm afraid we had to start without you.
A2 Counter-relief: Gosh, I should hope so.

Goffman's remedial exchanges allow for the management of the defilement of sacred or proper territory, and for the violation of spatial and temporal boundaries. But *how* is being late or early a violation of a time boundary? If early you are present in a socially distinct period which, for example, may be a preparatory period for the action to come, and a great deal of back-stage equipment may still litter the scene (the cooking utensils or the baby's toys have not yet been put away). The style of the action may be inappropriate to the presence of a stranger. Under these conditions a remedial interchange is required to maintain the social order. The equilibration of civility may even require the early arrival to join the home team, and pay the penalty by tidying up the sitting room. In short, times as well as spaces may be distinguished as front and back stage.

To be late is equally the breaching of a boundary, since you were not there for some temporal sections of the action, though expected, and you did not arrive through the time portal provided just before the beginning of the action. A remedial interchange is required. We need no special theory to account for the fact that late arrivals are very much more common than early. Of course late arrival may be part of a presentational sequence, susceptible of dramaturgical analysis, as when someone conspicuously arrives late in order to be noticed.

Another important category of plots concerns the scenarios for the preservation or restoration of social identity and dignity in the face of actual or potential threats. I instance here, for illustrative purposes, one such plot, which may take somewhat different forms in an actual production — the plot Goffman (1972) has called 'face-work'. He defines 'face' as 'the positive social value a person effectively claims for himself by the lines others assume he has taken during a particular contact', for example, he may have been supposed to have knowledge and experience of mountaineering. Since it is usually demeaning to everyone in the group if any one member, previously in good standing, loses face by some contradiction to his right to take the line emerging, it is in each person's interest to support every other person's line. By saving the face of others, each person saves his own and vice versa. Goffman calls activities directed to this end 'face-work'.

The sequence of the actions by which actual and potential loss of face can be dealt with are sufficiently standardized for them to be treated as ritual. The sequence begins with a challenge in which the actual or potential offence is 'noticed'. The offender is then given the chance to re-establish the expressive order either by redefining the action as of another social-type, 'not this act but that', or by making some form of compensation, or by punishment with a 'silly me!', or something of the sort. The offering is then usually accepted and the offender's gratitude made known in a terminal move. Though these actions are conventionally called for as expressions of ritually correct social attitudes and relations, they serve to illustrate the sensibility and personhood of the actor.

A more subtle form of ritual remedies can be found during 'trouble in school'. The point of 'trouble' is reached by a growing feeling in some schoolchildren that the school system and the school teachers do not value them. They experience the efforts to make them study pointless subjects ('getting at me') as well as abandonment of those very efforts ('writing me off') as degrading. A gap opens up between how they would like to be evaluated, the dignity they would like to have ritually ratified, and how they interpret themselves as presently conceived by others. Children create systematic remedial exchanges in which dignity between teachers as representations (as opposed to representatives) of the educational system and pupils as persons supposedly in good standing, is equilibrated.

To understand these cycles of remedial exchange two categories of insult have to be distinguished. There are the results which preserve dignity such as a teacher swearing at a child or even hitting it. At least in principle, the child is noticed as someone of consequence as the recipient of that act. Insults of that sort are reciprocated in kind. But children distinguish another category of insults — those that demean. Demeaning insults include two main sub-categories. Failing to know a child's name or treating it like a sibling are construed as wounding to personal esteem. They indicate both a lack of care from the teacher and a lack of reputation for the child. But worse, according to the childish interpretation, is some form of 'writing off'. This is illustrated most vividly for them by their discovering that some of their teachers are both weak and frightened. 'If we were being taken seriously', they reason, 'we would not be given such feeble teachers.'

How to restore the dignity they have lost? Many such children have devised a double cycle of retribution to balance the dignity equation. The first cycle tests the teacher by 'playing up' or 'dossing about'. If the teacher is strong and seriously concerned with them this will be shown in firmness and there is no disequilibrium in the equation to balance. The balance is already there. But if the teacher fails, this is read as an insult, a demeaning not of the teacher but of the class itself, and each member of it. An imbalance exists. The second phase of the cycle involves a retesting of the teacher aimed at reducing him or her to a condition of indignity — breakdown or retreat. When that has been achieved the pupils withdraw, amplifying their own dignity by completely ignoring the impotent rage and posturing of their official mentor. But, we have found, if the teacher tries to break the cycle by an attempt at strength and fails a second time, the results are likely to be violent. Physical punishment of the teacher may be meted out, so high a state of excitement is reached.

1.2. Resolutions

My second category of examples of scenarios are resolutions. Situations arise where private and personal attitudes shadow forth or anticipate a relationship not yet publicly or socially ratified. I take friendship and its complement as examples. (Since there is no word for 'enemyship' in English, I shall distinguish the formally ratified states as *Bruderschaft* and *Feindschaft,* borrowing a convenient German distinction.) I take it that the individuals who proceed to these ritual ratifications whose scenarios I will describe are psychologically prepared for transformation, and are aware of each others' attitudes; though misreadings may show up in the unfolding of the scenarios.

We can look at the action sequences which one might have to carry out for the creation or maintenance of a friendship or a state of enmity as the performing of a ritual, a *Bruderschaft* ceremony in the first case and a *Feindschaft* ceremony in the latter.

It seems to me that there are two quite separate aspects that can be studied. There are ways of speaking to friends, styles of speech manifestly regardless of content. I notice myself adopting a peculiar half-jocular style of speech with friends. I certainly use such a style for indicating and maintaining friendship. The emergence of this style could be explained by seeing it as a kind of test. If one uses language which if it were taken literally by someone would be insulting, and then make it jocular, this could be a test of friendship of the other person. As a friend he does not take offence at being called an 'old bastard' or whatever happens to be the local expression. There are uses of speech and action which are strictly ceremonial in character. Alan Cook has pointed out to me, for example, the reference to fighting as a *Bruderschaft* ritual in the works of D. H. Lawrence. I have outlined earlier Mary Douglas's idea that some ritual resolutions and markings of social relations can be done with food and drink. Peter Marsh has noticed that in working-class communities, except for close relatives, there is not much inviting of people to meals. Intimacy may be politely ratified by 'going out' together. We need another project, complementary to Mary Douglas's, to look at how intimacy is 'done' amongst cultures other than the British professional classes. Inversion is frequent, I believe, in anomic communities. People who are not very intimate tend to invite each other to a meal. Mary Douglas's hypothesis that passage through more structured rituals, in passing a person through degrees of greater intimacy, needs two further dimensions. When a relation goes beyond a certain degree of intimacy, then ratification rituals become less and less structured. When one wishes to express a purely formal relation, though a close one, the rhetoric of friendship may be used metaphorically and without irony.

The maintenance of hostility among intimates involves verbal rituals I shall call 'needlings'. An apparently harmless opening pair of remarks are made by A and B. But B's remark contains two aspects: it has a literal, primary sense, but it also has a performative or secondary sense in which it could be used as a needle. If A foiled the initial needle by taking B's speech as a provocation and replying to it, then B has a further move open, namely, 'I meant it literally'. B's opening remark means *q* literally and *r* performatively. If A takes it performatively, as *r,* and B says 'No, I meant q', B opens up an option of condemning A for implying that B is the kind of person to needle A. Such dialogue opens the way for B to trap A into seeming to denigrate the relationship which A and B ought to have. This structure is probably very general.

One must not suppose that Feindschaft-sustaining ritualized needlings are disruptive, that is necessarily lead to a break up of relationships. In marriage, where people are forced into intimacy, a complex relationship exists which involves some *Bruderschaft* and some *Feindschaft.* The relative quality of the marriage does not depend on whether *Feindschaft* is totally absent but how it is managed and how sustaining it is.

As Berne (1970) seems to imply, some *Feindschaft* rituals turn out to be highly sustaining to a relationship. He has some convincing examples of action-sequences which, if taken literally, are rather vicious, but since they are played out in a ritual way, simply tend to keep the formal relationship going.

Little is known about contemporary ritual ways of transforming private feelings of enmity into a stable, hostile and publicly realized relationship. Social psychologists have tended to look at attraction and altruism rather than their opposites, and even Goffman has concentrated more on person supportive rituals such as 'face work' than the ritual maintenance of hostility. A great deal remains to be done, both in the collection of examples and in their analysis and classification and in the investigation of how they work.

In the past there have been much more publicly visible forms of *Feindschaft* rituals. I distinguish these into two broad categories. There are negative rituals where a point of the ritual is a ceremonial and stylistic display of lack of interaction. The obsolete practice of the 'cut', the stylized refusal to acknowledge or greet someone, was an example of a negative ritual. Positive rituals, on the other hand, involve hostile but stylized interaction. The hostile actions are strictly controlled by rule.[t]A duel does not fit this pattern since it is not a sustaining of the *Feindschaft* relationship, rather it is a formalized way of resolving it. Feuds, on the other hand, are sequences of interaction by which enmity is sustained according to rule. Two avenues of historical research suggest themselves. Feuds seem to have been conducted quite non-violently at times, particularly amongst village women in rural communities. Anthropologists have studied the violent or blood feud and there should be little difficulty in abstracting their material to reveal formal structures useful in the social psychological study of *Feindschaft.* Finally, a comparative study of violent and nonviolent forms of feuding could usefully be undertaken to try to discern any formal parallels in the initiation, maintenance and resolution of the state of enmity.

1.3. Monodrama

Throughout this work I have stressed the need to comprehend both action and talk about action (accounting) in our registration of social life. The distinction between practical and expressive aspects of social activity is, I claim, unevenly distributed between action and accounts, since action has both practical and expressive aspects, while accounting is primarily, though not exclusively, expressive.

1.3.1. Retrospective resolution of personal/social tension in a dramaturgical mode

In this section I describe a way of resolving a kind of situation where an actor feels a disparity between how they could be taken to be and how they want to seem to appear. The solution is to speak in a way that not only defines the actor's part in the unfolding action of a drama, but also treats him or her as a cast of characters.

One common device for the drawing of other people into playing parts is the use of syntactical forms which result in what I shall call, following Torode, (1977) 'the conjuring up of Voices and Realms'. A repeated pronoun, for example, is not accepted at its face value as having identical reference, but scrutinized for its 'voice'. The structure of the discourse is revealed by linking 'voices', not instances of lexically identical pronouns. Thus,

'You never know, do you?'

addressed to another involves two voices — 'You', the voice of abstract humanity (Voice 1), and 'you' (Voice 2), that of the addressee; and via this separation of voices we can understand why the proper response is,

'No, you don't', rather than, 'No, I don't',

since the 'You' who doesn't, is Voice 1. In this way the structure of the stanza comprising the two speeches becomes clear.

Realms are the characteristic territories of Voices and may be more or less well defined in the presentation of monodrama.

The retrospective reconstruction of psychological reality I want to illustrate works by a purely internal constituting of Voices and Realms. As might be expected, monodramatic presentations of social psychological matters involving the self are a prominent feature of accounts. A very common accounting technique involves the separation in speech of 'I' from 'me'. Typically, the account involves a scenario in which the 'I' is represented as losing control of the 'me', who then as an independent being, performs the action for which the account is being prepared. In some scenarios the 'I' is a helpless spectator of the unleashed 'me'; in others the 'I' fails to attend, or loses consciousness, or in some other way is prevented from knowing anything about what the 'me' has been doing. In the former scenario, the 'I' loses control and releases the 'beast within'. In the other, the 'I' in losing consciousness reveals a mere 'automaton within'.

What are the monodramas conjured up in the use of such expressions? Their plots are based upon social vignettes, drawing upon common-sense understandings of commonplace multidrama. By virtue of their origin they have an explanatory function, for example I represent myself as using the same technique of self control as I use to control others when, for example, I say 'I made myself do something which I was reluctant to undertake': or at least, that is how I represent the matter monodramatically.

The self-justifying aspects of the resort to a monodramatic rep-resentation of the reasons for setting about reformatory self-work appear clearly when we notice that *my* failings are transformed into personal characteristics of the characters of the monodramatic presentations conjured up in 'I talked myself into it' and so on, thus my reluctance to act or my weakness of character is masked in part by attributing it to a separated and in the plot rather feeble-willed quasi fellow, the 'me' who can be brought round by the eloquence of 'I'. 'I' as primary self-mover, can hog all the *Herrschaft* ('mastery') available in the little drama. Thus monodrama is not just presentative of the dynamics of self-management, but is also technique, a way of talking that facilitates self-mastery by separating, as into another person, situation-relative undesirable personal characteristics. Sometimes self-congratulation can also be emphatically expressed by separating off and claiming desirable characteristics for *all* the members of my self-colony of selves as in the little monodrama 'Myself, I did it!'

1.3.2. Projective casting of others into roles adumbrated in the forms of speech

But the same technique can be put to work in trapping others in a self-constituted monodrama. Torode has provided a beautiful example in which a Calvinistic world of exclusion and election is conjured up by a form of speech. I take this example and its general method of analysis from Torode's study of teachers' speech, though the analysis I shall propose is somewhat more elaborate than his.

'We don't have any talking when we do compositions. I hope that is clear.' The first person plural appears here in two voices. The first voice speaks from a transcendent world, the seat of authority and the source of order. The inhabitants of this realm are strict — they 'don't have any talking'. The second occurrence of the first person plural pronoun 'we' denotes a different set of voices, those of the members of the mundane world of the classroom — the subjects, those who 'do'.

Mr. Crimond presents himself as a member of both realms — a status to which we shall return — and also as a separated individual able to look at them both from an external standpoint in his character or voice as 'I'. Mr. Crimond is the only person in the classroom who is a member of the populations of both realms. He is benevolent towards the citizens of the mundane realm, and he hopes that the message from the transcendent world is clear.

At the same time he is the channel of mediation and interpretation between realms. His hoping is directed to the possibility of his making clear to the members of the mundane realm the authoritarian wishes of the Voices of the transcendent realm. Furthermore, as a member of the transcendent realm, he is elect, while the members of the mundane realm are unable to address the issue of order except through his mediation. However, they are shown the possibility of election. One of them, namely Mr. Crimond, is a member of both realms. However, aspiration to membership of the transcendent realm is matched by the possibility of being cast out of 'Heaven' altogether into what Torode calls 'Hell', an act in the monodrama expressed by such phrases as 'you boys...' The members of the mundane realm are trapped in Mr. Crimond's monodrama. In particular they are unable to address questions concerning the issue of order directly to the source of that order. If they do query these matters, Mr. Crimond replies in such phrases as 'We'll have to see', conjuring up an image of lofty deliberation among the immortals and of reserved judgements which may or may not be handed down. It should surprise no one that Mr. Crimond maintains a high degree of discipline without recourse to anything other than speech.

Torode also raises the question of different Voices of the 'I', particularly the 'I' of concern (that above which 'hopes') and the 'I' of action and authority, for Mr. Crimond occasionally speaks in the person of that Voice, as when he says 'I will not have that sort of thing'.

2. ACTOR ANALYSIS

In this section I shall be considering a person in action strictly in accordance with the dramaturgical model, that is as analogous to an actor in a staged performance of a traditional scripted play, or an improvised happening.

2.1. Social identity vs. personal identity

The most trite yet important distinction to bring to understanding life on the dramaturgical model is that between an actor and his or her parts. As a human being for whom acting is work or even a hobby, the person as actor in the theatre or in a film has an identity distinct from the parts they play. There might be problems for individual actors in keeping the distinction sharp. But clearly, ontologically, stage or film actors are primarily themselves, and stage parts have to be adopted.

When an episode of ordinary life is looked at in accordance with the dramaturgical model, this ontological relation is reversed. Except for machiavellian and socio-pathological individuals, people are primarily the parts they play, and the attitude of detachment that would allow them see their actions as performances of parts is a frame of mind which has to be consciously adopted and may induce a stultifying self-consciousness inhibiting convincing performance.

The psychological distinction between personal and social identity allows for the detachment of the actor as a person from the part, that is public self-presentations or personas in which one is usually almost wholly immersed. Detachment admits the possibility of control. As one detached from the action one can be an agent. But what can the actor control? To return to the source-model: on stage an actor must keep fairly close to the script, or in an improvised drama, the scenario, otherwise they will lose the presentation as a part. In very advanced experimental theatre that indeed may be the very effect aimed at. But usually the actor as creating the part is almost without effective agency. Only the ultimate agency remains — to stalk off and abandon the performance altogether. But while playing the part the actor can put a personal stamp on it — make it his Hamlet or her Ophelia. This is done by control of style, of the way the part is performed. As we noticed earlier, this allows an actor considerable expressive power, the power needed to illustrate the sort of person he or she is in the way the actions are performed. Actors allow their personal identities to show through the social identity they are forced to adopt. In this aspect of performance the psychological condition and the ontological category of both stage actor and the performer of daily life are identical. They must both monitor and control the style of the performance without becoming wooden or self-conscious. They must both be agents with respect to these matters, that is fulfilling projects of their own devising, free of promptings and controlling influences from other people and the scenes of the action. But adopting the stance of an actor and bringing action to explicit consciousness at certain times and moments in one's life can lead to the perception of disparities between presentations, personas, and our conception of character. A particularly prominent example of this phenomenon is the fits of seif-consiousness that can overcome an inexperienced or uncertain social actor, so that the reflection of oneself in the eyes of others can become so dazzling as practically to stultify action altogether. I take the form of consciousness which children experience in what we call 'showing off to be very similar. Struck by the disparity between the presented self and inner being they force embarrassingly over-presented personas on their audience. These are examples of the condition one might describe as 'over-awareness' of the management of one's actions. In the condition I have just described, the managing self and our conception of how we want to be, and the way we believe we are presenting ourselves, become the focus of explicit attention. But most people act in the social world in an unselfconscious and often unreflective manner, lost in their activities and intent upon their goals. This can lead to an under-awareness of the actor and personal aspects of one's life in the social world. The difficulty of sustaining an adequate conception of the complexity of oneself as a social person has been beautifully described by Doris Lessing: 'This is what it must feel like to be an actor, an actress — how very taxing that must be, a sense of self kept burning behind so many different phantoms.'

To grasp the complexity of the relationship between, and differentiation from an actor and his part, I offer as an example an event which occurred recently in Denmark. A man went into a chocolate shop to buy some confectionary. There was a customer ahead of him—a lady with a little dog. The shop keeper offered the dog a chocolate. The dog refused and left shortly afterwards with its mistress. The man turned to the shop keeper and went, 'Woof, woof!' and was given a piece of chocolate by the proprietor, who remarked, 'You ought to have begged as well'.

The first point to notice is that the legitimacy or propriety of social acts are related to the part in which they are occurring. The man in the social or dramaturgical part of 'Dog' could do things he could not do in one of his other parts, for example university professor. The shop woman was acting with perfect propriety in rebuking the man as 'dog' for leaving out part of the ritual proper to that part; as 'dog' he should have begged as well. But who was she rebuking? It seems to me clear that she was rebuking the individual as managing or controlling self, neither 'dog' nor the 'professor'. It is the managing or controlling self that is the proper object of rebukes of that nature, and, to put the matter more grandly, is the object of moral praise or blame. One might add that the shop keeper had let down the man as managing self rather lightly since as 'dog' he should have wagged his tail after receiving the chocolate. The important point to notice is that social failures occur relative to the parts being undertaken and the personas being presented, and are part of 'drama criticism', but rebukes occur in the moral world and are directed to the man that lies behind the 'dog'.

2.2. Techniques for the presentation of social selves

For the most part, the presentations of self as this or that persona proper to a certain kind of social event and amongst people of a particular sort, are achieved not so much in the instrumental activities in bringing off practical tasks such as counting money, driving cars, eating peas, making legal judgements, delivering lectures, screwing nuts on bolts, and so on, but in the style in which those activities are performed. Self-presentation is described in adverbs such as 'reluctantly', 'churlishly', 'gloomily', 'cheerfully', 'carefully', and so on, rather than in verbs of action. Impression-management, as Goffman calls it, consists largely in the control of style. Attributions of character are made to a person by others pretty much on the basis of how they see the style in which he or she performs the actions which are called for on particular occasions. Explicit statements or illustrations of personal qualities are usually unacceptable. They can be criticized as boasting or coming on.

However, the control of style leading to the attribution of character, however effective, takes time, so that to be seen to have authority or strength of character, or to be weak and easily led, excitable or withdrawn, are reputations which may take months or even years to achieve. But the practical purposes of society require certain people to be seen to have personal characteristics such as authority, sympathy, or wisdom directly and immediately. The solution to this practical problem is found in the use of regalia, uniforms, and so on, where the regalia suggest a specific character by framing and determining the persona that can be presented. A glance at the uniforms of the police of various nations is enough to establish the point, but priests and professors, radicals and air line stewardesses are all dependent on the same device. In accordance with the dramaturgical framework I shall call this 'costume'.

The most general form of regalia is clothes, illustrating and commenting upon the body, which they emphasize by concealing. This point has been made most elegantly in the following passage from *The Ogre of Kaltenborg:*

> I observed eagerly how their personalities altered with each [costume]. It is not that they came through the clothes as a voice does through a wall, more or less distinctly, according to the thickness. No — each time a new version of their personality is put forward, altogether new and unexpected, but as complete as the previous one, as complete as nakedness, it is like a poem translated into one language after another which never loses any of its magic but each time puts on new and surprising charms. On the most trivial level, clothes are so many keys to the human body. At that degree of indistinctness key and grid are more or less the same. Clothes are keys because they are *carried* by the body, but they are related to the grid because they cover the body, sometimes entirely like a translation *in extenso* or a long-winded commentary that takes up more room than the text, but they are merely a prosaic gloss, garulous and trivial, without emblematic significance.
>
> More even than a key or a grid, a garment is a *framing*. The face is framed and thus commented on and interpreted by the hat above and the collar below. Arms alter according to whether sleeves are long or short, close-fitting or loose, or whether there aren't any sleeves at all. A short, tight sleeve follows the shape of the arm, brings out the contours of the biceps, the soft swelling of the triceps, the plump roundness of the shoulder, but without any attempt to please, without any invitation to touch. A loose sleeve hides the roundness of the arm and makes it seem slimmer, but its welcoming ampleness invokes a caress which will take possession of the arm and go right up the shoulder if need be. Shorts and socks frame the knee and interpret it differently according to how low the first come and how high the second.

A socially symbolic object, like any other symbol, is partly defined by syntagmatic relationships with all the possible structures into which it might fit without loss of intelligibility. Sometimes these relationships may be very narrow. For example, roundlensed steel-rimmed spectacles cannot be replaced by any other form of spectacles, say *pince-nez* or horn-rims in a sartorial context defined by Afro-hair, Indian beads and flared jeans, without loss of intelligibility. So the paradigmatic dimension of round-lensed steel-rimmed spectacles is severely restricted. Equally we cannot insert round-lensed steel-rimmed spectacles into the context defined by low-cut shiny black shoes, white shirt, blue tie, grey suit, without loss of intelligibility. 'Just what sort of guy is that?' (Wolfe, 1968) But we can insert either *pince-nez* or horn-rims in the latter context, while horn-rims, though not *pince-nez,* can go into the context, suede desert boots, cavalry twill trousers, viyella shirts and knitted tie, and thick medium-length dark or black hair. Thus syntactically, round-lensed steel-rimmed spectacles are also very circumscribed, which horn-rims have a broader syntagmatic dimension.

Secondly, many symbolic objects do not have a meaning in isolation from their opposites. They mean only as contrasting pairs of symbols. Hair length has once again, in recent times, come into use as a social and political symbol. Historically it has not been the length as such, but the long/short contrast that has had political significance, that is the semantic unit 'long hair' is embedded in the structure 'long/short'. This explains how 'short hair' could be radical in 1640 and 1780, and reactionary in 1965. This example illustrates the way something which may appear at first sight to be a semantic unit in itself is, on more careful analysis, seen to be significant only as a member of a pair. Long hair is currently (or more accurately, was recently) used as part of a heraldic display manifesting a symbol for a radical political orientation. This went with round-lensed, steel-rimmed spectacles, flowing clothes and the like, the semantic unit comprised by the hair length is, for this total object, a diachronic entity '— as opposed to —' which is an opposition over time, that is long hair is worn, not just in opposition to the short hair of the squares, but as opposed in time to short, that is 'long, formerly short'. And of course either length can be a realization of either formula of opposition, the synchronic or the diachronic. The same explanation is available for the apparent contradiction between the role of a brassiere as a radical garment in the late nineteenth century and its discarding by certain radical women in the sixties. It was both in synchronic relation to those who continued to wear it, and in diachronic relation to the previous state of the radicals. And in some lexicons it has a meaning in itself, as a way of emphasizing basic femaleness, and of course, in its absence inhibiting physical actions which are deemed proper to the male, such as running or chopping wood.

However, as a general principle the basic form 'x as opposed to y' has no particular temporal order built into it. One could choose an instantiation of the relation now, in anticipation of the appearance of its contrary later. Though peculiarly appropriate to radical heraldry the basic form 'x as opposed to y' has been a very common form for the conveying of social meaning. For example, women have used the up/down contrast in hair style for expressing socio-sexual status, for example to put up the hair showed that childhood had ended and that the woman was marriageable. In a somewhat similar way the contrast clean-shaven/bearded expressed social distinctions amongst the Romans. In early days slaves were clean-shaven and their masters bearded, but in the reign of Hadrian a technological revolution in shaving techniques brought by Sicilian barbers to Rome made shaving much less disagreeable, with the consequence that Hadrian decreed that slaves be bearded, now that their masters were not.

However, as Cooper (1971) has pointed out, both long hair and beards have had a persistent standard meaning despite their frequent appearance as members of contrasting pairs. Long hair has generally been associated with romanticism, femininity and so on, while beards have usually been associated with intellectual and moral status as opposed to political, but *if* political, with conservatism, with the authoritarian father and the like.

The design of people then forms a very striking feature of the person as social actor. It is clear that such heraldic matters as hair-length and type of eye-decoration are parts of a more complex structure, the whole ensemble including clothes and shoes, and ways of walking and holding the arms, and so on. In the American West 'cowboy' is done by some cowhands even in their Sunday suits. It is predominantly marked by a way of walking.

But far the most important structural element in the design of people is clothes. There have been one or two inconclusive studies relating skirt lengths to economic factors, but they have paid little attention to the expressive features even of such correlations, if they could be established. In this chapter I can only draw attention to two features of the clothes in person design, basing my remarks on little more than impressionistic evidence.

The first point to remark about clothes as structured entities concerns their role in socially marking sexual differences, so that one can tell at a glance whether one is going to meet a member of the opposite sex, of the same sex, or a homosexual. Traditionally these have been marked by differential markers in all three possible modes, primary, secondary or tertiary differentia. By primary differentia I mean anatomical differentia based upon genitalia; by secondary I mean anatomical differentia, such as relative hairiness, bone formation, general outline of the limbs, face and so on; and by tertiary differentia I mean markings by different forms of clothes, or by differential regalia, such as different forms of decoration as among the Australian Aboriginals, or by such matters as the length of hair. In societies where 'unisex' fashions in the basic structure of clothing are predominant, such as among Western university students, or in the Muslim world, recourse may have to be made to secondary or primary differentia. Muslim men and women are differentiated by subtle stylistic differences in their *shalwa* and *kemis,* and by the use by women of various forms of face concealment, elaborations on a tertiary theme. Students generally have had recourse to both secondary and primary differentia. When the current fashion was for long hair in both sexes it was accompanied by the growth of beards amongst men, marking them off at a glance. There are, of course, certain surviving subtle modifications of the 'unisex' style so that some tertiary differentia do remain. But in general social marking is by secondary or primary characteristics. The survival of these markers through the transition to different forms of clothing discloses what I should like to identify as a 'social universal' or 'equilibrating principle' that requires that certain differentia be preserved through formal transformations. The way such principles are recognized, learned and promulgated is a much neglected branch of social psychology. Finally one should remark that the contrast heterosexual/homosexual, which *a fortiori* is incapable of being marked by primary or secondary differentia, is usually presented in both sexes by modifications of the basic form of the biologically appropriate clothing and regalia so that they are styled in the predominant stylistic mode of the sex with which the homosexual is identifying. Thus women might continue to wear jacket and skirt, but have them made in mannish materials and styled in a mannish way. The point may be underlined by the choice of accessories from the repertoire of the opposite sex, such as a handbag by a homosexual, and a collar and tie by a lesbian. By these public displays the differentia are the markers of the social dichotomies that go with various sociosexual categories.

But there are some structural differences in clothing which are manifestly but mysteriously related to the expression of social matters. So far as I can tell, so little is known about these phenomena that I can do little but describe them. Both men's and women's clothing is modified diachronically along a number of dimensions, long/short, loose/tight, what I can only call 'apex up'/'apex down', elaborated! nonelaborated, and there are no doubt others. To illustrate, the 'zoot suit' of the nineteen forties had a very long jacket, while the predominant jacket length in the fifties and early sixties was short. Trousers, which were tight in the Edwardian period, were styled in a loose manner up until the late fifties. In the forties and fifties men's clothes were designed as a triangle with the apex down, wide shoulders and narrow hips, but the introduction of flared jackets and trousers created a silhouette with apex up. And of course elaboration with more buttons, waist coats with lapels, turnups and so on, has come and gone. I would like to put these changes to the social scientist as problematic, through and through, both as to their genesis and their spread through the population of clothes. I believe that in the iconography of clothing there is a ready-made model for all forms of social change, and recommend it for the closest possible study, relatively neglected as it has been hitherto.

3. KEY CHANGE: ACTORS IN RELATION TO THE ACTION

Subtle differences in the musical meaning of a melody can be discerned when it is presented in one key rather than another, particularly if the key change is from major to minor. Goffman's (1974) metaphor of the social psychological 'key' draws our attention to the not-so-subtle differences that exist between the way a stage actor is related to his or her portrayal of a simulacrum of some slice of life and the way a person engaged in daily living of the 'same' slice of life is related to their actions. We could express the contrast as that between playing the melody in the 'dramaturgical key' and playing it in the 'reality key'.

What is the difference between play-acting a quarrel and quarrelling? Between play-acting the signing of a contract and really signing one? It cannot be in any of the actions performed.

They may be qualitatively identical. The example of the contract gives us a clue. The difference seems to be a matter of commitment. It is not anything in what is done. That is, the difference does not lie in the action. The whole of the difference must reside in the act, since it is by acts that we are committed. The play acting does not create acts. What is done on the stage has no illocutionary force, outside the boundaries of the play itself. The actors know this of course, so therein, I propose, we see the distinction between the seriousness of everyday life and the 'frivolity' of the stage. Those who treat everyday life like a stage play are importing just that feature, namely that their actions have no force as acts, and hence their performance involves no commitments.

Taking an action seriously, in the 'reality key', is to take myself as committed, and this is to take the future of the relationships forged and reforged in the episodes of everyday life, as preempted by the acts jointly performed by the relevant parties to an episode. A making and accepting of a proposal of marriage by the actors on the stage is not expected by either actors or audience to eventuate in a marriage. Why? It can only be because the act-force of the actions is contextually relative. There is a Burkian 'ratio' between scenes and acts. It follows that the exposition of the rules and conventions of how social life is to be done is not a complete social psychology. It falls short by the crucial issue of commitment. And yet the discovery of the rules and conventions, through which the correctness and propriety of life events is sustained is an essential part of any social psychology.

It remains to remark here that a key change is a change in how things are taken. What does that amount to psychologically? An act of commitment must involve at least the acquisition of some beliefs. Since beliefs are dispositions, acts of commitment involve changes in the dispositions of an actor, as one of the parties to the joint creation of a commitment. The beliefs involved are tied in with and possibly constitutive of attitudes. Again we find ourselves confronting the psychological phenomenon of dispositions. Since attitudes are interactively displayed, and in real cases, drawn from a repertoire of possible attitude displays, acts of commitment must be understood contextually. Much remains to be worked out both conceptually and empirically before an adequate account of the psychology of commitment as a discursive practice is fully understood. . . .

4. THE REPRESENTATION OF ORDER: ROLES AND RULES

4.1. 'Rule' as a descriptive and as an explanatory concept

The upshot of the application of the methods of enquiry sketched [above] can be expressed in the form of sets of rules. This is a natural way of writing out a description of the norms that seem to be adhered to by some cultural group, in their daily activities. The task of the social psychologist, as we see it, is the explicit presentation of the tacit knowledge of a group of people. Or to express the point another way, social psychologists look at the world of social interactions in the 'dramaturgical key', while for most of our lives and for most people life is lived with others in the 'reality key'. We find it natural to express what we think we know about the normative conventions of a society as sets of rules. But there are some misunderstandings to which this procedure is particularly liable.

It is tempting to think that if these are the rules, then orderly slices of life are produced by following them. This is the picture that springs to mind when we think of ceremonies and other formal episodes. Indeed, in such cases as the conferment of military honours, the participants may be consciously following explicit rules. But 'conformity to rule' may have a quite different interpretation. We can draw on Wittgenstein's (1953) insights into the matter. He devoted considerable effort to clarifying the role of rules in all sorts of human activities, including calculating. To put the matter bluntly, should we take it that rules are the causes, or among the causes, of orderly behaviour? Or should we say that the only active entities in the human world are people, who use rules for a variety of purposes? Wittgenstein argued, through the use of a great many examples, that only the latter view made sense. In outline his argument is very simple. If rules were the causes of behaviour then they would, so to say, fix the future. He imagined a boy being taught to do arithmetic. Adding is a skill, or even a habit. After using the 'add 2' rule to create the set of even numbers as far as 1000, the boy then goes on '1004', '1008' and so on. Of course that is a way of using the 'add 2' rule! But not the way the teacher meant. If the rule were causing the boy's answers then *ceteris paribus* he could not offer those as correct answers. It is not that we are never sure what rule we are using. Rather there is always the question of how we are to apply the rule. Any rule could be applied in countless ways. As Wittgenstein points out there could not be an infinite regress of rules for applying rules for applying rules. We could never act at all if that were so. People use rules, rules do not use people.

The conclusion we must hold on to is that the job of the psychologist is to describe the ways that people do indeed use rules. These range from the use of explicit rules by the actors in formal episodes as templates for action, to the use of a 'rule' metaphor by psychologists to describe the orderly behaviour of pedestrians in the Street. Included in this spectrum there is the important case of the use of rules to remind, instruct and admonish people about what they should or could be expected to do. Given this diversity the opportunities for confusion are many.

As a concept for social psychology, 'rule' and its natural partner 'role' are terms of art. The 'rules' that social psychologists assemble are of essentially the same status as those that people use to comment on the behaviour of others and to instruct them in proper conduct. We can say that in order to behave consistently with one another people must have tacit knowledge of the rules. But that is not to be taken as an invitation to imagine a hidden information processing mechanism programmed with these rules and causing people to behave in an orderly fashion. Once again one must remind oneself that while the question 'What caused X to exhale violently?' is a proper scientific question, with a proper procedure for finding an answer, the further question 'What caused X to blow his/her whistle?' may have no proper answer. 'Referees should blow their whistles when a player is offside' is not a causal law. Where the citation of rules is germane to under standing what is happening it is best to say that the dichotomy 'caused/not caused' does not apply. If we express these insights in the form of the statement 'rules do not cause behaviour' it sounds as if the statement 'rules do cause behaviour' makes sense and just happens to be false on the whole. The point is rather that these statements do not make sense. They are an illegitimate mix of incompatible concepts.

CHAPTER FIFTEEN

CONVERSATION AS SOURCE MODEL FOR HUMAN SCIENCES

1. THE ORIGINAL ROLE-RULE MODEL

1.1. Towards a new psychology

The first major step away from the primitive social psychology of the American experimental tradition began with a double shift in the foundations of this branch of human studies. Human beings were no longer to be taken as automata responding to stimuli, but as active agents trying to realize long term projects and to achieve short term goals. What people did in social encounters was no longer to be taken as mere responses, but as monitored and controlled role performances in accordance with socially given sets of rules. Given this new view of human beings and their collective and co-ordinated conduct the goal of empirical enquiry had perforce also to change. This was the second step in the creation of social psychology that could make some claim to be taken seriously and so be counted worthy of the accolade of being counted as a genuine human science. Participants' own accounts of their goals and the means they were adopting to realize them displaced the statistics of experimental manipulations as the prime source of data. Instead of constructing set-ups with dependent and independent variables in imitation of what we had naively taken to be the methods of the natural sciences, a new generation of social psychologists developed methods of eliciting participants' accounts. The new approach, as set out in detail by Secord and myself (1973), became known as the role-rule model.

1.2. Rules and cognition

In this paper I propose to explain why I am no longer satisfied with that model. In so doing I hope to be able to suggest where studies of mean-end structures in the control of action can be reinterpreted and elaborated in ways that are still coordinate with the latest version of the ethnogenic point of view in psychology, a version which has brought it close to general social constructionism (Coulter 1989). But before I can undertake that explanation I must spell out the details of the original proposal more carefully. When a description of a role-rule model for some characteristic social episode was written out as a scientific paper, what exactly had been recorded? For example, what interpretation was to be placed on the content of the paper in which J.P. de Waele and I (1977) described the various role-rule models for the ways in which strangers were brought into a social relation to the members of some existing group through introduction ceremony? In the original programme such writings served two functions. They constituted an anthropological report of the norms of a certain class of social interactions routinely undertaken by the members of a certain tribe or tribes. Whatever the minor variations, actual introduction ceremonies converged on something like structure we had described. Our report could also be read as a set of hypotheses as the cognitive resources that were necessary to whatever information processing lay behind the overt actions of the participants. As such possession of these cognitive resources explained the ability of each member to act correctly in the circumstances of the given interaction type. In our model, the actor was assumed to be able to recognize the kind of interaction that was under way and thus to draw on the appropriate of role-rule norms to regulate his or her actions. This interpretation was used by Rosser (1979) in her work on the production of orderly disruptions of official school room routines by disgruntled adolescents. It was also assumed that the propositional format of a rule-expression was adequate to capture the psychological reality of those norms. Rules as known to actors could then be assimilated to the category of beliefs and so be made to fit in with the developing theories of cognitive science as the testing of hypothetical desire/belief systems.

1.3. Acts and Actions

How were the episodes of coordinated and collective interaction into which human life could be analytically divided to be analyzed? A major contribution from the new social psychology to the study of both human and primate behaviour came from the attempt to answer this questions. Goffman (1981) used the term "strip" to refer to short, self-contained episodes. Of what units was a strip composed? This is essentially the same question as that posed by those whose unit of macroanalysis was the episode. A clue came from Austin's (1962) speech act theory. To understand the way language is used in everyday life, particularly when moral issues are relevant, one must pay attention to the social force of the utterance of a sentence rather than to its literal meaning. Speech act theory drew the attention of social scientists to the ways acts are accomplished linguistically. In particular it enabled investigators like Clarke (1983) to account for the structure of conversations by reference to the sequencing rules for utterances considered as speech acts. Also see Harré (1979). Speech act analysis requires a three level scheme. Articulated sounds are heard by those who know the relevant language as meaningful utterances, intended by the speakers. These utterances are effective in conversation just in so far as they are intended by the speakers and understood by the hearers, who are themselves also speakers, as having certain act forces. So there is mere vocalizing behaviour, which is taken as a speech action, by which a speech act is accomplished in the appropriate social group. Marga Kreckel (1981) found that there are some speech act conventions that are peculiar to a single family.

In 1981 this scheme was generalized by Reynolds, myself and others as an analytical tool for all forms of coordinated collective human interaction, as the act/action/behaviour distinction. This three level scheme fits in very well with von Cranach's ideas about goal setting and the means/end structure of action. Acts are the goals, actions are the means. So actions are at once intended behaviours and the accomplishment of intended acts. I mean to move my hand up and down in order to bid you farewell. It is important to emphasize that neither Reynolds and myself nor von Cranach thought that actors would necessarily be consciously aware of goal setting or of the choice of means to accomplish the goal once set. Some aspects of this scheme have not been very fully followed up by British social psychologists.

I shall call the distinction between acts and actions the "upper boundary" of the scheme and that between actions and behaviours the "lower boundary". At both boundaries the link between acts and actions and between actions and behaviours is created by cultural conventions. It is a cultural fact that closing one eyelid is taken as an intended actions, while it is also a cultural fact, though a different one that in some places that action is a sign of complicity. The same physical movement may be taken as intended action in one culture and ignored or treated as an inadvertence in another. Similarly at the upper boundary, the same action may serve to accomplish an act of a certain type in one social circle and of quite another type elsewhere. For instance the making of an appointment for a certain hour is taken as an expression of firm intention to be there by British and Americans, but is meant as at most a loosely defined hope by many Greeks. Joint Anglo-Greek academic meetings can be made quite difficult by reason of the inevitable clash of conventions.

Another important aspect of this important threefold distinction has recently come to light. Some feminist social psychologists have tried to give body to the idea that there are some people who live out contradictory episodes. Clearly some hierarchical analytical scheme would be necessary in order for the idea to make any sense at all since behaviours, as phenomena in the physical world cannot be "contradictory". By that I mean that a simultaneous movement of the hand both up and down is impossible. This is not because there are some adamantine laws of nature which forbid it, in the way that physics forbids flying to the moon by waving one's arms, but that the purported description of what is to be done is self-contradictory and therefore makes no sense. The intuition that some people are called upon to live out contradictory strips of life can be made sense of by reference to the hierarchical structure of episodes when analyzed according to the threefold distinction between acts, actions and behaviours. Since behaviours must form a linear stream of non-contradictory physical movements, the "contradiction" must be in the higher levels of the structure. Acts are the conventional meanings of behaviours intended as actions so there is no ontological bar to one and the same behaviour having two incompatible conventional meanings. This can be most readily seen if we locate one member of the act pair in the practical order, say as a request, and the other in some currently valid expressive order, say as an act of condescension. Incompatible acts can be performed simultaneously by the same person. What of the next action relative to the action sustaining the two acts. In general, one would expect there to be a complex interweaving of two or more action streams each of which the double act in question would have a place. Of course, it is theoretically possible for there to be indefinitely many distinct acts performed by means of same action/behaviour, "same" here defined by reference to physical properties also. And extending that idea, it is also possible for there to be many incompatible episodes considered with respect to acts, accomplished by the same sequence of behaviours. Of course this is only one kind of contradictory human action. There is also the case which someone acts contrary to their declared intentions. Then there are those people who display great variations in personality not only when they interact with different people something we would expect on general theoretical grounds but also when they are in the company of the same person. That is one of the cases in which people accuse each other of inconsistency.

2. PROBLEMS WITH THE ROLE-RULE APPROACH

2.1 An over-simplified cognitive theory

In recapitulating the early forms of action theory in psychology I used the current way of describing the role-rule approach, namely as a model. There is a persisting source of confusion in psychology in that the word "model" is used equivocally. In the physical sciences, a model is an analogue or ideal form of an entity, structure or process. The thought behind this use of the word is that for certain purposes it is often easier to study something that resembles what one is really studying, rather than tackle the real thing. In psychology this use does occur though it is rare. The usual meaning of "model" is "theory". So the role-rule model is best read as the role-rule theory of human social action. It was not intended as an ideal or analogue of human action, though it involved a hierarchy of models, from the dramaturgical to the ludic. The concepts of "role" and "rule" were intended literally. But the theory can be weakened and at the same time be given greater power by actually being offered as if it were really a model in the sense in which that term is meant in the physical sciences, namely as an analogue. We can now ask in what respects is what people do like acting out roles. We can also ask whether there is anything in their ways of talking and thinking that is like having recourse to rules.

In the original formulation of the role-rule model it was assumed that there were only two kinds of episodes relative to the applicability of the methodology of account analysis. Either the model did apply or people were simply improvising a moment civility which could break down at any time (Harré 1979). This dichotomizing of episodes is purely mistaken. For me it came about because of fascination with the power of the liturgical model. It seemed that a great many episodes of ordinary social life were very like ceremonials. In ceremonials there are clearly assigned and accepted roles and publicly formulated rules according to which an actor is to regulate his or her behaviour.

Work, such as that by Douglas (1975), in which she showed that there was a powerful liturgical element in the progression of entertainments by which intimacy is established, seemed to confirm the methodological wisdom of a ubiquitous application of the model. It became clear, however, that there was another model, in the sense of that term in the physical sciences, which was at least as important in understanding the sources of and means of maintenance of social order. This was the idea of the lived narrative.

In the work reported in von Cranach and Harré (1982) there was an almost universal assumption that for each person engaged in producing a strip of life, even when that strip involved essential contributions from others, there existed only one means-end hierarchy, though which that person's actions were guided. But more recent reflection on the complexity of act/action behaviour hierarchies has shown that not only in principle but also in actual lived episodes, people can and do perform more than one act in the doing of an action which is physically realized in an identifiably singular piece of behaviour. Multiple goals may be set and yet a singular means for their achievement may be selected.

2.2. The essential role of narrative conventions

In addition to roles and rules the analyst of social action must also pay attention to characters and story lines. While the liturgical model emphasizes the analogy of certain strips of life to fixed and formal routines, the narratological analogy emphasizes their dynamic and negotiable characteristics. Both analogues impose a radically different framework from the experimental tradition since neither involves any reference to causes. Each analogue can be read as a source for hypotheses about the norms to which members subscribe, and which for one reason or another they will try to realize in action. The two analogues are complementary in that they allow an investigator of a form of life to focus on orderlinesses which comprehend different time spans. The dramaturgical and liturgical analogues illuminate the structure of relatively short and self-enclosed episodes, while the narratological analogy enables one to get a feeling for the orderliness of a whole life. Character takes on a double meaning for the social psychologist. It is at once a concept referring to the moral qualities of an actor and also a concept that can be used to refer to the exemplary person of a favoured narrative. But how then does it differ from role? Role is fixed, preordained by the existing social order. Character is something that can be built, adopted, transformed. In short while one's roles are determined by the impersonal forces of collective hegemony, character, in both senses, is a personal creation. One writes one's own story. Of what use then is the narratological model? I think the idea of genre is the key. It is not so much that someone lives out another version of the life of Anna Karenina, through the *représentalion sociale* of that actual novel. Rather there is a genre of tales in which a neglected wife embarks on a love affair with a romantic other, who eventually tires of her by reason of the very qualities of character that led her husband to become indifferent in the first place. How do people know how to conduct business negotiations, love affairs, heroic exploits and so on? Narratology is based on the thesis that there are socially maintained and publicly displayed exemplars of these life forms, and that people draw upon them. They are not used as rigid templates with which to conform, but as genre models to suggest ways of doing things that make a kind of sense. When one opens a book called something like *Myths and legends of the Maori*, is one making contact with the narratology of those people? Did the tale of Maui show young Maoris how to be heroes? The anthropological and historical literature is silent on the question of the relation between tales that are told and lives that are lived.

It is worth remarking that the act/action/behaviour distinction applies to human activities at both poles of the spectrum of normative controls, the role-rule model and the narratological analogy. The difference in practice is confined to the "place" to which one goes to find the sources of the taxonomies of acts through which people try to make what they do intelligible and with which they grasp the sense of what others are doing. Quite early in the development of the ethogenic point of view it was realized that the creation of acts is an interactive process, best described by another analogy, that of negotiation. Acts are not what people individually intend by what they do, nor are they just what others take the actions of some individual to be. Acts are created by the joint production of meaning. At this point I am stating the main principle of social constructionism quite dogmatically. But if it is taken as a regulative principle of enquiry there are some well-defined research targets in sight—namely the discovery of the normative structure of these negotiations themselves, and subjecting that structure to just the same polarity of possibilities between role-rule model and narratological analogy as served to reveal the existence of normative order at the level of action in the first place. In short is there a metalife which we must live collectively so that first order social life should be possible? Are processes of goal-setting and means—selecting essentially social processes after all—and how are goals set and means selected for them? These questions raise serious ontological issues.

3. ONTOLOGICAL PROBLEMS: WHAT IS THE STATUS OF RULES?

3.1. Rules are not causes of action

I have written freely about rules, conventions and norms in what I have already discussed. I have assumed that these entities can appear in propositional form in accounts and function as guides to action in the cognitive hierarchies of means-end structures discussed in von Cranach and Harré (1982). They might be considered to be the same kind of entities that served as major premises in Aristotle's practical syllogisms. But how exactly do they exist? I used to think that the metaphor of "templates" for action would be adequate as a model for role-rule conventions and act/action/behaviour rules. A rule would be like a plan or template, which pre-existed the action and determined the form that action took, because the actor simply followed it. But "following" is itself a metaphor and needs to be spelled out. What is it that ensures that some strip of life will be sufficiently similar to a preceding strip to count as the doing of the same again?

Wittgenstein realized an exceedingly important fact about the way rules actually function in real life. It is evident that rules do not cause people to act in regular ways. If they did then there would never be any question but that a competent actor would continue repeating the same steps over and over again into the indefinite future. But a mere *rule* cannot preempt the future. We, the actors, have to be determined to apply it in precisely the same way over and over again. Must there be a rule which determines how the original rule is to be applied? Here a second problems appears. If we admit that rules for the application of rules are needed, are we not then committed to an indefinite hierarchy of rules for the application of rules for the application of rules, and so on indefinitely? There are now two problems to deal with. Do rules stand outside action, preexisting what people do and determining it? How can we ever be sure that rules are applied correctly if yet another rule is required for the application of the first rule? Both problems can be solved by attending more closely to the ways rules are actually used. Rules do not determine what will happen, that is what anyone will do, but they do determine whether what has been done is correct or incorrect. Rules express norms of intelligible and warrantable conduct. They are not the causes of regularities in behaviour. According to Wittgenstein this regulative use of rules depends on there already being regularities in human conduct. These could be either the natural regularities the predispositions for which we inherit along with the rest of our biological nature. Or they could be the patterns of regular action into which we have been trained as infants. Rules, as expressions of norms, obviously have a part to play in teaching someone the routine patterns of everyday life. "What you are doing is not correct. Try to do it this way!" says the teacher, and then demonstrates the correct way to drink one's soup.

3.2. Immanent rules and normative imitation

Further light is cast on the matter by drawing on yet another metaphor, the distinction in theology between immanence and transcendence. Do rules exist only in their use and application or do they have an independent existence? Well there *are* rules which exist independently of their use. There are all sorts of written or memorized codes which plainly exist independently. The social psychological concept of "rule" is drawn from these examples by analogy. To argue for a ubiquitous transcendental use of the rule metaphor in explanatory theories requires the invention of a special category of rules, tailor-made for the ordinary case where we just know how to do something without explicitly calling a rule to mind. These are the implicit or unconscious rules that "lie behind" those regularities which are clearly normative but for which there are no public or private rule consultations. I would like to argue that it is more economical and psychologically more plausible to suppose that there are no such entities as implicit rules. Instead we should look for immanent norms in actual, concrete strips of life. How could these support immanent norms? One way could be as cases that an actor remembers and tries to copy. Or it might be that there are concrete cases used in training someone in a useful or expressively proper routine, which then becomes inculcated as a habit. The model here would shift from the literal rule analogy towards something like a skill or competence.

Perhaps the difference between immanence and transcendence is most telling when we try to formulate an ontology for narratology. Told narratives are not transcendental templates for lived narratives. In real life, unlike those pale simulacra of life that old- fashioned social psychologists create in their laboratories, people frame and underpin their actions with snippets of autobiography. They are telling their lives as they are living them. And what is being told turns out to be some of the material necessary interpreting what has been done, is being done and will be done.

4. A NEW MODEL FOR THE ANALYSIS AND EXPLANATION OF SOCIAL ENCOUNTERS

4.1. Reflexive or "Inner" modelling in theory construction

In constructing his theory of organic evolution on the model of stock and plant breeding, Darwin used a methodology of very great sophistication. Human beings breeding their domestic animals and plants are not only a model for nature but also part of nature. Darwin used his analysis of the processes that occurred in one region of biology to illuminate the whole. Recently a new model or analogue has appeared in social studies though as yet it has been largely ignored by social psychologists. It makes use of the trick of reflexive modelling that Darwin used with such power. The idea is to take conversation as a model of social encounter and to use it as an analogue for other kinds of social actions (Mair 1989). Hints of this development have been about for some time, for instance in some of the opaque writings of post-structuralists. The notion of life as text goes some way towards it. The conversation idea has sometimes been taken further than a working analogue. I have myself proposed the idea that the concept of conversation might serve as a universal basis for an ontology for the whole of social life, including practical and economic activity. Just as in accordance with speech act theory we suppose every utterance to have some illocutionary force and to be significant to conversational exchanges only by reason of its meaning as a social act, so every human encounter is to be thought of as the accomplishment or failure to accomplish some social act or other through the deployment of actions which are literally conversational, that is have to be understood in terms of their illocutionary and perlocutionary forces. Pearce and Cronen (1981) have called this the "coordinated management of meaning". When analysts speak of "grammar of action" the conversational ontologist bids us to take this talk literally. Since there is nothing but conversation and some conversations exist only as commentaries on others, all strips of life are, in principle, nested hierarchies of conversations. What becomes of the questions concerning the maintenance of social order with which I was concerned in the last section?

I propose that any attempt at explanation of order should begin with a hypothesis of immanence, that is that order is created by reference to the exigencies of the conversation itself and not by reference to some transcendent template. This idea began life in ethnomethodology, where it was used to draw attention to the way that conversations proceeded by realizing the demands of certain adjacency pairs, such as question/answer. Unfortunately it became corrupted by a strict positivism. Only facts about adjacency pairs and others intratextual phenomena were accepted as genuine material for conversational analysis (Schegloff 1984). This development has been most unfortunate since it has meant that one of the great insights of early ethnomethodology has been lost, namely that the macroproperties of multiple sets of encounters must be immanent in the most intimate moments of everyday life. Of course there are social classes and social changes and so on, but they are not to be found in the numbers generated by large-scale surveys. They must be right there in the way people interact with one another in the most mundane ways. The social world does not happen. It is constructed.

4.2. The development of "conversation" as an "Inner" model

If the social world is nothing but a conversation, how are individual human beings to be understood in relation to it? In early forms of cognitive social psychology the whole of the cognitive processes of goal setting etc. for each person were conceived to be located in that person, consciously or unconsciously. But our new thought is that these processes are actually themselves also conversational. Many of them, and certainly the prototypes, will occur in interpersonal discussions of various kinds, organized according to a variety of possible power relations between those who take part. Looking at the possibilities this way forces us to take account of *another* important distinction, that between discursive and procedural knowledge. If we think of what individual people know as discursive knowledge we are in the realm of the common sense concept of belief. But if we think of what people know as procedural knowledge we are in the realm of the common sense concept of skill. On the skills view the cognitive account of the management of action should be interpreted as a private conversation, modelled on public discussions of what should be done. "What shall I do?" "How shall ***I*** do it?" "What will count as achieving it?". This is a conversational *model* for the grounding of a skill. It is a model of the prototypical case of thinking out what I must do to carry out a proper performance. But it is itself the exercising of a skill, one that has been learnt in practice, a conversational skill.

There is a further, even more revolutionary step to be taken if one adopts a conversational ontology. It has been assumed by all those who have adopted the general idea that people are active agents, that what anyone does is determinate. Even when, as above, I pointed out that acts only exist as joint products of intentions and interpretations, it still seemed as if each contributed to a definite outcome. But as Pearce and Cronen (1981) and others have shown records of strips of real life do not support that assumption. People make the indeterminate "matter" of the social world as determinate as need be for the purpose in hand. A great deal of the time we do not bother to sharpen either our own performance nor our apprehension of that of the others with whom we are then engaged. The concept of "umwelt" can usefully be borrowed from biology. By making the behaviour of ourselves and others determinate we jointly create a social living space, an "umwelt" peculiar to our local culture. It may be quite impenetrable to someone from elsewhere. This idea fits with Kreckel's observations about the idiosyncratic speech act interpretations that are used by each family to construct their private and unique social worlds. The idea of making something relatively determinate in the course of an unfolding conversation can be applied not only to the process by which utterances become determinate as speech acts but also to the way that bodily feelings become determinate as emotions in much the same way. Our skills as actors then are directed to making what happens determinate in acceptable ways.

In summary then, I believe that progress over the last ten or so years has been patchy. While the role-rule model and the cognitive theory of goal setting and means selection that went with it were adopted for a while by traditional social psychologists such as Argyle (Argyle, Furnham, and Graham, 1981) many have slipped back into the bad old ways, producing work of little permanent value. An ironic aspect of this backsliding has been that during this period social constructionism has been greatly elaborated both in breadth and sophistication. Whether the interaction between advanced theory and conventional practice that proved so fruitful in the nineteen seventies will ever recur is a matter on which I am unable to speculate.

CHAPTER SIXTEEN

AI AS HYBRID MODEL MAKING

I am far from wanting to mount an attack on the general project of using the computer and its programs as a resource in the furtherance of psychological research. My concern in this essay is not at all Luddite. Instead, my aim is to develop an analysis of a problem field for a possible psychology so that a certain version of the artificial intelligence/cognitive science program (hereafter AI/CS) can actually be productive. So far as I can see, the promise of AI/CS has far exceeded any of its deliverances. I believe that the trouble lies in the use by AI/CS practitioners of a distorted version of what psychology could be and what language is. This has encouraged unfortunate trends in AI/CS that have led the field to the kind of impasse chronicled by Winograd and Flores (1988). Some authors, notably Fodor (1981) and Dennett (1978) perhaps more than others, represent the point of view that is responsible for the current crisis. By reworking our conceptions of what a discipline of psychology might be, free of assumptions about what it would take for it to be a science, I hope to revive the AI/CS approach in a more promising way.

I take the techniques of the AI/CS to be essentially modes of modeling. In this respect they are very similar to the modes of modeling in use in general engineering, so I shall be drawing on the kinds of analyses of iconic and homeomorphic modeling that are familiar in the philosophy of science. Models have served two main roles in physical science and engineering. Homeomorphic models like the pictures of blood vascular systems to be found in textbooks have been fruitful as devices for the schematic representation and so for the analysis of complex phenomenon. Paramorphic models like lines of force have been fruitful as devices for controlling the formulation of plausible theories in a realistic style. These are theories which purport to describe plausible but so far unobserved causal mechanisms productive of the phenomena in question. I hope to show which of these aims AI/CS as a modeling technique might fulfill and to which phenomena it has a proper application.

Preliminary philosophical work is needed to define the proper field of psychology because, despite the short history of AI/CS, there are already tendencies for this very important development to go off the rails—largely, I believe, because it has been hooked to the wrong ontological locomotive. Like much academic psychology of recent provenance, it assumes the unexamined thesis of individualism in psychology, whether this thesis is taken mentalistically or physicalistically. The divide between the defenders of alleged folk psychology and those who would like to replace it with a descriptive and explanatory rhetoric derived wholly from neurophysiology has already been rendered empty by Wittgenstein's insights into the role of words like belief, want, and so on and their place in the universe of interpersonal action and intrapersonal reflection. Beliefs, etc., are not hypothetical entities in an alleged folk psychology of a hypothetically deductive kind. The Hertzian model of explanation is proper only to the physical sciences as Wittgenstein realized. There is no false or inadequate folk psychology on the Hertzian model to be supplanted by something better of the same sort. Such an idea is at least partly a consequence of the 'individualism' assumption. I believe that the most fundamental error reproduced in such cognitive psychology has been an assumption about the very nature of psychological reality itself, and it is to a repair of this fundamental error that I shall be devoting most of this essay.

I shall set out five theses, all, to some extent, controversial. Using these theses as a touchstone, I propose "to check out" how far a cognitive science could be developed by being made compatible with them.

By AI/CS, I understand the following double-sided project: (1) to develop an analysis of all forms of human activity, including perception, judgment, action, and so on, relative to some conceptual system, the content of which is yet to be settled; (2) to invent programs, in the technical sense, which could be run on some conceivable but not necessarily actual computer. In running these programs the output of the machine would simulate, to any preassigned degree, the human activity as originally analyzed. In AI/CS there is a thorough-going interaction between, or involvement of, the mode of analysis of human activity and the form and status of the output of the programmed computer.

The metaphysical or ontological element, which has been present in cognitive science since the very beginning and which has proved attractive to scientific realists, is the assumption that structures of successful programs will, in some way, serve as models of the structure of the mind. Furthermore it is hoped that the structure of the computer, when so programmed, can serve as a model of the structure of the brain. At least these two modeling assumptions are built into the project. These assumptions are not in question in this essay. At issue will be: What is the psychological reality for the representation and understanding of which the modeling enterprise is to be undertaken? In particular, what is the nature of the mind?

1. THE FIVE THESES

1.1. Hacking's Thesis: That Mind Is Not a Trans-historical Category

According to Hacking, there is no such thing as *the mind,* of which cognitive science could be true or false. The word "mind" does not refer to some genus of entities, innumerable examples of which have appeared and whose nature is open to investigation. It is not a word like "virus." "The mind" might figure in a theory of what it is for people to think. Such a theory might be modeled on the kind of theory exemplified by the idea of the molecule, which figures in a theory of gases expanding. An entity of this sort provides an ontological basis for an account of the genesis of a certain kind of process. According to Hacking, there are only 'local' concepts of "the mind" arising out of whatever psychological "science" is currently in favor. Thus, apropos of AI/CS, he says, epigrammatically, "The mental life' is not modelled by the programme, it becomes the programme." So, in assessing the standing of AI/CS, we must compare the mental life as it emerges in cognitive science studies with the mental life as it emerges in other kinds of study, for example, socio-linguistic studies. The mental life is what we are trying to give an account of, but which aspects of our mental lives are in focus is determined by the *form* of the kinds of study that we are currently favoring. For instance, is it the kind of study whose form involves hypothetical entities? To make comparisons of mental life pictures properly we need to clear up certain muddles about the role of ordinary language in the production of minds. In particular we must clear up the folk psychology muddle that has been at the heart of much of the debate about the apparently bizarre proposals by the Churchlands (cf. P. Churchland 1986) for a neuropsychology which is to exclude all mentalistic concepts by relexicalizing all psychological terms as physiological technical expressions. It is worth remarking at this point, to forestall misunderstandings, that only a Luddite would think that the comparison between the mental life as it emerges in AI/CS and the mental life as it emerges in social linguistics leads to some kind of competition. It is not that one of these is going to stand in place of the other.

According to Hacking, AI/CS picks out certain aspects of the mental life, particularly those involving tasks and achievements, by virtue of the kinds of studies undertaken. It is not that something else could be studied by current AI/CS. This restriction is built into the very language in which the AI/CS project is presented. So, AI/CS contributes to an understanding of an important part of human life as it is currently lived, namely that kind of life that is particularly associated with entrepreneurial capitalism. But that is by no means the whole of life, even life as lived by modern people. AI/CS as it is currently presented, that is, through a rhetoric by which its projects and results are told as narratives of a certain kind (celebrating success rather than salvation), runs the danger of reintroducing Lockean ideas with the use of the tendentious concept of mental representation and of reinventing Cartesian mind-stuff with the use of the tendentious concept of mental states. I shall return to the issue of mental states it later. Since it is an essential part of what I shall call the Level One AI/CS rhetorical apparatus, its reevaluation will play an important part in defining Level Two AI/CS.

My problem is this: Can we inoculate AI/CS in its proper role as a way of contributing to the understanding of a certain part of contemporary life in such a way that it will not continually lead its practitioners to slip back into the reconstitution of the ontology of a seventeenth-century psychology that we ought long since to have abandoned? Hacking's thesis is an important contribution to this inoculation. If mind is not a transhistorical category, identifying a transhistorical ontological species that has been continuously reproduced as individual "minds," then the seventeenth-century thesis that minds are sets of ideas or immaterial substances is not something that needs to engage us at all in our construction of the psychology for current modes of everyday living.

1.2. Vygotsky's Thesis: "Mind" is a Collective Production Largely Mediated by Language

According to this point of view, "minds" are continually being produced, partly publicly and partly privately, in the discursive practices of communities of speakers. Mentality consists, in part, of the products of these practices, like rememberings, emotings, etc. But an adequate psychology must also include in its field the powers and skills needed to produce them. In this way of thinking of the mind there is no substance. The mind products are activities or actions or processes. Nouns like "memories" are to be taken as substantive metaphors, convenient shorthand, hypostatizations for complex social acts or the conditions for their possibility, and not as the names of mysterious entities which it is the job of the psychologist to track down and study. Beliefs, memories, wants, and so on should not be taken as entities on the model of HIV viruses which it is the job of the virologist to track down and study. Emotions are abstractions from the everyday activity of emoting, of angry, penitent, cheerful *displays,* which are part of the ordinary conversational interaction of groups of human beings. Emotions are abstractions. At best, words like "anger," "sorrow," "happiness," and so on are stylistic devices for a rhetoric and not the names of mysterious entities. This is not the place to argue the Vygotsky thesis. I shall assume it as part of my resources for the project of revamping AI/CS.

In all of the cases of mind production that will be referred to in this essay, the human ground is primarily collective. Minds are produced in the joint activities of psychologically symbiotic diads, triads, and so forth of people engaging in conversation and other conversation-like symbolic activities. The grounding of individual skills is only secondarily personal, as individual and private versions of fragments and genres of the public and collective conversation.

I want to illustrate this point with the example of remembering as a human practice. I shall argue that characteristically remembering is a cognitive process that is accomplished in conversation distributed through the relevant collection. I shall also argue that, far from being an individual achievement, remembering is socially structured. One must distinguish between the personal and private phenomenon of recollection (the current presentation of some record of past events to oneself as such, the study of which was inaugurated by Ebbinghaus) and the full-blown human activity of remembering in which private recollections are certified as verisimilitudinous representations or presentations of some past occasion, state, or event. Clearly the psychology of remembering, in the full sense, cannot possibly be reduced to some feature of recollection, since there is nothing, and could be nothing, in recollections themselves that could enable anyone to determine whether his or her "memorial experiences" are verisimilitudinous or not, accurate or inaccurate, and so on. Something other than subjective conviction must enter into the process by which recollections are certified as memories.

One suggestion might be that certification takes place through some kind of testing procedure, a kind of archaeology of everyday life. Recollections are checked out against the traces that have been left by the events in question. For instance, a recollection might be checked out against an old photograph or it might be checked against a diary entry. One hardly needs reminding that checkings-out of this kind are really very rare. In most cases, recollection is not transformed into memory in that fashion.

If one thinks about how a committee organizes its memorial activities, it becomes clear that the transformation of a recollection into a memory is achieved very largely through a public and conversational process. It is often a matter of negotiation between the various parties with an interest in what is to count as the past of the institution or group of people in question. Recollections are set against one another, not against archaeological remains of everyday life. The minutes of a committee are constructed by the chairman and the secretary from the rough notes that have been taken during the meeting. With the certification of those minutes at the next meeting, a verisimilitudinous record is created there and then. There is no further question as to the true and accurate "record of the past." In this formal procedure, we have a kind of model of much of the memory work that goes on in everyday life (Haug 1987). The conversations that are common in families, in which members propose recollections and in which these are discussed and worked over to determine the family past, are very much the same sort of activity as the taking and certification of the minutes of the meeting. In each case recollections are juxtaposed to recollections and social power plays an important role in the outcome of negotiations of conflicting claims. However, in the case of those discussions in which the memories of families are created, the overt exercise of memorial power is much clearer than in the behind-the-scenes work of chairman and secretary in the preparation of the minutes of the meeting of an ordinary institution. Marga Kreckel, in her study of the conversations of the British Broadcasting Corporation's series "The Family" noticed on several occasions that discussions about the past were finally settled by the exercising of memorial rights by the mother, who simply declared what the past had been. Other research has shown a similar sort of power structure and the formation of memories in accordance with it.

There are plenty of other cases where we must free ourselves from the idea that something private and individual is behind a psychological process that occurs publicly and collectively. For example, if we consider emotions like indignation and regret, we have no need to say that a hidden moral assessment is behind the display of these emotions. Emotions, as displayed, are the form that the moral assessment sometimes takes. To display one's indignation is to make a moral assessment of some event that has occurred. Similarly to display one's regret is retrospectively to assess one of one's own actions unfavorably. Indeed, in some cases, one person's display may be defined as indignation, not by any intentional states of that person but by what the other people who will take part in the conversation do and say.

It was Vygotsky (1962) who first drew our attention to the fact that conversation is typical of the means of the collective production of mentality and also, one might add, the best model for other such production processes.

At this point we can begin to unravel the tangle of confusions about folk psychology. The main error, compounding their individualism, into which both the critics and defenders of folk psychology have fallen is to assume that the English words for all sorts of aspects of mentality work for the folk that use the language, as if they played a similar role to the theoretical concepts that are used in the terminology of physics to denote real or imaginary hypothetical entities. It was exactly this position that Wittgenstein pilloried in the psychological sections of Wittgenstein's *Philosophical Investigations* (1953). Both defenders and critics of "folk psychology" suppose that, for example, the folk treat beliefs as the mental states which purportedly lie behind declarations. Wittgenstein enabled us to see what is in front of our very eyes. Beliefs are the content of actual or possible declarations. A statement like "I believed so and so all along" is not to be interpreted as "somewhere in me was an entity, 'the belief in so and so' that, though, at that time unobserved, persisted in existence during the period in question?" Rather it means something like "I would have been prepared to say or to declare all along that so and so." What is the temporally continuous grounding of this disposition? There need be nothing *behind* a declaration but physiology. But a declaration is not a disguised physiological phenomenon. That is the mistake which the Churchlands make. A declaration exists only in a collectivity. It is incorporated, and only exists as such, in a moral order in which people as speakers stand in certain relations of mutual trust and commitment.

Belief is a conversational phenomenon. So the idea that folk psychology is a kind of diaphanous physics of the mind with beliefs and desires as its theoretical entities is an error common to both sides of the debate. It is not that folk psychology can be defended against critics such as the Churchlands by advocates such as Dennett. The whole idea is radically misconceived. There is no folk psychology that could be displaced by neurophysiology. The collective production of mentality by symbolic means does not stand over against neurophysiology. On the 'Vygotskian' view a reformed psychology would be dominated by the study of the "grammars" that is, the norms of order of those conversations in which mental products, like rememberings, emotings, declarations of belief, etc., are produced. This is not exhaustive of psychology, only of mutation. There is the whole field of manual skill to be investigated, the conditions of acquisition and use, the physiological groundings and so on.

The remaining three principles could be thought of as special cases of the two general principles with which I have begun this discussion.

1.3. Winograd's Thesis: Linguistic Productions Are More or Less Indeterminate, Their Meanings Are Essentially Contestable, and Retrospectively Revisable

The ground for this principle or thesis is that we must acknowledge three particularizing aspects of acts of speaking: (1) their historicity, the way in which the meaning of an utterance depends on the actual history of past uses and which is specific to the present moment of its production; (2) contextuality, the way in which the meaning of an utterance is made relatively determinate in the immediate social and conversational context of its utterance; (3) indexicality, the way the meaning of an utterance is dependent on the identity of the engaged speakers as embodied beings and persons of a certain moral standing. The meaning of all linguistic items is subject to historicity, contextuality, and indexicality and, thus, so are the meanings of all linguistic acts. This is an important thesis, indeed, perhaps the most important thesis for the critical reconstruction of cognitive science. One of the striking consequences of accepting it is that now the existence of norms of order become a problem. Given the historicity, contextuality, and indexicality of all utterances, how then can we say that there are right and wrong things to say, correct and incorrect, proper and improper contributions to a conversation?

Since there are no such things as constant transhistorical, transsituational, and transpersonal units of language as produced, all linguistic structures are indeterminate to some degree. Yet we are able to construct continuous conversations. This is achieved, I believe, by virtue of the general acceptance of some sort of principle of charity. For example, we overlook the minute contextual and indexical variance in the terms of a traditional syllogism. The difference in the context provided by the major and minor premises, together with the conclusion, are not sufficiently great to invoke serious equivocation. There will be equivocation but it may be of such little moment that the argument can be allowed to run through. However, it would be wise to assume that formal argument forms, if realized as conversation, are equivocal unless shown not to be, and so any kind of model using a formal calculus is to some extent suspect.

1.4. Davies's Thesis: Insofar as Mentality is Produced Conversationally, Contradiction is Sometimes To Be Tolerated

This thesis has been mostly developed in feminist linguistics to deal with the problems that are created, particularly for women, by the existence of a variety of discursive practices in which they must engage in order to complete, so to say, a day of ordinary life. In this context, a discursive practice is to be understood as a locally recognized way of producing a proper conversation and thus of producing some aspect of one's mentality, in particular oneself as a 'singularity'—a 'self'.

Many people organize their reflexive reasoning, their sense of themselves, around apparently formally contradictory declarations. So for instance, we might have the declaration "I am a mother of small children" framed within one discursive practice and "I am an academic sociologist" framed within another, both declarations made by the same publicly identified person (Smith 1987). Considered relative to each distinctive discursive practice, there is microconsistency. But since to live one's daily life requires the use of both discursive practices, there is no macroconsistency. There would be no matter of interest here unless there was some way in which the discursive practices interacted with one another and, indeed, that is exactly what happens.

Complications arise when there is a leakage between the constituent conversations. An interesting example (Davies and Harré 1990) is the Alwyn Peter case. Alwyn Peter was an aboriginal Australian who was acquitted of murdering his wife on the grounds that his culture had been destroyed. In a brief autobiography, Alwyn Peter claims both that he did kill his wife, and that he did not. The former declaration is made sense of within the discursive practices of the European system of legality and psychology in which intentional acts are ascribed to the person who performed the action. Alwyn Peter does not deny that it was by his hand that the fatal knife-thrust was made. However, the latter declaration makes sense within the discursive practices of aboriginal Australians, in which Peter's failure to conform to tribal law by marrying the woman he murdered was a punishable offense. But since the traditional penalty of spearing is no longer possible, Alwyn Peter believed that a collective tribal punishment was meted out to him through a magical impulsion to kill his wife when he was drunk. The suffering he has subsequently endured is the punishment for his violation of tribal custom. The leakage between the two discursive practices occurs in his account of the events that followed his second marriage. The second wife asked him to leave on grounds defined within European discursive practice, namely that he was dangerous (that is, not in control of his actions) when drunk. He himself understands this action as rational in its own terms, but *also* as embedded within the aboriginal discourse. The second onset of misery, subsequent on his expulsion by the second wife, is the indication that he is still being punished for his original violation.

Even if we restrict the application of Davies's thesis to only some discursive productions of self, the necessity to tolerate contradiction is fatal to the idea of a cognitive science program in which the whole of a person's language in use is formally mapped onto some logical calculus. The contradiction condition would be fatal to any such enterprise.

Theses three and four therefore impose constraints on the language models which would be necessary to represent linguistically mediated mental processes. They certainly run counter to naive cognitive science, but as I shall argue, they do provide opportunities for a more sophisticated version of that project. Theses three and four show, if they are correct, that standard formal logics cannot be adequate tools for representing mental processes or structures if the mentation is conceived according to the Vygotskian model, that is, conversationally. Thus we reach a disturbing intermediate conclusion: naive cognitive science is not an empirical descriptive project, but it is normative and prescriptive. It presumes individualism and logicism as essential conditions or perhaps even premises in its theoretical constructions. Sociologists and historians will recognize these for the political assumptions that they are.

1. 5. Coulter's Thesis: Mentation involves Whole Person Skills and Performances

Coulter has argued in several places (Coulter 1989) that it is a mistake to assign, either to the brain or to an information-processing module, aspects of human mentation that can only be ascribed to a whole embodied social being, a person. His argument is essentially based upon usage and borrowed consciously from Wittgenstein. For instance, everyday uses of 'remember', 'forget', and so on determine what, for this culture, at this moment, remembering and forgetting are, and these are always attributed to whole persons. We do not say, for instance, that part of ourselves forgets or part of our brains remembers. This is not just an accident of our ignorance. This is a feature of the logic of the concept itself, and it reflects the fact that remembering and forgetting are morally accountable and socially constructed phenomena, in the conversational production of which whole persons are engaged.

But there is a second argument, a moral argument. Declarations (for example, memories as claims to recollect a past event), displays of anger which involve moral commitments and moral stances, or displays of indignation and of regret are interjections into conversations which can only be made on behalf of moral beings. It is the whole person that is a moral agent, not some part of the person, and in particular it is not the brain that stands in moral relations to others.

Thesis five, if correct, would restrict the use of modularity treatments and support the project of AI/CS as a modeling of whole conversations rather than of mysterious goings-on in individual speakers.

2. THE TWO LEVELS OF COGNITIVE SCIENCE

2.1. Level One

I have tried to represent in Figure 1 the overall structure of the naive form of all CS that I believe must be abandoned for research into most psychological topics. I shall call that AI/CS of Level One. In this scheme the tendentious concept of "mental state" is defined in terms of structural isomorphisms that can be set up between the structure of an abstract program, the structure of a brain, and the structure of a computer whose microswitches have been organized by the taking up of the program. The one thing such a "mental state" is for sure is not mental! In terms of the scheme (Figure 1), the upshot of the argument involving the five theses is that only *p"* could be a mental state. *S"* is a brain state under a mentalistic description.

There is a sharp clash between the technical expression "mental state" as it is defined in Level One AI/CS and the concept of "mental state" as it emerges from the application of the five theses about language and the production of mind that I have been developing in this essay. I believe that the clash is so severe that the project of a Level One AI/CS is actually impossible. What "p" is as a mental state (for instance, as an act of recollection made *relatively and for the moment* determinate as a memory through the developing structures of conversation and other

The Structure of Contemporary Cognitive Science

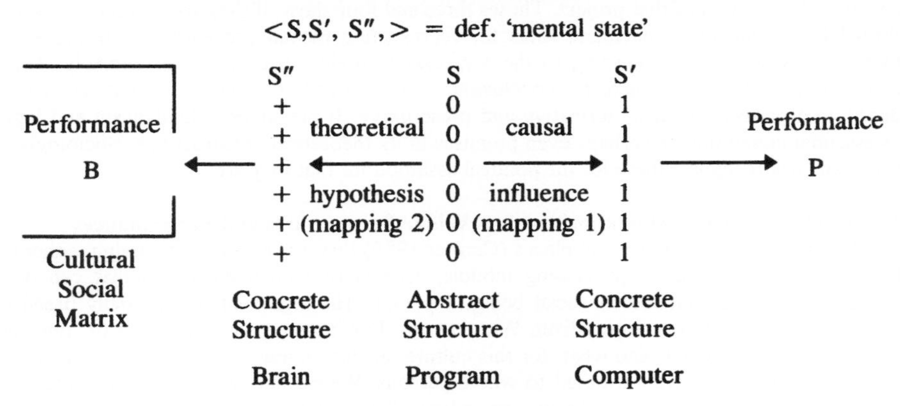

Figure 1

social actions within which it has a place) exists only in a collective of persons. There is no finite determinate entity that is *the* mental state "p". As a memory claim, it is essentially contestable, and this is the case for all such mental products of conversational activity. However, there could be a Level Two AI/CS in which the very same modeling techniques could be put to proper use, but it would be focused on something very different. It would involve the modeling of the very processes in which mental states of the kind I have argued for in this essay, would have their place. The pair <S', S">, that is, the pair which is formed by the mapping of the structure of a program onto the structure of a brain, would be an incomplete but necessary part of the explanation of the ability of the group of people so trained as to produce a conversation, which would in its turn engender such and such a 'real' mental state, say, a certified act of remembering. But, of course, the pair <S', S"> would be a formal model in neurophysiology. It would not, it could not, be a stand-in for a mental state. There is a project for cognitive science and an important one, but it is not at all the project that has been undertaken under the influence of the mistaken idea that in the total model expressed in Figure 1, the computer is to the program as the brain is to the mind.

3. CONCLUSION

I would like to see AI/CS developed as a cluster of formal theories of conversation and, in this respect, it could become very deep. If this program were followed, it could become a theory of the mind in the only sense I can take seriously, that is, a theory of the discursive production of mentation. Just as quantum mechanics and relativity cannot be combined consistently into one overall physics of the universe, it may be that no consistent overall theory could be developed which incorporates all the individual theories of the clusters of formal representations of the procedures and conventions that we find necessary to develop our indexical, historical, and contextually specific conversations.

REFERENCES

REFERENCES

Ackemann, R.J. (1985), *Data, Instruments, Theory*, Princeton, Princeton University Press.

Ackermann, R. (1988), *Wittgenstein's City*, Amherst, University of Massachusetts Press.

Aitchinson, I. J. R. (1985), "Nothing's Plenty: the Vacuum in Modern Quantum Theory", *Contemporary Physics*, vol. 26, pp. 331-391.

Aitchison, I. J. R. and Hey, A. J. C. (1982), *Gauge Theories in Particle Physics*, Bristol, Adam Hilger.

Argyle, M. (1972), "Non-Verbal Communication in Human Social Interaction", in Hinde, R. A. (ed.), *Non-verbal Communication*, Cambridge, Cambridge University Press, pp. 243- 269.

Argyle, M., Furnham, A. and Graham, J.A. (1981), *Social Situations*, Cambridge: Cambridge University Press.

Aronson, J. L. (1988), "Testing for Convergent Realism" in Fine, A. and Leplin, J. (eds.), *Proceedings of the 1988 Biennial Meeting of the Philosophy of Science Association*, vol. 1, pp. 188-193.

Aronson, J. L. (1991), "Versimilitude and Type-Hierarchies", *Philosophical Topics*, vol. 18, pp. 5-28.

Aronson, J. L., Harré, R., and Way, E. (1994), *Realism Rescued*, London, Duckworth.

Austin, J.L. (1962), *How to do things with words*, Oxford, Clarendon Press.

Bachelard, G. (1934), *Le Vouvel Esprit Scientifique*, Paris.

Bannister, D. and Mair, J. M. M. (1968), *The Evaluation of Personal Constructs*, London, Academic Press.

Berkeley, G. (1710) (1988), *Treatise Concerning the Principles of Human Knowledge*, London, Penguin Books.

Berne, E. 1970, *Games People Play*, Harmondsworth: Penguin.

Berger, P. L. and Luckmann, T. (1966), *The Social Construction of Reality*, New York, Doubleday.

Bernstein, B. (1965), "A Socio-Linguistic Approach to Social Learning", in Gould, J. (ed.) *penguin Survey of the Social Sciences*, Baltimore, Penguin.

Bhaskar, R. (1978), *Realist Theory of Science*, Sussex, England, Harvester.

Bhaskar, R. (1989), *Reclaiming Reality*, London, Verso.

Billig, M. (1987), *Arguing and Thinking*, Cambridge, Cambridge University Press.

Black, M. (1962), "Metaphor", in *Models and Metaphor*, Ithaca, New York, Cornell University Press.

Black, M. (1977), "More about Metaphor", *Dialectica*, vol. 31, pp. 431-457.

Boscovich, R. J. (1763/1966), *A Theory of Natural Philosophy*, Cambridge, Massachusetts, MIT Press.

Bohr, N. (1958), *Atomic Physics and Human Knowledge*, New York, Wiley.

Boyd, R. (1979), "Metaphor and Theory Change" in Ortony, A. (ed.), *Metaphor and Thought*, Cambridge, Cambridge University Press.

Boyle, Hon. R. (1666), *The Origins of Forms and Qualities According to the Corpuscular Philosophy*, Oxford, Davis

Bridge, J. (1977), *Beginning Model Theory*, Oxford, Oxford University Press.

Brooke, J. (1980), "The Chemistry of the Organic and the Inorganic", *Kagakushi (Journal of the Japanese Society for the History of Chemistry)*, vol. 7, pp. 37-60.

Brooke, J. (1987), "Methods and Methodology in the Development of Organic Chemistry", *Ambix*, vol. 34, pp. 147-155.

Brown, H. R. and Harré, R , (eds.) (1990), *Philosophical foundations of Quantum Field Theory*, Oxford, Oxford University Press.
Bunge, M. (1973), *Method, Model and Matter*, Dordrecht and Boston, Reidel.
Burke, K. (1969), *A Grammar of Motives*, Berkeley and Los Angeles, University of California Press.
Canguilhem, G. (1963), "The Role of Analogies and Models in Biological Discovery", in Crombie, A. (ed.), *Scientific Change*, New York, Basic Books, pp. 507-520.
Cartwright, N. (1983), *How the Laws of Nature Lie*. Oxford: Clarendon Press.
Cartwright, N (1989), *Nature's Capacities and their Measurement*, Oxford, Clarendon Press.
Cartwright, N. (1999), *The Dappled World: A Study of the Boundaries of Science*, Cambridge, Cambridge University Press.
Chomsky, N. (1973), *For Reasons of State*, New York, Pantheon Books.
Churchland, P. (1986), *Neurophilosophy*, Cambridge, Massachusetts, MIT Press.
Clagett, M. (1959), *The Science of Mechanics in the Middle Ages*, Madison, University of Wisconsin Press.
Clarke, D.D. (1983), *Language and action*, Oxford, Pergamon Press.
Cooper, W. (1971), *Hair, sex, society, symbolism*, London, Alden.
Coulter, J. (1989), *Mind in action*, Cambridge, Cambridge University Press.
Cranach, M. von and Harré R. (1982), *The analysis of action*, Cambridge: Cambridge University Press.
Culler, J. (1975), *Structuralist Poetics*, Ithaca, Cornell University Press.
Darwin, C. (1859), *The origin of Species, or, the Preservation of Favoured Races*, London, John Murray.
Davies, B., and R. Harré (1990), "Positioning", *Journal for the Theory of Social Behaviour*, vol. 20, no. 1, pp. 43-63.
Del Rey, G. (1974), "Current Problems and Perspectives in MO-LCAO Theory of Molecules", *Advances in Quantum Chemistry*, vol. 8, pp. 95-136.
Dennett, D. C. (1978), *Brainstorms: Philosophical Essays on Mind and Psychology*, Cambridge, Massachusetts, MIT Press.
Descartes, R. (1954), *Descartes' Philosophical Writings*, Anscombe, E. and Geach, P. T., (eds.), London, Nelson.
Devall, B. and Sessions, G. (1985), *Deep Ecology*, Salt Lake City, Utah, Gibbs Smith.
Devitt, M. (1984), *Realism and Truth*, Oxford, Basil Blackwell.
De Waele, J-P. and Harré, R. (1976), "Rituals for the incorporation of a stranger", in Harré, R. (ed.), *Life sentences*. Chichester, John Wiley & Sons, pp. 76-86.
De Waele, J. P. and Harré, R. (1979), "Autobiography as a Psychological Method", in Ginsberg, G. P. (ed.), *Emerging Strategies in Social Psychological Research*, Chichester, Wiley.
Diedrich. W. (1989), "The development of Marx's economic theory", *Erkenntnis*, vol. 30, pp. 147-164.
Douglas, M. (1970), *Natural Symbols, New York*, Pantheon Books.
Douglas, M. (1975), *"The meaning of meals"*, Daedalus Autumn.
Einstein, A. (1905), "On the electrodynamics of Moving Bodies" *Annalen de Physik*, vol. 17.
Emirbayer, M. and Mische, A., (1998), "What Is Agency?" *American Journal of Sociology*, vol. 103, no. 4, pp. 962-1023.
Filmer, P., Phillipson, M., Silverman, D., and Walsh, D. (1972), *New Directions in Sociological Theory*, Cambridge, Massachusetts, MIT Press.
Fisk, M. (1974), *Nature and Necessity*, Bloomington, Indiana, Indiana University Press.

Fodor, J. A. (1981), *Representations: Philosophical Essays on the Foundations of Cognitive Science*, Cambridge, Massachusetts, MIT Press.

Foucault, M. (1972), *The Archaeology of Knowledge* (transl. By A. M. S. Smith), London, Dorset Press.

Galileo, G. (1623/1957) *Il Saggiatore* in Drake, G. S. (ed.), *The Discoveries and Opinions of Galileo*, New York, Doubleday.

Gergen, K. J. and Gergen, M. (1991), "Towards reflexive methodology" in Steiner, F.(ed.), *Research and Reflexivity*, London, Sage Publications, pp. 76-95.

Gibson, J. J. (1968), *The Senses Considered as Perceptual Systems*, Boston, Houghton Mifflin.

Gibson, J. J. (1979), *The Ecological Approach to Visual Perception*, London, Houghton Mifflin.

Giere, R. L. (1988), *Explaining Science: A Cognitive Approach*, Chicago, University of Chicago Press.

Goffman, E. (1959), *The Presentation of Self in Everyday Life*, Garden City, New York, Doubleday.

Goffman, E. (1961), *Asylums*, Garden City, New York, Anchor Books.

Goffman, E. (1963), *Stigma*, Englewood Cliffs, NJ: Prentice-Hall.

Goffman, E. (1972), *Relations in Public*, New York, Harper and Row.

Goffman, E. (1974), *Frame Analysis*, Cambridge, Massachusetts: Harvard University Press.

Goffman, E. (1981), *Forms of Talk*, Philadelphia, University of Pennsylvania Press.

Goldsmith, T. (1992, May 30), "the roads to Rio: No, the real global threat is the relentless demand for growth" *Sunday [London] Times*, p. 14.

Gooding, D. (1990), *Experiments and the Making of Meaning*, Dordrecht, Kluwer.

Goodman, N. (1978), *Ways of Worldmaking*, Indianapolis, Hackett.

Goodman, N. (1996), "Conditional plurality of pluralisms", *Dialectica*, vol. 3, pp. 69-80.

Hacking, I. (1983), *Representing and Intervening: Introductory Topics in the Philosophy of Natural Science*, Cambridge, Cambridge University Press.

Hacking, 1. (1988), "Making Up the Mind", *London Review of Books*, 1 September, pp. 15-16.

Hacking, I., (1992), "The Self-Vindication of the Laboratory Sciences", in Pickering, A. (ed.) *Science as Practice and Culture*, Chicago, University of Chicago Press, pp. 29-64.

Hall, T. (1968), "On Biological Analogs of Newtonian Paradigms", *Philosophy of Science*, vol. 35, pp. 6-27.

Hallam, A. (1973), *A Revolution in the Earth Science*, Oxford, Clarendon Press.

Hanson, N. R. (1963), *The concept of the Positron: A Philosophical Analysis*, Cambridge, Cambridge University Press

Harré, R. (1961). *Theories and Things*, London: Sheed and Ward.

Harré, R. (1970), *The Principles of Scientific Thinking*, Chicago: University of Chicago Press.

Harré, R (1971), "Johnson's Dilemma", *Bulletin of the British Psychological Society*, vol. 24, pp. 115-119.

Harré, R. (1972), "Surrogates for Necessity", *Mind*, vol. 82, pp. 358-380.

Harré, R. (1976), "The Constructive Role of Models", in Collins, E. (ed.), *The Use of Models in the Social Sciences*, London, Tavistock, pp. 16-43.

Harré, R. (1979), *Social Being*, Oxford, Blackwell.

Harré, R. (1981), *Great Scientific Experiments*, Oxford, Phaidon.

Harré, R. (1986), *Varieties of Realism*, Oxford, Basil Blackwell.

Harré, R (1988), "Where models and analogies really count", *International Studies in the Philosophy of Science*, vol. 2, pp. 118-133.

Harré. R. (1990) 'Parsing the amplitudes', in Brown, H. and Harré, R. (eds..) *Philosophical Foundations of Quantum Field Theory*, Oxford, Clarendon Press, pp. 59-71.

Harré, R. (1996), "There is no time like the present", in Copeland, B. J. (ed.), *Logic, Modality and Time*, Oxford, Oxford University Press, pp. 389-409.
Harré, R., Brockmeier, J., Mühlhäusler, P. (1999), *Greenspeak: A Study of Environmental Discourse*, Thousand Oaks, California, Sage Publications.
Harré, R. and Madden, E. H. (1975), *Causal Powers*, Oxford, Basic Blackwell.
Harré, R. and Reynolds, V., eds., (1984), *The meaning of primate signals*, Cambridge: Cambridge University Press.
Harré, R. and Secord, P. F. (1972), *The Explanation of Social Behavior*, Oxford, Basil Blackwell.
Haug, F. (1987), *Female Sexualisation,* translated by E. Carter. London, Verso.
Hertz, H. (1894), *The Principles of Mechanics*, (reprint) New York, Macmillan, reissued by Dover, 1956.
Hesse, M. (1974), *The Structure of Scientific Inference*, Berkeley, University of California Press.
Honner, R, J. (1987) *The Description of Nature*. Oxford, Clarendon Press.
Hume, D. (1788/1978), *A Treatise of Human Nature,* ed. Selby-Bigge, L.A., Oxford, Clarendon Press.
Husserl, E. G. A. (1960), *Cartesian Meditations,* V, Trans. By P. Cairns, The Hague, Nijhoff.
Kahn, M. (1992), "The Passive voice of science: Language absent in the wild-life profession", *Trumpeter*, vol. 9, pp. 152-154.
Kant, I. (1781/1950), *The Critique of Pure Reason,* trans. Meikeljohn, J. M. Dent, London.
Kant, I. (1786/1970) *Metaphysical Foundations of Natural Science*, trans. J. Ellington, Indianapolis, Indiana, University of Indiana Press.
Katz. J. J. (1966), *The Philosophy of Language*, New York and London, Harper and Row.
Kim, K. M. (1994), *Explaining Scientific consensus,* New York and London, Guildford Press.
Kreckel, M. (1981), *Communicative acts and shared knowledge*, London, Academic Press.
Knorr-Cetina, K. (1981), *The Manufacture of Knowledge*, Oxford, Pergamon Press.
Koestler, A. (1959), *The Sleepwalkers* , London, Hutchinson.Butterfield..
Kuhn, T. S. (1962), *The Structure of Scientific Revolutions,* Chicago, University of Chicago Press.
Lakatos, I. (1970), "Falsification and the Methodology of Scientific Research Programmes", in
Lakatos, I. And Musgrave, A. (eds.), *Criticism and the Growth of Knowledge*, Cambridge, Cambridge University Press, pp. 91-196.
Latour, B. (1987), *Science in Action*, Cambridge, Massachusetts, Harvard University Press.
Latour, B., and Woolgar, S. (1979), *Laboratory Life*, Beverly Hills, California, Sage.
Laudan, L. (1984), "A Confutation of Convergent Realism" in Leplin, J. (ed), *Scientific Realism*, Berkeley, California, University of California Press, pp. 218-249.
Lehner, A. (1969), "Semantic Cuisine", *Journal of Linguistics*, vol. 5, pp. 39-55.
Lelas, S. 1993. “Science as Technology, ” *British Journal for the Philosophy of Science*, vol. 44, pp. 423-442.
Lévi-Strauss, C. (1966), *The Savage Mind*, London, Weidenfeld and Nicolson
Levi-Strauss, C. 1973), *Triste Tropiques,* trans. J. and D. Weightman, New York, Pocket Books.
Leibniz, G. (1717) (1956), *The Leibniz-Clarke Correspondence*, Alexander, H. G., Manchester, Manchester University Press.
Lipton, P. (1991), *Inference to the Best Explanation*, London, Routledge.
Lipton, P. (1985), *Explanation and Evidence*, Oxford, Doctoral Dissertation.
Locke, J. (1690/1947), *An Essay Concerning Human Understanding*, Wilburn, R., London, Dent. Dutton.
Lovelock, J. E. (1987), *Gaia: A new look at life on Earth*, Oxford, Oxford University Press.
Lucas, J. R., (1973), *A Treatise on Time and Space*, London, Methuen.

Lucas, J. R. (1985), *Space, Time and Causality*, Oxford, Oxford University Press.
Lucas, J. R and Hudgson, P. E. (1990), *Spacetime and electromagnetism*, Oxford, Clarendon Press.
Lyman, S. M. and Scorr, M. B. (1975), *The Drama of Social Reality*, Oxford, Oxford University Press.
Mach, E. (1914), *The Analysis of Sensations*, trans. T. J. McCormack and C. M. Williams, Chicago, Open Court.
Madden, E. H. and Harré, R. (1971), "The Powers that Be", *Dialogue*, vol. 10, pp. 12-30.
Mair, M. (1989), *Between psychology and psychotherapy: A poetics of experience*, New York Routledge, Chapman & Hall
Mannheim, K. (1936), *Ideology and Utopia*, trans. L. Wirth and E. Shils, New York, Harcourt Brace.
Mason, J. (1992), "The greenhouse effect and global warming," in Cartledge, B. (ed.) *Monitoring the Environment*, Oxford, Oxford University Press, pp. 55-92.
McGuire, J. (1970), "Atoms and the 'Analogy of Nature': Newton's Third Rule of Philosophizing", *Studies in the History and Philosophy of Science*, vol. 1, pp. 3-58.
Mill, J. S. (1872), *A System of Logic*, London: Routledge.
Miller, A. I. (1984), *Imagery in Scientific Thought*, Boston, Massachusetts, Birkhauser.
Mixon, D. (1972), "Instead of Deception", *Journal for the Theory of Social Behavior*, vol. 2, pp. 145-176.
Moken, G. (1988), *Particles and Ideas*, Oxford, Clarendon Press.
Monod. J. (1972), *Chance and Necessity*, London, Collins.
Mühlhäusler, P. and Harré, R (1990), *Pronouns and people*, Oxford, Blackwell.
Munn, A. M. (1960), *Freewill and Determinism*, London, MacGibbon & Kee.
Nagel, E. (1961), *The Structure of Science*, New York, Harcourt, Brace.
Noaken, G. R. (1977), *Intermediate Physics*, Dondon, Methuen.
North, J. (1980), "Science and Analogy", in Grmek, M., Cohen, R., Cimino, G., (eds.), *On Scientific Discoveries*, Vol. 34, Boston Studies in Philosophy of Science, Dordrecht: Holland, D. Reidel, pp. 115-140.
Oddie, B. (1986), "The Poverty of the Popperian Programme for Truth-Likeness", *Philosophy of Science*, vol. 53, pp. 163-178.
Ogden, C. K. and Richards, I. A. (1923/1972), *The meaning of Meaning*, New York, Harcourt, Brace & World.
Pearce, W.B. and Cronen, V. (1980), *Communication, action and meaning*, New York, Vernon E. Cronen.
Pickering, A. (1981), "Constraints and Controversy: the Case of the Magnetic Monopole", *Social Studies of Science*, vol. 11, pp. 63-93.
Pickering, A. (1984), *Constructing Quarks: A Sociological History of Particle Physics*, Edinburgh, Edinburgh University Press.
Platts, M. (1979), *Ways of Meaning*, London, Routledge & Kegan Paul.
Popper, K. R. (1959), *The Logic of Scientific Discovery*, London, Hutchinson.
Popper, K. R. (1969), *Conjectures and Refutations*, London, Routledge and Kegan Paul.
Popper, K. R., (1972), *Objective Knowledge, an Evolutionary Approach*, Oxford, Clarendon Press.
Quine, W. V. O. (1953), *From a Logical Point of View*, Cambridge, Massachusetts, Harvard University Press, Second Edition.
Redhead, M. (1980), "The Use of Models in Physics", *British Journal for the Philosophy of Science*, vol. 31, pp. 145-163.
Reid, T. (1788/1969), *Essays on the Active Powers of the Human Mind*, Cambridge, Massachusetts,

MIT Press.
Richards, A. I. (1965), *The Philosophy of Rhetoric*, New York, Oxford University Press.
Rorty, R. (1979), *Philosophy and the Mirror of Nature*, Princeton, New Jersey, Princeton University Press.
Rorty, R. (1989), *Contingency, Irony, and Solidarity*, Cambridge, Cambridge University Press.
Rosser, E. (1979), "Trouble in school", in Hammersley, M. (ed.), *The process of schooling*, Milton Keynes, Open University.
Rothbart, D. (1997), *Metaphors, Models and Meanings: Explaining the Growth of Scientific Knowledge*, Lewiston, New York, The Edwin Mellen Press.
Ryle, G. (1949). *The Concept of Mind*, London, Hutchison.
Schegloff, E.A. (1984), "On some questions and ambiguities in conversation", in Atkinson, M. and Heritage, J. (eds.), *Structures of social action. Studies in conversational analysis*, Cambridge, Cambridge University Press, pp. 28-52.
Schön, D. (1963), *The Displacement of Concepts*, London, Tavistock.
Schutz, A. (1971), *Collected Papers*, 3rd ed., The Hague, Nijhoff.
Schutz, A. (1972), *The Phenomenology of the Social World*, G. Walsh and F. Lehrert, trans, London, Heinemann. (1st edition; 1932 Vienna.)
Scott, M. B. and Lyman, S. M. (1968), "Accounts", *American Sociological Review*, vol. 133, pp. 46-62.
Scott, M. B. and Lyman, S. M. (1970), *A Sociology of the Absurd*, New York: Appleton-Century-Croft.
Secord, P. and Backman, C. (1964), *Social Psychology*, New York, McGraw Hill.
Shotter, J. (1973), "The transformation of natural into personal powers", *Journal for the Theory of Social Behaviour,* vol. 3, pp. 141-156.
Shotter, J. (1975), *Images of Man in Psychological Research*, London, Methuen.
Shotter, J. (1984), *Social Accountability and Selfhood*, Oxford, Blackwell.
Skinner, B. F. (1957), *Verbal Behavior,* New York, Appleton-Century-Crofts.
Skinner, B. F. (1971), *Beyond freedom and Dignity*, New York, Knopf.
Sklar, L. (1992), *Philosophy and Spacetime Physics*, Berkeley, University of California Press.
Smith, D. E. (1987), "Women's Perspective as a Radical Critique of Sociology", in Harding, S. (ed.), *Feminism and Methodology,* Indianapolis, Indiana University Press.
Sneed, T. (1971), *The Logical Structure of mathematical Physics*, Dordrecht, Reidel.
Sokal, A. (1997), "Correcting the excesses of post-modernism", *Symposium on Consciousness,* May 2, New York: New School.
Southwood, R. (1992), "The Environment: Problems and Prospects", in Cartledge, B. (ed.) *Monitoring the Environment*, Oxford, Oxford University Press, pp. 5-41.
Stegmüller, W. (1979), *The Structure and Dynamics of Theories*, Berlin and NewYork, Springer Verlag.
Sutherland, N. S. (1963), "Visual Discrimination and Orientation of Shape by the Octopus," *Nature*, vol., 179, pp. 11-13.
Thagard, P. (1997), "Scientific Change: the Discovery and Acceptance of the Bacterial Theory of Ulcers", Address to *The British Society for the Philosophy of Science,* London: 10 March, 1997.
Tiger, L. and Fox, R. (1971), *The Imperial Animal*, New York, Holt, Rinehart and Winston.
Toolan, M. (2001), *Narratives: A Critical Linguistic Introduction,* 2nd edn, London, Routledge.
Torode, T. 1977, "The revelation of the theory of the social world as grammar," in Harré, R. (ed.), *Life sentences.* Chichester, John Wiley & Sons, pp. 87-97.

Toulmin, S. E. (1953), *The Philosophy of Science*, London, Hutchinson.
Toulmin, S. E. (1961), *Foresight and Understanding*, London, Hutchinson.
Toulmin, S. E. (1972), *Human Understanding*, vol. 1, Princeton, New Jersey, Princeton University Press.
Urban, G. (1989), "The 'I' of Discourse", in Lee, B. and Urban, G. (eds.), *Semiotics, Self and Society* New York, Mouton de Gruyter.
Turner, R. (1974), *Ethnomethodology*, Harmondsworth, Penguin.
Ursus, N. R. (1588), *Fundamentum Astronomium*, Vienna.
Van Fraassen, B. (1980), *The Scientific Image*, Oxford: Clarendon Press
Von Uexküll, J. (1909), *Umwelt und Innenwelt der Tiere*, Berlin, Springer.
Vygotsky, L. S. (1962). *Thought and Language*, Trans. and edited, E. Hanfmann and G. Vakar, Cambridge, Massachusetts, MIT Press.
Wallace, W. A. (1996), *The Modeling of Nature*, Washington, D. C., The Catholic University of America Press.
Way, E. C. (1991), *Knowledge Representation and Metaphor*, Dordrecht, Kluwer.
Whewell, W. (1967), *The Philosophy of the Inductive Sciences*, 3rd edition, London, Cass.
Wigner, E. P. (1970), *Symmetries and Reflections*, Moore, W.J. and Scriven, M (eds.), Second Edition, Cambridge, Massachusetts, MIT Press.
Winograd, P., and Flores, F. (1987), *Understanding Computers and Cognition.* Reading, Massachusetts, Addison-Wesley.
Wittgenstein, L. (1922), *Tractatus Logico-Philosophicus*, trans. D. F. Pears and B. McGuiness, London and New York, Routledge and Kegan Paul..
Wittgenstein, L. (1953), *Philosophical Investigations*, Oxford, Blackwell.
Wittgenstein, L. (1956), *Remarks on the Foundations of Mathematics*, von Wright, G. H. and Rhees, R., eds., Oxford, Blackwell.
Wittgenstein, L. (1969), *On Certainty*, Trans. D. Paul and G.E.M. Anscombe, Oxford, Blackwell.
Woolf, V. (1977), *To the Lighthouse*, New York, Harcourt, Brace Jovanovich

SUBJECT INDEX

Subject Index